Medicinal Chemistry for Pharmacy Students

(Volume 1)

Fundamentals of Medicinal Chemistry and Drug Metabolism

Edited by

M. O. Faruk Khan

Department of Pharmaceutical Sciences and Research, Marshall University School of Pharmacy, Huntington, WV, USA

&

Ashok Philip

Department of Pharmaceutical Sciences, Union University College of Pharmacy, Jackson, TN, USA

Medicinal Chemistry for Pharmacy Students

Volume # 1

Fundamentals of Medicinal Chemistry and Drug Metabolism

Editors: M. O. Faruk Khan and Ashok Philip

ISSN (Online): 2589-6989

ISSN (Print): 2589-6997

ISBN (Online): 978-1-68108-687-3

ISBN (Print): 978-1-68108-688-0

General:

1. Any dispute or claim arising out of or in connection with this License Agreement or the Work (including non-contractual disputes or claims) will be governed by and construed in accordance with the laws of the U.A.E. as applied in the Emirate of Dubai. Each party agrees that the courts of the Emirate of Dubai shall have exclusive jurisdiction to settle any dispute or claim arising out of or in connection with this License Agreement or the Work (including non-contractual disputes or claims).
2. Your rights under this License Agreement will automatically terminate without notice and without the need for a court order if at any point you breach any terms of this License Agreement. In no event will any delay or failure by Bentham Science Publishers in enforcing your compliance with this License Agreement constitute a waiver of any of its rights.
3. You acknowledge that you have read this License Agreement, and agree to be bound by its terms and conditions. To the extent that any other terms and conditions presented on any website of Bentham Science Publishers conflict with, or are inconsistent with, the terms and conditions set out in this License Agreement, you acknowledge that the terms and conditions set out in this License Agreement shall prevail.

Bentham Science Publishers Ltd.
Executive Suite Y - 2
PO Box 7917, Saif Zone
Sharjah, U.A.E.
Email: subscriptions@benthamscience.org

BENTHAM SCIENCE

CONTENTS

FOREWORD

For a pharmacist to be a successful member of the health care team, a foundation of medicinal chemistry knowledge is one of the essential elements to developing an adequate knowledge base, and critical thinking and problem solving skills. This e-book provides the fundamental principles of medicinal chemistry including the functional groups occurring in medicinal agents, the acidity and basicity of drugs, and their water and lipid solubility as well as drug-receptor interactions. The physicochemical principles, isosterism and spatial characteristics of drugs are prerequisites to understanding drug pharmacodynamics, pharmacokinetics, biopharmaceutics, formulations and pharmacotherapeutics. Another important aspect crucial to comprehending the mechanism of drug action is the knowledge of important biosynthetic pathways frequently encountered in the pharmaceutical interventions. A comprehensive approach has been taken to explaining the phases of drug metabolism, modifications of drug chemical structures and their effects on drug pharmacokinetics, safety and efficacy.

These authors have carefully integrated the key aspect of Doctor of Pharmacy curriculum in the design and organization of the contents in this eBook series; its novel and innovative layout includes 4 volumes in three distinct areas - the fundamental concepts, detailed structure activity relationships of different drug classes, and recent developments in the area of medicinal chemistry and drug discovery. It offers students the opportunity to learn the principles of drug action in a logical stepwise manner. The case studies, student's self-study guide and self-assessments at the end of each chapter are unique features of this book that would be beneficial to both students, faculty members and clinicians. Although there are several medicinal chemistry textbooks available in the market, to my understanding, this is the first textbook of its kind focusing on the integration of medicinal chemistry into the foundations for a pharmacy curriculum.

As a pharmacy educator and pharmaceutical scientist, I am pleased to testify and endorse this e-book to the pharmacy educators, and learners as a novel and innovative addition to the available literature. One of the key advantages of this e-book is that it focuses its approach on a student-friendly manner that incorporates the appropriate illustrative diagrams and study guides as well as self-assessments to enable students to enhance their skills as self-learners. The increasing emphasis on clinical and management focuses in our pharmacy curriculum makes it challenging for students with limited time available to learn and internalize concepts in the area of medicinal chemistry. This e-book series brings those memories of my early years as a pharmacy student and the evolution of the pharmacy education, and gives me the confidence that it will pave the way of future medicinal chemistry education for pharmacists and other health professionals.

Gayle Brazeau
Dean School of Pharmacy
Marshall University
Huntington, WV
USA

Preface

This is the first volume of the 4-volume eBook series, "*Medicinal Chemistry for Pharmacy Students*". The primary objective of this e-Book series is to educate Pharm-D students in the area of medicinal chemistry and serve as a reference guide to pharmacists on the aspects of chemical basis of drug action. A thorough discussion of key physicochemical parameters of therapeutic agents and how they affect the biochemical, pharmacological, and pharmacokinetic processes and clinical use of these agents is the primary focus of the whole book. The rationale for putting together an e-Book of this nature is to equip Pharm-D students with the scientific basis to competently evaluate, recommend and counsel patients and health care professionals regarding the safe, appropriate, and cost-effective use of medications.

This first volume of the series titled, "*Fundamentals of Medicinal Chemistry and Drug Metabolism*", is comprised of 8 chapters focusing on basic background information to build a firm knowledge base of medicinal chemistry. It takes a succinct and conceptual approach to introduce important fundamental chemical concepts, required for a clear understanding of various facets of pharmacotherapeutic agents, drug metabolism and important biosynthetic pathways that are relevant to drug action.

Chapter 1 is designed to ensure that the students learn about the scope and importance of medicinal chemistry, in addition to some important definitions. This chapter is an introduction to the overall role of a pharmacist and to the significance of medicinal chemistry in pharmacy education. It discusses the role of the pharmacist, history of medicinal chemistry and intellectual domains of medicinal chemistry.

Chapter 2 includes a comprehensive discussion of the four major biomolecules: proteins, carbohydrates, lipids, nucleic acids and key heterocyclic ring systems. Organic functional groups present in drugs, and biomolecules are reviewed in this chapter. Additionally, heterocycles present in drugs and biomolecules are reviewed in this chapter.

Chapter 3 focuses on acid base chemistry and salt formation; a brief review of the concepts of acid-base chemistry including the Arrhenius, Brønsted-Lowry, and Lewis concepts of acids and bases. It also highlights the significance of salt formation in pharmaceutical products, factors that determine ionization, and acid-base strengths. The application of acid-conjugate base, base-conjugate acid and Henderson-Hasselbalch equation in pharmacy and drug action and bioavailability are also discussed. The interpretation of pH partition theory, its significance in drug pharmacokinetics, the purpose of salt formation with drug molecules and the acidity or basicity of the salts are illustrated in this chapter.

Chapter 4 covers solubility and lipid-water partition coefficient (LWPC) concepts in detail with specific drug examples. Hydrophilicity, hydrophobicity and lipophilicity of drugs, and their effect on solubility are discussed to enable the readers to predict the water or lipid solubility of drugs based on their chemical composition. Additionally, the chapter includes a discussion of the effects of partition coefficient on drug bioavailability and action.

Chapter 5 reviews the concepts of isosterism, stereochemical principles, and their application. An explanation of drugs' spatial factors, and their influence on drug action including definitions of important stereochemical parameters are also included.

Chapter 6 is a brief review of the mechanisms of drug action and discusses drug receptor interactions critical for pharmacological responses of drugs. This chapter also discusses the theories of drug action that include: occupancy theory, rate theory, induced-fit theory, macromolecular perturbation theory, and occupation-activation theory of "two-state" model.

Chapter 7 provides a detailed account of drug metabolism, prodrugs and related terminology. It provides a comprehensive account of the fundamental concepts of drug metabolism, describes the significance of drug metabolism, key enzymes involved in the sites of drug metabolism, phase I and phase II metabolic pathways that include oxidation reduction hydrolysis, glucuronic acid conjugation, sulfate conjugation, conjugations with glycine and other amino acids, glutathione or mercapturic acid, acetylation, and methylation. This chapter also defines and differentiates between prodrug, soft drug and antedrugs and discusses their clinical significance.

Chapter 8 provides a brief review of biosynthetic pathways frequently targeted by pharmaceutical interventions. The biosynthetic pathways discussed in this chapter include: eicosanoid biosynthesis (prostaglandins, prostacyclins and leukotrienes), epinephrine and norepinephrine biosynthesis, folic acid biosynthesis, steroid biosynthesis (cholesterol, adrenocorticoids and sex hormones) and nucleic acid biosynthesis (purines and pyrimidines anabolism, catabolism and salvages).

The chapters in this volume are designed to guide the reader to review, integrate and apply medicinal chemistry concepts to the study of therapeutic agents that are the focus of subsequent volumes. All concepts are illustrated with diagrams or figures, with the keywords highlighted, bulleted or numbered. Wherever needed, special boxes and case studies are included. In addition, each chapter is reinforced with practice problems and answer sets. Special notations are highlighted using call-out boxes for visual effect. Tables and figures are used to augment the text as needed.

We would like to express our sincere gratitude to the contributing authors for their time and effort in completing this volume. We would also like to thank Bentham Science Publishers, particularly Mr. Shehzad Naqvi (Senior Manager Publication), and Ms. Fariya Zulfiqar (Assistant Manager Publications) for their support. We are confident that this volume of the eBook series will guide students and educators of pharmacy and related health professions worldwide.

M. O. Faruk Khan, Ph.D.
Department of Pharmaceutical Sciences and Research,
Marshall University School of Pharmacy,
Huntington, WV,
USA

&

Ashok Philip, Ph.D.
Department of Pharmaceutical Sciences,
Union University College of Pharmacy,
Jackson, TN,
USA

List of Contributors

M. O. Faruk Khan	Department of Pharmaceutical Sciences and Research, Marshall University School of Pharmacy, Huntington, WV, USA
Ashok Philip	Department of Pharmaceutical Sciences, Union University School of Pharmacy, Jackson, TN, USA
Ashim Malhotra	Department of Pharmaceutical and Biomedical Sciences, College of Pharmacy, California Northstate University, Elk Grove, California, CA, USA
Hardeep Singh Saluja	Southwestern Oklahoma State University College of Pharmacy, Weatherford, Oklahoma, USA
Timothy J. Hubin	Department of Chemistry, Southwestern Oklahoma State University, Weatherford, Oklahoma, USA
Taufiq Rahman	Department of Pharmacology, University of Cambridge, Cambridge, UK
Rahmat Talukder	Department of Pharmaceutical Sciences, University of Texas at Tyler College of Pharmacy, Tyler, TX, USA
Jason L. Johnson	Department of Chemistry and Physics, Southwestern Oklahoma State University, Weatherford, Oklahoma, USA

CHAPTER 1

Introduction

M. O. Faruk Khan[1,*] and **Ashok Philip**[2]

[1] *Department of Pharmaceutical Sciences and Research, Marshall University School of Pharmacy, Huntington, WV, USA*

[2] *Department of Pharmaceutical Sciences, Union University School of Pharmacy, Jackson, TN, USA*

Abstract: This chapter is an introduction to the overall role of pharmacist and the significance of medicinal chemistry in pharmacy education. After study of this chapter students will be able to:

• Discuss the role of the pharmacist
• Highlight the history of medicinal chemistry
• Illustrate the relevance of medicinal chemistry in pharmacy education
• Recognize the significance of medicinal chemistry in pharmacy education
• State important definitions in medicinal chemistry and pharmacy.

Keywords: Common terms in medicinal chemistry, Domains of medicinal chemistry, History of medicinal chemistry, Pharmacy education, Role of pharmacist.

BRIEF HISTORY AND ROLE OF PHARMACISTS

Although the history of pharmacy goes back a long way, the profession of pharmacy emerged as a separate entity in America in the mid-19th century. In 1869, William Proctor Jr. defined pharmacy as the "art of preparing and dispensing medicine", which "embodies the knowledge and skill requisite to carry them out to practice." Philadelphia College of Pharmacy was the first to start the journey of institutional level pharmacy education in America in 1821 by establishing night school for apprentices and discussion groups for scientific pharmacy. The professional credentials for American pharmacists were strengthened with the initiation of the 4-year B.Sc. degree requirement for pharmacy licensure in 1932, which also gave the formal recognition of medicinal chemistry in academic pharmacy education [1, 2].

* **Corresponding author M.O. Faruk Khan**: Department of Pharmaceutical Sciences and Research, Marshall University School of Pharmacy, Huntington, WV, USA; Tel: 304-696-3094; Fax: 304-696-7309; E-mail: khanmo@marshall.edu

Since the beginning, both preparation and dispensing of medicine have been the scope of pharmacy practice; over decades, the role of a pharmacist has gradually shifted to a service and information provider and eventually to a patient care provider. Pharmacists are now required to provide counseling to Medicaid patients and to participate in drug utilization review programs. Soon after the Omnibus Budget Reconciliation Act of 1990 (OBRA 1990) was approved, all schools of pharmacy in America shifted to the 6-year (2 + 4) Pharm.D. program [1]. This trend has now continued to grow globally.

BRIEF HISTORY OF MEDICINAL CHEMISTRY

The origin of medicinal chemistry may be traced back to the early history of mankind when procedures such as refining sugar, fermenting wine, extracting vegetable oils, rendition of vegetable fats and saponification were known. The Chinese scholar-emperor Shen Nung, recommended the use of *ch'ang shang* for the treatment of malaria in 2735 B.C. The ancient Babylonian-Assyrian culture used 250 vegetable drugs and a few drugs from mineral and animal sources. Egyptian medicine, developed from a vegetable origin, was fully developed before 1600 B.C. Later in the 5th century B.C., Hippocrates recommended metallic salts for medical treatments and, together with Discarides, Pliny, and Galen, had a strong influence on Western medicinal chemistry. *Materia Medica* by Discarides discussed medicinal products at length. Galen introduced the term 'Galenicals' and Pliny translated the Greek medical symbols into Latin, thus making the knowledge of medicinal chemistry available to the scientific Western World [3].

The alchemists in the 13th century contributed a great deal to the development of medicinal chemistry. The Arabian medical treatises described senna, camphor, rhubarb, tamarind, and nutmeg as natural drugs. In 1240, German emperor Frederick II paved the way to assigning professional status for medicinal chemistry practitioners by issuing the *Magna Charta* of medicinal chemists. However, it was not until the Renaissance which gave birth to independent thought that further development in this young science was achieved [3]. Paracelsus, in the first half of the 16th century, challenged alchemy by announcing that the object of chemistry is not to make gold but to prepare medicine [4]. He believed that sickness is a disturbance of body chemistry and thus the preparation of medicine is an important job of the chemist – the basic tenet of today's pharmaceutical chemistry was thus born [3].

Many drugs of plant origin emerged in South America, notably *ginseng*, *mandrake*, and *wild licorice,* about a century before North America started their search. The *United States Pharmacopoeia* and the *National Formulary* of North America in 1925 listed many more; Maya Indians alone had over 400 drugs.[2] The isolation of benzoic acid in 17th century guided the principles of careful exami-nation of plant and animal products for their medicinal values and the isolation of ephedrine by Nagai in 1887 is a milestone of such investigation. Subsequent efforts were extended toward the synthesis of chemical constituents of plant and animal origins. Wohler was the first chemist to synthesize the organic compound *urea* and began the new science of synthetic organic medicinal chemistry as an important tool for the synthesis of modern drugs. The Kolbe synthesis of salicylic acid in the mid-19th century, Perkin synthesis of dyes in 1856, Dreser synthesis of aspirin in 1889 and barbital synthesis in 1903 by Fischer and Mering are a few milestones of discovering drugs by organic syntheses [5].

Ehrlich's 'side chain theory' of drug action (1885) may be viewed as the starting point of modern medicinal chemistry [6], which led him in 1891 to coin the term 'chemotherapy'. A chemotherapy is defined as *"chemical entity exhibiting selective toxicities against particular infectious agent"* [3]. Cambridge physiologist Langley (mid-1890s) supported the 'side chain theory', and described it in his publications as *'receptive substances'*. Today's advancement of medicinal chemistry is largely indebted to this concept of receptors and its role in diverse biological processes [7]. A few key milestones concerning principles of drug action and modern medicinal chemistry are: Fischer's (1894) *lock-and-key theory*, and Henry's (1903) hypothesis on enzyme-substrate complex formation. Grimm's and Erlenmeyer's concepts of isosterism and bioisosterism (1929-1931) are also critical for understanding structure activity relationship (SAR, the relationship of chemical structure with biological activity) and drug discovery [5]. A few notable advancements in drug action and design in the mid to late 20th century include: (a) charge transfer concept of drug-receptor interaction (Kosower, 1955), (b) induced-fit theory of drug action (Koshland, 1958), (c) concepts of drug latentiation (Harper, 1959) and prodrug (Albert, 1960), (d) application of mathematical methods and transformation of SAR studies into quantitative SAR (QSAR) by Hansch and others (1960s), and (e) application of artificial intelligence to drug research by Chu (1974) [8].

Merrifield's Nobel winning (1984) discovery of solid phase synthesis (1960s) led to the peptide synthesis on pin-shaped solid supports and discovery of *combinatorial chemistry* in 1984 by Geysen of Glaxo Wellcome Inc [9, 10]. The technique received widespread application by industries since the 1990s; the deconvolution of libraries (see latter in vocabulary section) and their intrinsic problems were addressed by Hougten and Frier (1993 and 1995) [11, 12]. In the mid-nineties Schultz *et al.* pioneered the discovery of luminescent materials by applying the combinatorial technique [13].

DEFINITION OF MEDICINAL CHEMISTRY

Medicinal chemistry is an interdependent mature science that combines applied (medicine) and basic (chemistry) sciences. It focuses the discovery, development, identification and interpretation of the mode of action of therapeutic agents at the molecular level. The clear understanding of chemical and pharmacological principles of drug molecules is dependent on detail SAR study, which is main emphasis of medicinal chemistry. Medicinal chemistry plays a major role in drug research and development by taking advantage of newer techniques and increased knowledge of different branches of related sciences including all branches of chemistry and biology [14].

MEDICINAL CHEMISTRY IN PHARMACY EDUCATION

The discipline of medicinal chemistry is devoted with the invention, discovery, design, identification and synthesis of biologically active compounds. Drug design and development and ADMET (absorption, distribution, metabolism, excretion, and toxicity) assessments are most relevant to pharmacy education and training. The interpretation of mode of action at the molecular level and construction of SAR of drugs and other therapeutic agents are important knowledge base for future pharmacists. Pharmacists are required to be competent in therapeutic decision making, which is dependent on their ability to conduct detail ADMET assessments of therapeutic drug classes. The pharmaceutical care and counseling to patient and other healthcare professionals are dependent on clinical pharmacists' knowledge and training in therapeutic use of medications, evaluation of scientific literature, proper therapeutic evaluations of medications and development of evidence based patient-specific pharmacotherapy plans. They serve as primary resources for drug related facts and provide sound advice regarding the safe and cost-effective use of medications. Thus to become a competent pharmacist, the knowledge of the physicochemical principles, SAR, mechanism of action, pharmacology, toxicology and ADME of drugs are indispensable components of a professional Pharm.D. curriculum.

An Abridged Medicinal Chemistry Vocabulary [15]

- The **affinity** of a drug is defined as its ability to bind to receptor, enzyme, or other biological target. Mathematically, it is the reciprocal of the equilibrium dissociation constant of the drug-receptor complex (K_A), which is quantified by dividing the rate constant for offset (k_{-1}) by the rate constant for onset (k_1). **Efficacy** is the property by which drugs are able to elicit a response. Agonists may vary in the relative intensity of their response regardless of their affinity and number of the receptor sites they occupy. The maximal stimulatory response produced by a compound compared to that of a standard, or physiological

agonist is defined as **intrinsic activity (α)**. When the intrinsic activity of a compound is 1, it is categorized as full agonists, and when the value is zero, it is categorized as antagonists. Compounds with fractional values between 0-1 are known as partial agonists. A drug or an endogenous substance that interacts with a receptor to fully activate it and initiate a response is known as an **agonist**. An **antagonist** on the other hand opposes or blocks the physiological effects of the agonist by binding with the same receptor because of its lack of intrinsic activity. A **partial agonist** possessing fractional intrinsic activity for a particular receptor in a tissue is unable to produce maximal activation of that receptor regardless of dose. **Potency** of a drug is defined as the dose required to produce a specific intensity of response compared to a reference standard. A **receptor** is a polymer (macromolecule) present inside a cell or on the cell surface to specifically recognize and bind a drug molecule or any other compound acting as a molecular messenger to elicit a response. Biochemically it is a protein on the cell membrane or within the cytoplasm or the cell nucleus, and it binds to a physiological substrate such as a neurotransmitter, hormone, other substance, or a drug molecule to initiate the cellular response. Four kinds of regulatory proteins that serve as primary drug targets are: (i) enzymes, (ii) carrier molecules, (iii) ion channels, and (iv) receptors ('true receptors'). There are still other structural proteins that may produce important pharmacological effects, *e.g.*, tubulin is specific to colchicine, and drugs like taxoids bind to this colchicine binding site of tubulin to elicit the antineoplastic effects. When a messenger molecule (agonist) stimulates a specific receptor it may sometime increase or decrease another molecule or a metabolite or ion called a **secondary messenger**.

- **Allosteric enzymes** contain allosteric binding sites, which are separate from the substrate binding sites to which small molecules (other than substrate) may bind to enhance or reduce the effect of the enzymes by changing the conformation of the active site. **Allosteric binding sites** of enzymes and receptors may regulate (activate, or inhibit) the binding of the normal ligand by inducing conformational changes of the enzyme, or receptor – a process known as **allosteric regulation**.
- When a drug is structurally related to another drug with similar or different chemical and biological properties, it is known as an **analog**. A member of a series of compounds differing only in a repeating unit, such as a methylene group, or a peptide residue is termed **homolog**. A member of a series of compounds synthesized by similar chemical reactions and procedures are called the **congeners** (literally *con-* meaning with; *generated* meaning synthesized). Thus the term congener is often a synonym for homologue, but it is also frequently used interchangeably with the term analog in the literature.
- **Combinatorial chemistry**, or **combinatorial synthesis** is used to synthesize a

combinatorial library (a large number of compounds) by combining sets of building blocks. High throughput screening (HTS) of the combinatorial library helps to discover lead compounds from a set of combinatorial libraries, which is a complex process. However, for ease of outcomes the deconvolution method is used. **Deconvolution of libraries** is a process of backtracking, reanalyzing, and resynthesizing the subset of structures of the combinatorial library showing promising activity in the preliminary HTS screening with the goal of tracking down the active principle(s).

- **Drug targeting** is a drug delivery strategy to a particular tissue of the body, which is often the desired site of action of the drug, to achieve **drug selectivity** and safety. Drug targeting can be performed by altering the drug structure to show increased selectivity for the target receptor. Not only it will produce the desired pharmacological response, but will also reduce adverse effects. Another strategy often used in drug targeting is **site-specific delivery** of drugs to its target tissue, using prodrugs or antibody recognition systems to reduce their systemic side effects and/or increase potency.

- **Hansch analysis** is a well-known technique that quantitates relationship of the physicochemical parameters, *e.g.*, hydrophobic, electronic, steric and other characteristics, of a set of compounds with their pharmacodynamics and pharmacokinetic properties by using multiple regression analyses.

- **Isosteres** are molecules or ions containing the same number of atoms and valence electrons, *e.g.*, O^{2-}, F^-, and Ne. **Bioisosteres** (or **Non-classical isosteres**) are compounds resulting from the exchange of atoms or of a group of atoms. Isosteric and bioisosteric replacements are often performed based on physicochemical or topological characteristics of the parent compounds to obtain a new drug with similar pharmacological activities.

- In the field of drug discovery, new compounds (the leads) with interesting biological activities and potential to become new therapeutic agents are routinely identified by a process called **lead discovery**. The strategy to identify such lead compounds is known as **lead generation**. The leads are further modified by a process called **lead optimization** to fulfill all stereological, electronic, physicochemical, pharmacokinetic and toxicological requirements to translate into clinically useful drugs.

- A **pharmacophore** is defined by the IUPAC as the "ensemble of steric and electronic features necessary to ensure the optimal supramolecular interactions with a specific biological target structure and to trigger (or to block) its biological response". The concept of pharmacophore stems from the capacity of an imaginary group of compounds to interact with their receptor or an enzyme, at the molecular level. It does not necessarily represent a real molecule or a real association of functional groups, but is generally shared by a set of active molecules. The concept of pharmacophore is often mistakenly used to denote

simple structural skeletons such as sulfonamides, flavones, phenothiazines, prostaglandins or steroids. Most useful drugs bind through the use of multiple weak bonds with the aid of H-bonding, hydrophobic and electrostatic interaction sites present in the atoms, rings, and centers, which are commonly defined as **pharmacophoric descriptors**.

- An inactive derivative of a drug that exerts pharmacological effects only after bioactivation is known as **prodrug**. When a prodrug is deliberately modified by the aid of a transient carrier group, which is often an easily hydrolyzable ester group, is known as **carrier-linked prodrug**. The purposes of such modification is to produce improved physicochemical or pharmacokinetic properties. There are also another group of prodrugs known as **bioprecursor prodrugs** that possess no such carrier and undergo one or more metabolic transformations in the body through normal metabolic pathway to their pharmacologically active forms. A **soft drug** is a compound that is metabolized quickly after exerting its therapeutic action in a predictable manner *in vivo* to produce inactive metabolites. An **antedrug** is an active derivative of a drug that is intended for local use which, upon entry into the systemic circulation, quickly undergoes metabolic inactivation and elimination. A **hard drug** is a metabolically stable, highly lipophilic compound that accumulates in adipose tissues and organelles or highly polar compound that is excreted easily through urine due to high water solubility. Pharmacologically, a powerful drug of abuse such as cocaine or heroin is commonly termed as a "**hard drug**".

Medicinal Chemistry in Drug Discovery and Development

Lomberdino and Lowe extensively reviewed the scope of medicinal chemistry in highly sophisticated process of drug discovery from a historical perspective [16]. Technological advances over the past few decades including computational chemistry and combinatorial chemistry have dramatically expanded the scope of medicinal chemistry in drug discovery. It is a long, complex, and costly process taking 12-24 years to launch a drug from bench to the market and costing up to US$ 1.4 billion for a single drug. It has been estimated that about 1 drug per 10,000 active principles in the bench come to the market and ~1 out of 15-25 clinical candidates survive the safety and efficacy standards to be marketable. Medicinal chemists play a crucial role in the early phases of drug discovery with the goal of maximizing efficacy and minimizing side effects. Their knowledge in modern organic chemistry and medicinal chemistry, biology of disease, *in vitro* and *in vivo* pharmacological screening and pharmacokinetic characteristics are the driving forces in a drug discovery project. The medicinal chemist is also well aware of ADMET issues related to medicines currently in the market for a target disease, regulatory affairs for similar drugs, current competitors in market, drugs in the pipeline, other related scientific matters in literatures, and technological

advances. The study of drug design and discovery is an excellent source of knowledge base for pharmacists about drugs and diseases.

Medicinal Chemistry in Pharmacy Education

The sound knowledge of functional group chemistry of drug molecules, along with ADMET parameters, is fundamental to understanding routes of drug administration, selection of appropriate therapeutic agent and/or formulation, and the dosages [17]. Functional groups are critical to receptor binding, and thus influence the mode of drug action, determine drug potency and consequently their dosages. In this context, the ADMET intellectual domain of medicinal chemistry, especially the metabolism of drugs, is of value. Since metabolic reactions are dependent on the drugs' electronic and steric characteristics of functional groups, one can effectively predict the potential drug metabolic outcomes from the knowledge of functional group chemistry and biochemistry.

Structurally based therapeutic evaluation (SBTE) is an innovative concept utilized by medicinal chemistry courses within pharmacy curriculum, developed by Alsharif *et al.* SBTE uses the knowledge of drug structures in making therapeutic decisions and emphasizes the relevance of medicinal chemistry to pharmaceutical care. All seven criteria of therapeutic decision making *i.e.* drug history/drug response, patient compliance, current medical history, past medical history, side effects, biopharmaceutics and pharmacodynamics are addressed in this SBTE approach. Students apply this newly designed SBTE approach to solve therapeutic problems for each class of drugs [18]. It has been described that the SBTE approach is valuable in guiding different functions of pharmaceutical care that include participating in drug selection decision process, patient counselling, monitoring patients to prevent drug interactions, selecting appropriate dosage forms and maximizing patient compliance by appropriate case stories. SBTE has also been shown to be of importance in developing professional practice skills like problem solving and decision making, learning from problem solving experiences, communicating, teaching, educating and collaborating [19]. Most importantly, using cardiovascular drugs as an example, SBTE has been shown to be a valuable tool for curriculum integration and interdisciplinary teaching [20 - 22].

The application of problem based learning (PBL) in medicinal chemistry teaching by some educators has shown to be valuable in pharmacy education. Medicinal chemistry based case studies were developed by integrating medicinal chemistry and pharmacology courses to solve clinical problems through group discussions [23]. Consequently, the overall outcome of clinical problem solving skills and the confidence of the students markedly improved, reiterating the significance of

medicinal chemistry as a critical component of this pharmaceutical care directed learning [24, 25]. Roche and Zito developed computerized case studies emphasizing medicinal chemistry principles in the practice of pharmacy and evaluated the seven performance criteria with four of them showing positive results, specifically in: identifying relevant therapeutic problems, conducting thorough and mechanistic SAR analyses of the drug product choices provided, evaluating SAR findings in terms of patient needs and desired therapeutic outcomes, and solving patient related therapeutic problems [26, 27].

The design and discovery of drugs is the primary responsibility of a medicinal chemist, which is the source of pride to the pharmacist as the entrepreneur and innovator of the most important armor of health care, the medicine, and thus the leadership position in the healthcare sector. The subject areas that are fundamental to drug discovery also serve as the sources for a complete set of knowledge base of the diseases and their safe and economic treatments. By incorporating these into the pharmacy curriculum, pharmacists become invaluable to the healthcare community.

The uniqueness of the pharmacy profession primarily lies in the comprehensive expertise of medicines and other pharmaceutical products when compared to other healthcare professionals including doctors and nurses. Since medicines are primarily chemical entities, early histories of both pharmacy and medicinal chemistry overlap and are inherently bonded to each other. From the beginning of the academic pharmacy program in the United States, medicinal chemistry has been the indispensable component of its curriculum. Because of pharmacists' unique knowledge of a medicine's design, pharmacological action, manufacture, storage, use, supply and handling, they are in a suitable position in the health care sector. The legislative support (*e.g.* OBRA 1990) has increased the legal role of pharmacists in patient care; the product of which is today's "Pharmaceutical Care". Professionally, pharmacist cannot afford to ignore his or her identity as the medication safety expert if they want to successfully perform and hold on to this newly assigned additional responsibility on them. Thus, medicinal chemistry is an indispensable component for pharmacy profession. One cannot consider himself a pharmacist without a sound knowledge of all the components of medicinal chemistry. By embracing the discipline of medicinal chemistry, the pharmacy profession can reap manifold advantages.

SCOPE OF THIS BOOK SERIES

A drug is defined as any substance presented for treating, curing or preventing disease in human beings or in animals and can also be used for making a medical diagnosis or for restoring, correcting, or modifying physiological functions (*e.g.*,

the contraceptive pill). Chemically, a drug may be extremely complex or very simple. The study of medicinal chemistry and thus the scope of this book involves all about the drugs, simple or complex, that turns the mystery or fantasy regarding drugs' behavior in the body into rationality. As medication experts, pharmacists routinely provide medication therapy evaluations and recommendations and counseling to patients and health care professionals. Clinical pharmacists are the primary resource for evidence-based information and advice regarding the safe, appropriate, and cost-effective use of medications. Thus, to become a competent pharmacist, the knowledge of chemical basis of drug action, its stability, pharmacology and toxicology are indispensable. The study of medicinal agents or drugs that are in clinical use, their metabolism, physicochemical principles, SAR, mechanism of action and toxicity are the scopes of professional pharmacy degree, which guide the breadth of this book.

This book is divided into four volumes:

Volume 1: Fundamentals of Medicinal Chemistry and Drug Metabolism.

Volume 2: Medicinal Chemistry of Drugs Affecting Autonomic and Central Nervous System.

Volume 3: Medicinal Chemistry of Drugs Acting on Cardiovascular and Endocrine Systems.

Volume 4: The Analgesic (Pain Management), Anti-infective, Anticancer and Other Agents and Recent Advances.

The first volume comprising of 8 chapters, focuses on basic background information to build a firm knowledge base of medicinal chemistry. It is a succinct and conceptual initial approach that introduces important fundamental chemical concepts required for a clear understanding of various facets of pharmacotherapeutic agents. The following volumes provide an in depth discussion of topics, including but not limited to: pharmacological and chemical basis of drug action, ADMET outcomes, drug-interactions, and adverse effects, which is expanded into different concepts and practical information on specific drug classes. A thorough discussion of key physicochemical parameters of therapeutic agents and how they affect the biochemical, pharmacological, pharmacokinetic processes and clinical uses of these agents are the primary focus of these volumes. The medicinal chemistry concepts of each drug class are illustrated by appropriate drug structures and relevant case studies. Drugs widely prescribed, both generic and brand names (Top 200), and those widely used in a hospital setting in the last four to five years have been selected as examples to reinforce the concepts. The last chapter of volume 4 will also cover topics related

to advanced areas of medicinal chemistry such as drug discovery and development techniques. Overall, the goal of this book is to provide the Pharm. D. students with a comprehensive, student-friendly educational tool.

CONSENT FOR PUBLICATION

Not applicable.

CONFLICT OF INTEREST

This chapter is prepared based on an article published by the authors in American Journal of Pharmaceutical Education (Khan MOF, Deimling MJ, Philip A. Medicinal Chemistry and the Pharmacy Curriculum. Am J Pharm Educ 2011; 75(8): Article 161) with permission.

ACKNOWLEDGEMENT

None declared.

REFERENCES

[1]   Higby GJ. Evolution of pharmacy.Remington: The Science and Practice of Pharmacy. 21st ed. Philadelphia, PA: Lippincott Williams & Wilkins 2006; pp. 7-19.

[2]   Erhardt PW, Proudfoot JR. Drug discovery: historical perspective, current status, and outlook. Comprehensive Medicinal Chemistry II 2007; 1: 29-96.
        [http://dx.doi.org/10.1016/B0-08-045044-X/00002-X]

[3]   Ceresia GB, Brusch CA. An introduction to the history of medicinal chemistry. Am J Pharm Sci Support Public Health 1955; 127(11): 384-95.
        [PMID: 13283040]

[4]   Brown CE. Some relations of early chemistry in America to medicine. J Chem Educ 1926; 3(3): 267-79.
        [http://dx.doi.org/10.1021/ed003p267]

[5]   Burger A. History and economics of medicinal chemistry.Medicinal Chemistry. 3rd ed. John Wiley & Sons 1970; Vol. 1: pp. 4-19.

[6]   Witebsky E. Ehrlich's side-chain theory in the light of present immunology. Ann N Y Acad Sci 1954; 59(2): 168-81.
        [http://dx.doi.org/10.1111/j.1749-6632.1954.tb45929.x] [PMID: 13229205]

[7]   Maehle AH, Prüll CR, Halliwell RF. The emergence of the drug receptor theory. Nat Rev Drug Discov 2002; 1(8): 637-41.
        [http://dx.doi.org/10.1038/nrd875] [PMID: 12402503]

[8]   Korolkovas A. Essentials of Medicinal Chemistry. 2nd ed., New York, NY: John Wiley & Sons, Inc. 1988.

[9]   Merrifield RB. Solid phase peptide synthesis. I. The synthesis of a tetrapeptide. J Am Chem Soc 1963; 85: 2149-54.
        [http://dx.doi.org/10.1021/ja00897a025]

[10]   Geysen HM, Meloen RH, Barteling SJ. Use of peptide synthesis to probe viral antigens for epitopes to a resolution of a single amino acid. Proc Natl Acad Sci USA 1984; 81(13): 3998-4002.

[http://dx.doi.org/10.1073/pnas.81.13.3998] [PMID: 6204335]

[11]   Dooley CT, Houghten RA. The use of positional scanning synthetic peptide combinatorial libraries for the rapid determination of opioid receptor ligands. Life Sci 1993; 52(18): 1509-17.
[http://dx.doi.org/10.1016/0024-3205(93)90113-H] [PMID: 8387136]

[12]   Freier SM, Konings DAM, Wyatt JR, Ecker DJ. Deconvolution of combinatorial libraries for drug discovery: a model system. J Med Chem 1995; 38(2): 344-52.
[http://dx.doi.org/10.1021/jm00002a016] [PMID: 7830277]

[13]   Xiang X-D, Sun X, Briceño G, *et al.* A combinatorial approach to materials discovery. Science 1995; 268(5218): 1738-40.
[http://dx.doi.org/10.1126/science.268.5218.1738] [PMID: 17834993]

[14]   Timmerman H. Reflection of medicinal chemistry since 1950s. Comprehensive Medicinal Chemistry II 2007; 8: 7-15.
[http://dx.doi.org/10.1016/B0-08-045044-X/00310-2]

[15]   Wermuth CG, Ganellin CR, Lindberg P, Mitscher LA. Glossary of terms used in medicinal chemistry (IUPAC Recommendation 1998). 2008 January 15. Available from: http://www.chem.gmul.ac.uk/iupac/medchem/.

[16]   Lombardino JG, Lowe JA III. The role of the medicinal chemist in drug discovery--then and now. Nat Rev Drug Discov 2004; 3(10): 853-62.
[http://dx.doi.org/10.1038/nrd1523] [PMID: 15459676]

[17]   Harrold MW. Importance of functional group chemistry in the drug selection process: a case study. Am J Pharm Educ 1998; 62: 213-8.

[18]   Alsharif NZ, Theesen KA, Roche VF. Structurally based therapeutic evaluation: a therapeutic and practical approach to teaching medicinal chemistry. Am J Pharm Educ 1997; 61: 55-60.

[19]   Alsharif NZ, Destache CJ, Roche VF. Teaching medicinal chemistry to meet outcome objectives for pharmacy education. Am J Pharm Educ 1999; 63: 34-40.

[20]   Alsharif NZ, Shara MA, Roche VF. The structurally-based therapeutic evaluation (SBTE) concept: an opportunity for curriculum integration and interdisciplinary teaching. Am J Pharm Educ 2001; 65: 314-23.

[21]   Alsharif NZ, Galt KA, Mehanna A, Chapman R, Ogunbadeniyi AM. Instructional model to teach clinically relevant medicinal chemistry. Am J Pharm Educ 2006; 70(4): 91.
[http://dx.doi.org/10.5688/aj700491] [PMID: 17136210]

[22]   Alsharif NZ, Galt KA. Evaluation of an instructional model to teach clinically relevant medicinal chemistry in a campus and a distance pathway. Am J Pharm Educ 2008; 72(2): 31.
[http://dx.doi.org/10.5688/aj720231] [PMID: 18483599]

[23]   Currie BL, Roche VF, Zito SW. Medicinal Chemistry Case Story Workbook. Baltimore, MD: Williams & Wilkins 1996.

[24]   Herrier RN, Jackson TR, Consroe PF. The use of student-centered, problem based, clinical case discussions to enhance learning in pharmacology and medicinal chemistry. Am J Pharm Educ 1997; 61: 441-6.

[25]   Dimmock JR. Problem solving learning: applications in medicinal chemistry. Am J Pharm Educ 2000; 64: 44-9.

[26]   Roche VF, Zito SW. Computerized medicinal chemistry case studies. Am J Pharm Educ 1997; 61: 447-52.

[27]   Roche VF, Aitken MJ, Zito SW. Evaluation of computerized medicinal chemistry case study modules as tools to enhance student learning and clinical problem-solving skills. Am J Pharm Educ 1999; 61: 289-95.

Medicinal Chemistry for Pharmacy Students, 2018, Vol. 1, 13-75

CHAPTER 2

Review of Bioorganic Chemistry

M. O. Faruk Khan[1,*] and **Ashim Malhotra**[2]

[1] *Department of Pharmaceutical and Biomedical Sciences, College of Pharmacy, California Northstate University, Elk Grove, CA, USA*

[2] *Department of Pharmaceutical and Biomedical Sciences, College of Pharmacy, California Northstate University, Elk Grove, California, CA, USA*

Abstract: This chapter is a brief review of the important organic functional groups and biomolecules. After study of this chapter, students will be able to:

• Identify important organic functional groups in any drug structure
• Identify major biomolecules: proteins, carbohydrates, lipids, nucleic acids
• Apply chemical principles in all four classes of biomolecules
• Define monomer units of the biomolecules and their chemical properties
• Evaluate the structures of important heterocycles occurring in drugs and biomolecules
• Summarize the significance of of biomolecules

Keywords: Biomolecules, Carbohydrates, Eicosanoids, Fats, Functional groups, Nucleic acids, Proteins.

AN INTRODUCTION TO ORGANIC FUNCTIONAL GROUPS

The reactivity of organic compounds stems from an atom or a group of atoms referred to as a **functional group**. Based on the characteristic features of the functional groups, organic compounds are classified into a large number of groups. The most important ones are: alkanes, alkenes, haloalkanes, aromatic hydrocarbons, alcohols and phenols, ethers and thioethers, aldehydes and ketones, amines, carboxylic acids and their derivatives, *e.g.*, esters, amides, anhydrides, carbonates, carbamates, and ureas. Knowledge of functional groups is critical since they determine the physicochemical properties of organic compounds and drugs.

---

* **Corresponding author M. O. Faruk Khan**: Department of Pharmaceutical Sciences and Research, Marshall University School of Pharmacy, Huntington, WV, USA; Tel: 304-696-3094; Fax: 304-696-7309; E-mail: khanmo@marshall.edu

Alkanes, Alkenes and Cycloalkanes

Definition and Nomenclature of Alkanes, Alkenes and Cycloalkanes

Alkanes

The alkanes are hydrocarbons with the general molecular formula C_nH_{2n+2}. They are also referred to as **saturated hydrocarbons** (contain only single bonds), **aliphatic** or **alicyclic** (cyclic, but not aromatic) alkanes. This class of compounds possess tetrahedral atoms, which are sp^3 hybridized. The common names of this family have a suffix *ane* and a prefix highlighting the number of carbon atoms, *e.g.*, methane (1 carbon), ethane (2 carbons), propane (3 carbons), butane (4 carbons), pentane (5 carbons), hexane (6 carbons) and so on. However, as the compounds possess an increasing number of branched chains, resulting in multiple isomers, this system becomes less useful and the **IUPAC (International Union of Pure and Applied Chemistry) nomenclature** should be used.

Different compounds with the same molecular formula are referred to as **isomers**, which can be **structural** or **conformational isomers**. Two types of structural isomers exist, **continuous chain** and **branched-chain isomers**. For example, n-butane (a continuous chain structure) and isobutane (a branched chain structure) are structural isomers with the same molecular formula C_4H_{10} (4 carbons), but different structures. As the carbon number increases, the number of isomers also increases dramatically. Thus, 75 isomeric alkanes are possible for 10 carbons and 366,319 isomers are possible with 20 carbons, thereby making IUPAC nomenclature more appropriate for use.

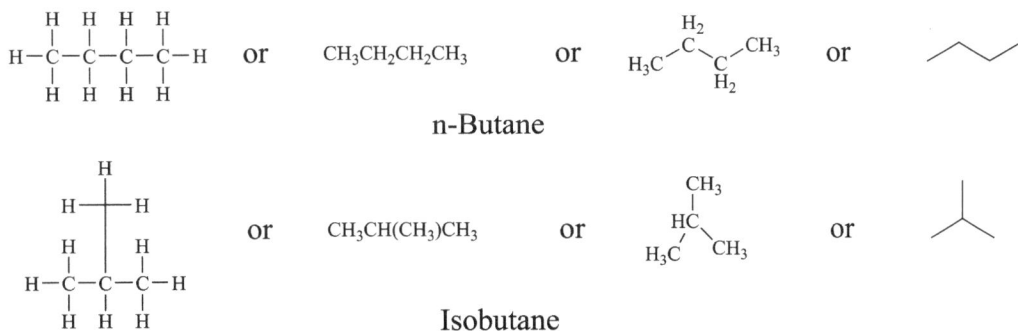

n-Butane

Isobutane

The general rule in the IUPAC system is to find the longest continuous chain alkane first, to assign the base name. The chain is then numbered in a way so that the substituents on the chain get the lowest possible number. For example, the IUPAC system of naming of one of the structural isomers of dodecane (12 carbons), 3,3-diethyl-2,5-dimethylhexane, is shown below.

$$CH_3$$
$$|$$
$$CH_3 \qquad CH_2 \quad CH_3$$
$$| \qquad\qquad | \quad\quad |$$
$$H_3C— CH — CH_2 —C— CH —CH_3$$
$$\quad 6 \quad\ 5 \qquad 4 \quad |\ 3 \quad 2 \qquad 1$$
$$CH_2$$
$$|$$
$$CH_3$$

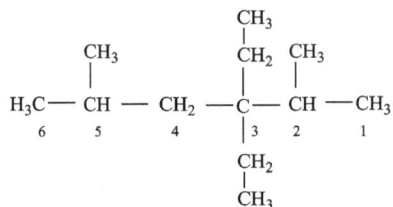

3,3-Diethyl-2,5-dimethylhexane

In the example shown above with 3,3-diethyl-2,5-dimethylhexane, the carbon atoms with 3 hydrogen atoms (C-1 and C-6) are called **primary carbons** and the hydrogens attached to it are referred to as **primary hydrogens**. The carbon atoms with 2 hydrogen atoms (C-4) are referred to as **secondary carbons** and the hydrogens attached to it are referred to as **secondary hydrogens**. The carbon atoms with 1 hydrogen atom (C-2 and C-5) are referred to as **tertiary carbons** and the hydrogens attached to it are referred to as **tertiary hydrogens**. The carbon atoms with no hydrogen atom (C-3) are referred to as **quaternary carbons**.

Alkenes

Alkenes, also known as olefins, are unsaturated hydrocarbons containing *carbon-carbon double* bonds (also known as *ethylenic double bond*) with a general molecular formula C_nH_{2n}. Ethylene (C_2H_4) is the simplest member of this series. The molecular formula for both 1-butylene and isobutylene is C_4H_8. The double bond consists of a σ (*sigma*) bond, which is similar to that of an alkane, and a π (*pi*) bond, which is relatively more reactive. The valence electrons of the double bonds are sp^2 hybridized. The sp^2 carbon is known as a vinylic carbon and the sp^3 carbon attached to the sp^2 carbon is known as an allylic carbon.

Table 1. Common names and IUPAC names of a few alkenes.

Structure	Common Name	IUPAC Name
$H_2C{=}CH_2$	Ethylene	Ethene
$\underset{H}{\overset{H}{>}}C{=}C\underset{H}{\overset{CH_2CH_3}{<}}$	α-Butylene	1-Butene
$\underset{H}{\overset{H}{>}}C{=}C\underset{CH_3}{\overset{CH_3}{<}}$	Isoutylene	2-Methyl-1-propene
$\underset{H}{\overset{H}{>}}C{=}C\underset{H}{<}$	Vinyl-	Substituted-ethene
$\underset{H}{\overset{H}{>}}C{=}C\underset{H}{\overset{CH_2{-}}{<}}$	Allyl-	3- Substituted-1-propene

The prefix of the names of alkenes stems from the radical names of the corresponding alkanes, ending in with the suffix *ene*. Thus, the ethylene is a combination of the 'ethyl' radical of ethane and *ene* and so forth (Table 1). A similar IUPAC rule is applied with slight modifications. The longest chain containing the double bond is the parent, and the numbering of the carbon atoms starts from the end, nearest to the double bond. If there are more than 3 carbon atoms in the chain, the position of the double bond is indicated by the lowest carbon number of the double bonded carbons, before the alkene name, *e.g.*, 1-butene. For lower molecular weight alkenes, the common names are often used (Table 1), but for higher molecular weight alkenes, the IUPAC names should be used, *e.g.*, 3,3-diethyl-2,5-dimethyl-2-hexene.

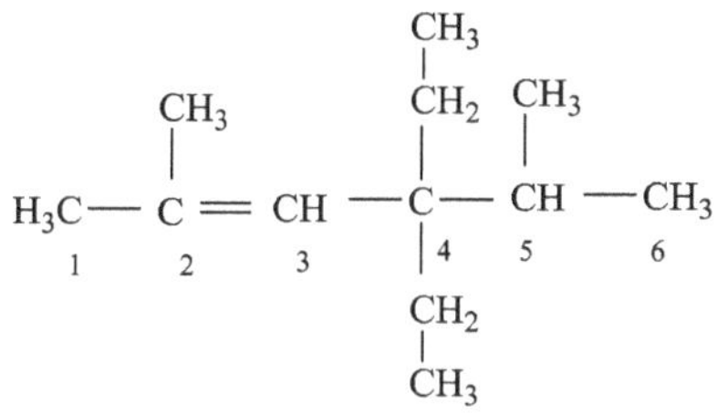

3,3-Diethyl-2,5-dimethyl-2-hexene

In addition to the structural isomers, alkenes also display geometric isomerism. The **geometric isomers** are configurational isomers (or stereoisomers) arising from the position of the groups with respect to the rigid double bond. They are also known as *cis-trans* isomers or *E-Z* isomers as shown below.

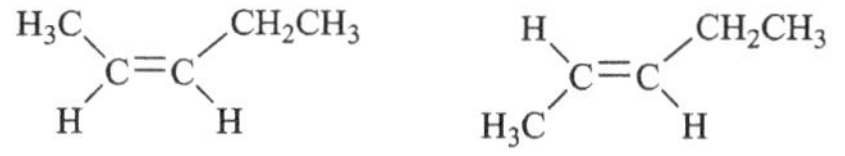

cis-2-Pentene   *trans*-2-Pentene

Z-2-Pentene   E-2-Pentene

Cycloalkanes

Cycloalkenes, also known as *alicyclic* compounds, are the structural isomers of alkenes with same molecular formula (C_nH_{2n}); however, instead of a double bond, they possess a ring structure. The same IUPAC rules as for alkanes in naming these compounds, with the addition of a prefix *cyclo-* to the names of the corresponding alkanes. Thus, the cycloalkane of propane is referred to as cyclopropane (3 carbons; C_3H_6), which is a structural isomer of propene. Other pharmaceutically important cyclic alkanes include cyclobutane (4 carbons; C_4H_8), cyclopentane (5 carbons; C_5H_{10}) and cyclohexane (6 carbons; C_6H_{12}) which are structural isomers of butene, pentene, and hexene, respectively. When the ring

contains a double bond, the suffix *ene* is added, instead of the suffix *ane* (*e.g.*, cyclohexene).

Cyclopropane   Cyclobutane   Cyclopentane   Cyclohexane   Cyclohexene

Physicochemical Properties of Alkanes, Alkenes and Cyclic Alkanes

Alkanes

The weak van der Waals forces are the only intermolecular attractions possible with alkanes. van der Waals forces increase with an increase in the surface area; the larger the molecule, the stronger the intermolecular attraction. The alkanes with 1-4 carbon atoms (methane through n-butane) are gaseous, whereas ones with 5-17 carbon atoms (n-pentane through n-heptadecane) are liquids, and the higher homologs are solids at room temperature. Since these compounds do not interact *via* hydrogen bonds, or through ionic interactions, they are nonpolar in nature. Thus, alkanes in general are hydrophobic or lipid-soluble, and when attached to a drug molecule, exhibit high lipid-water partition coefficients. Drugs intended for action in the brain, for example narcotic analgesics, general anesthetics, and antipsychotics contain larger alkyl chains for higher lipophilicity (ability to dissolve in fats, oils, lipids, and non-polar solvents) and better brain delivery.

Alkanes are chemically inert at room temperature and therefore are referred to as paraffin hydrocarbons (Latin *param affinis*, meaning little affinity). Such non-reactivity, and thus chemical stability makes them metabolically stable. Drugs containing alkane functional groups are excreted mostly in their unchanged form. However, exceptions exist with some drugs containing long chain alkanes, metabolized (hydroxylated) by mixed function oxidase at ω (terminal) and ω-1 carbon atoms.

Alkenes

The physical properties of alkenes are similar to those of alkanes. The alkenes with 2-4 carbon atoms are gaseous, 5-17 carbon atoms are liquids, and the higher homologs are solids at room temperature. Like alkanes, alkenes are also hydrophobic and lipid-soluble. When alkenes are attached to a drug molecule, they exhibit high lipid-water partition coefficients.

Presence of the double bond makes alkenes relatively more reactive than alkanes. The double bond is electron rich (nucleophile) with a loosely held π bond, which

attacks an electrophile. Electrophilic addition to the double bond is more common in olefins. The common reactions include hydrogenation (addition of hydrogen), also known as reduction, hydration (addition of water), hydroxylation (formation of diol), and peroxide and epoxide formations (oxidative reactions). Drugs containing double bonds are relatively more prone to metabolism than alkanes. The mixed function oxidases can cause hydration, epoxidation and peroxidation. These may also undergo reductive metabolism, such as the reduction of 3-ketone group and 4-alkene group of glucocorticoids to saturated alcohol metabolite.

Cycloalkanes

Like alkanes, the cycloalkanes are also nonpolar and thus lipid-soluble compounds, and often occur in drug molecules. Due to ring strains, the smaller cycloalkanes, cyclopropane and cyclobutane, are relatively more reactive like alkenes, but the larger cycloalkanes, *e.g.*, cyclopentane and cyclohexane, are relatively stable like the alkanes. In general, the larger the ring size, the lesser the ring strain and lower the reactivity. The cycloalkanes behave like alkenes in terms of chemical reactivity.

Alkyl Halides

Definition and Nomenclature of Alkyl Halides

When a halogen atom (-F, -Cl, -Br, -I; commonly designated by -X) is attached to a tetrahedral carbon (sp3) atom, it is known an alkyl halide. Some examples are: methyl chloride, or chloromethane (CH_3Cl), ethyl bromide, or bromoethane (CH_3CH_2Br), methylene chloride, or dichloromethane (CH_2Cl_2), chloroform, or trichloromethane ($CHCl_3$) and tetrachloromethane or carbon tetrachloride (CCl_4). All the above alkyl halides are primary (1°) alkyl halides, because the carbon containing halogen atoms are attached to only one alkyl group. The alkyl halides with 2 and 3 alkyl substitutions on the halogenated carbon atom are known as secondary (2°) and tertiary (3°) alkyl halides, respectively. The examples of 2° and 3° alkyl halides are shown below.

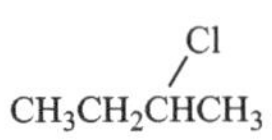

$$CH_3CH_2\overset{\displaystyle Cl}{\overset{\displaystyle /}{C}}HCH_3 \qquad CH_3\overset{\displaystyle Cl}{\overset{\displaystyle /}{\underset{\displaystyle \underset{CH_3}{\backslash}}{C}}}CH_3$$

2-Chlorobutane	*tert*-Butylchloride
Isobutyl chloride	2-Chloro-2-methylpropane
(2° Alkyl halide)	(3° Alkyl halide)

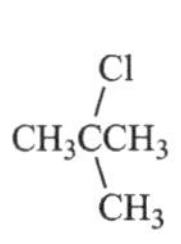

According to the IUPAC system of nomenclature, the halogen prefix (*fluoro*, *chloro*, *bromo*, *iodo*) is used to the corresponding alkane structure to name a

haloalkane. Thus, chloromethane, bromoethane, 2-chlorobutane, 2-chloro-2-methylpropane are the IUPAC names of the examples shown above. Compounds with two halogens on the same carbon atom are commonly known as methylene halides (*e.g.*, methylene chloride), three halides are called haloforms (*e.g.*, chloroform) and four halides are called carbon tetrahalides (*e.g.*, carbon tetrachloride).

Physicochemical Properties of Alkyl Halides

The alkyl halides have a permanent dipole due to the presence of halide, but are not hydrogen bonded. Like alkanes they also interact *via* van der Waals forces, which are slightly stronger than the corresponding alkanes. All the methyl halides, except methyl iodide, are gases and all other alkyl halides are liquids at room temperature. The boiling point increases with the increasing chain length of the alkyl group as well as the molecular weight of the halide group. It is thought that the halogenation increases the lipid solubility of the alkanes or other organic compounds.

The halogens are electron withdrawing in nature and act as "leaving groups". Thus, the alkyl halides may undergo *nucleophilic substitution* and *elimination* reactions. However, *in vivo* halogenation tends to increase the metabolic stability of drugs and thus prolong biological half-life.

Aromatic Compounds

Definition and Nomenclature of Aromatic Compounds

As defined by Erick Hückel (1931), an aromatic compound is one that contains a cyclic structure with $(4n + 2)$ π electrons ($n = 1, 2, 3$ *etc.*) and forms a cyclic delocalized π cloud above and below the plane of the molecule. A few examples are shown below. Benzene is **monocyclic**, naphthalene, anthracene and phenanthrene are **polycyclic benzoids**, azulene is a **non-benzenoid**, and annulene is a **macrocyclic** aromatic compound.

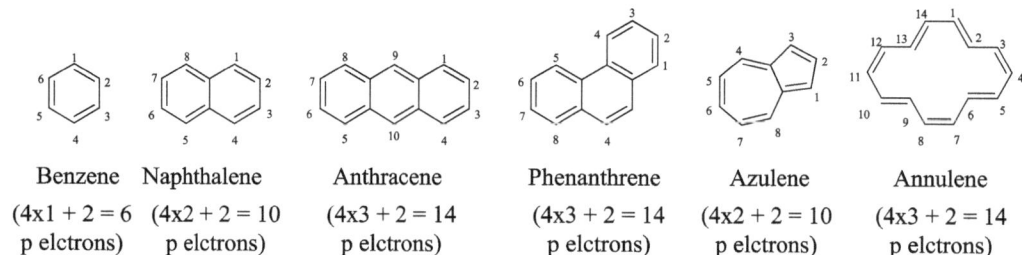

Benzene	Naphthalene	Anthracene	Phenanthrene	Azulene	Annulene
(4x1 + 2 = 6 p elctrons)	(4x2 + 2 = 10 p elctrons)	(4x3 + 2 = 14 p elctrons)	(4x3 + 2 = 14 p elctrons)	(4x2 + 2 = 10 p elctrons)	(4x3 + 2 = 14 p elctrons)

The monosubstituted benzenes are named by adding a prefix to benzene, *e.g.,* *chloro*benzene and *nitro*benzene. Sometimes trivial names are used, *e.g.,* phenol, toluene, and aniline.

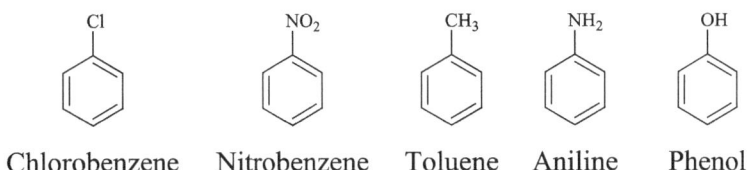

Chlorobenzene   Nitrobenzene   Toluene   Aniline   Phenol

Disubstituted benzenes are named by using *ortho, para,* and *meta* (*o-, p-* and *m-,* respectively) identifiers. If there are more than two substitutions, then the positions are specified to name these. See the examples below.

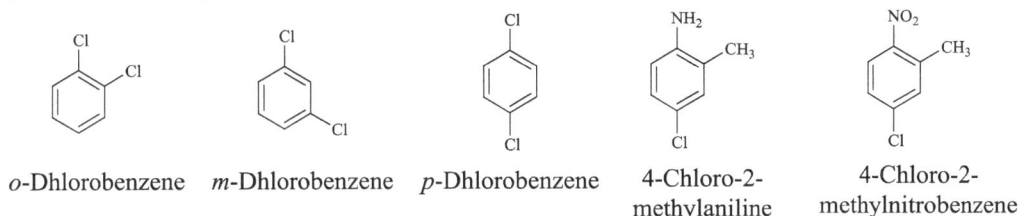

| *o*-Dhlorobenzene | *m*-Dhlorobenzene | *p*-Dhlorobenzene | 4-Chloro-2-methylaniline | 4-Chloro-2-methylnitrobenzene |

Physicochemical Properties of Aromatic Compounds

The aromatic ring is a flat structure with electron clouds above and below the plane, which are not readily available for chemical reaction. However, they contribute to better van der Waals interactions and are very important for binding to receptors in the biological systems. Most of the therapeutic agents contain aromatic ring systems, enhancing their lipophilicity. An unhindered aromatic ring can contribute about 4.2 Kcal/mol van der Waals force plus 4.2 Kcal/mol of hydrophobic interaction towards receptor binding.

Electron dense aromatic systems are less reactive than alkenes and are prone to attack by electrophiles to undergo a variety of electrophilic substitutions. Some of the examples include alkylation, acylation, halogenation, nitration, and sulfonation. A pharmaceutically important electrophilic reaction that occurs *in vivo* is *hydroxylation*, a common metabolic reaction of drugs containing the aromatic system. When the aromatic ring is substituted with electron donating substituents, *e.g.,* $-CH_3$, -OH and $-NH_2$, the reactivity including hydroxylation increases, but if it is substituted with an electron withdrawing group like the halides, $-NO_2$, $-SO_3H$, -COOH, the reactivity decreases. Thus, the electron withdrawing groups increase the metabolic stability and half-life of drugs containing aromatic ring, while the electron donors will enhance metabolism and decrease half-life.

Alcohols and Phenols

Definition and Nomenclature of Alcohols and Phenols

Alcohols

When a hydroxyl (–OH) group is attached to an aliphatic carbon atom, it is known as an alcohol (ROH). Depending on the type of carbon atom the –OH group attached to an alcohol can be *primary* (1°), *secondary* (2°), or *tertiary* (3°). According to the IUPAC system, an alcohol is named by replacing the *e* ending of the corresponding alkane name with *ol*, and the position of the hydroxyl group is also indicated by the corresponding number. If there is more than one hydroxyl group, then *diol* (2 hydroxyl), *triol* (3 hydroxyl), *etc.* are used. If an alcohol is unsaturated, the hydroxyl group gets the lower number, not the point of unsaturation. **Thiols** (RSH) are analogous to alcohols with –SH group in place of –OH group.

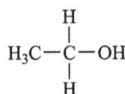 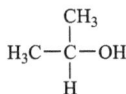 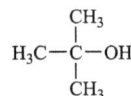

Methanol	Ethanol	Isopropanol	*tert*-Butanol
(Methyl alcohol)	(Ethyl alcohol)	(2-Propanol)	(2-Methyl-2-propanol)
1° Alcohols		2° Alcohol	3° Alcohol

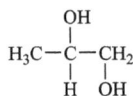 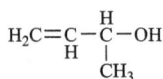

1,2-Propanediol	3-Buten-2-ol

Phenols

When a hydroxyl group is directly attached to an aromatic ring, the resulting compounds are referred to as phenols. Phenols differ markedly from alcohols in their properties such as hybridization, acidity, solubility in mild alkali and so forth. Phenol is the simplest member of the series, which is the basis for the nomenclature of phenols. The carbon bearing the OH is assigned number 1, and based on this, other substituents are numbered. Some phenols are known by trivial names, *e.g.*, cresol, hydroquinone, catechol and resorcinol. Sometimes the OH is prefixed by *hydroxy*, if other functional groups of higher precedence (*e.g.*, COOH, -CHO, -SO$_2$NH$_2$ according to *chemical abstract*) are also present.

Phenol   *o*-Bromophenol   *o*-Cresol
         (2-Bromophenol)

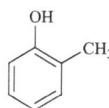

Hydroquinone   Catechol   Resorcinol   *m*-Hydroxybenzoic acid

Physicochemical Properties of Alcohols and Phenols

Alcohols

Alcohols can form hydrogen bonds with the aid of their hydroxyl groups and thus are water soluble. Their hydroxyl groups are polarized due to a high electronegativity difference between hydrogen and oxygen and are known as hydrophilic groups. The aliphatic chain is known as the hydrophobic group. The larger the chain length of the alcohol, the lower the water solubility and the more the more its lipid solubility. Because they can form intermolecular hydrogen bonds, alcohols are liquid and exhibit higher boiling points. The thiols are less polar than alcohols due to less electronegativity of sulfur than oxygen atom. Thus thiols have weaker hydrogen bonds and lower boiling points.

In a favorable environment, as in the presence of a strong base (*e.g.,* NaOH) alcohols can behave like acids. Overall, alcohols have a pKa of 16-19, which is essentially neutral. Thiols on the other hand are stronger acids (pKa = 10). The polar hydroxyl group can also react as a base or as a nucleophile in nucleophilic substitutions. Other major reaction of alcohols includes oxidation. The primary alcohols can be oxidized to aldehyde and carboxylic acids, the secondary alcohols to ketones while the tertiary alcohols are not oxidized. Alcohols also react with acids to form esters. Thiols, on the other hand, can be oxidized to disulfides and sulfonic acids.

Drugs containing hydroxyl functional groups undergo oxidative metabolism, catalyzed by a variety of oxidase enzymes, resulting in formation of carboxylic acids and ketones from primary and secondary alcohols, respectively. Other important metabolic reactions include conjugations with glucuronic acid (glucuronidation) and sulfuric acid (sulfonate conjugation). Thiol containing drugs are primarily metabolized by oxidation to disulfides as well as

glucuronidation.

Phenols

Phenols are liquids, or sometimes solids with low melting points. These compounds have high boiling points due to strong intermolecular bonding through hydrogen bonding and possibly some ionic bonding, as well as dipole-dipole and ion-dipole interactions. The hydroxyl group of phenols is more polar than that of alcohols and thus, phenols are more water soluble when compared with alcohols with an equivalent number of carbon chains. The ionic nature of this bond is derived from their inherent acidic character (pKa ~10). The acidity increases with the electron withdrawing substituents and decreases with the electron donating substituents. The electron withdrawing substituents as well as the lipophilic substituents will also decrease water solubility due to less ionizability and an increased hydrophobic chain length.

Phenols are somewhat reactive and can undergo electrophilic substitutions as well as other reactions including salt formation with a strong base like NaOH and KOH. It can also undergo ether and ester formation with alkyl halides and acyl halides, respectively. If exposed to air, it undergoes oxidation to *o*- and *p*-quinones.

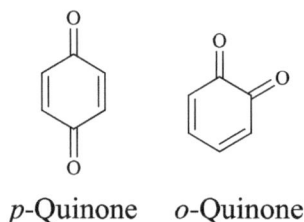

p-Quinone   *o*-Quinone

Phenolic drugs undergo aromatic hydroxylation as well as conjugation reactions which enhances water solubility and elimination through urine. However, some phenols are prone to methyl conjugation, which results in decreased water solubility and enhanced cell penetration.

Ethers and Thioethers

Definition and Nomenclature of Ethers

When an oxygen atom is attached to two "R" groups, aliphatic and/or aromatic, it is known as an ether. Thus, an ether has a general structure R-O-R (R = aliphatic or aromatic group). A few common ethers include, diethyl ether (commonly known as ether), tetrahydrofuran (cyclic ether) and anisole (alkylarylether). The thioethers are the sulfur analogs of ethers, *e.g.,* ethylthioethane.

Diethylether
(Ether)

Tetrahydrofuran
(THF)

Anisole

Ethylthioethane
(Ethylsulfide)

Ethypropylether (Common)
1-Ethoxypropane (IUPAC)

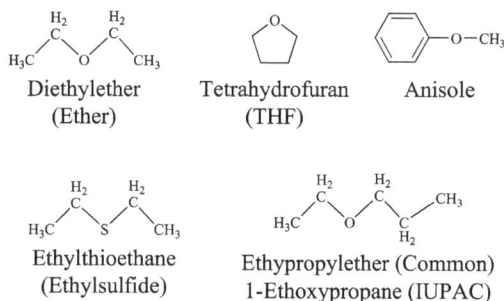

The commonly used nomenclature of ethers uses the radical names of the two alkyl groups suffixed by *ether*. However, as the ethers become larger and more complex, this system is not very useful, therefore, the IUPAC system is more appropriate. According to the IUPAC system, ether is considered as an *alkoxy* (RO-) derivative of hydrocarbon. Thus, the longest continuous alkane chain that gives the *suffix* is considered the parent. The chain is numbered in a way so that the alkoxy group that gives the *prefix* gets the lowest number.

Physicochemical Properties of Ethers

Ethers form weak hydrogen bonds with water, with van der Waals being the primary intermolecular attractions. In ether, the hydroxyl group of alcohols and phenols is masked (alkylated) making it poorly water soluble and a lower boiling liquid. The water solubility of ethers decreases dramatically as the alkyl chain length increases.

The masking of the hydroxyl function in ether also makes it almost chemically inert. The major metabolic reaction that occurs *in vivo* with ether containing drugs is their dealkylation to form an alcohol and a carbonyl compound. Thioethers undergo similar metabolic reactions forming a thiol and a carbonyl compound, in addition to S-oxidations to form sulfoxides and sulfones.

Aldehydes and Ketones

Definition and Nomenclature of Aldehydes and Ketones

Aldehydes and ketones are classified as carbonyl compounds (those containing C=O group). If the carbon atom of the C=O group is attached to at least one hydrogen, it is known as an **aldehyde**, while if it is attached to two carbon atoms (alkyl or aryl), then it is known as a **ketone**. Formaldehyde is the simplest aldehyde (1 carbon) where the carbonyl carbon is attached to two hydrogen atoms and acetone is the simplest ketone where the carbonyl carbon is attached to two methyl groups (3 carbons).

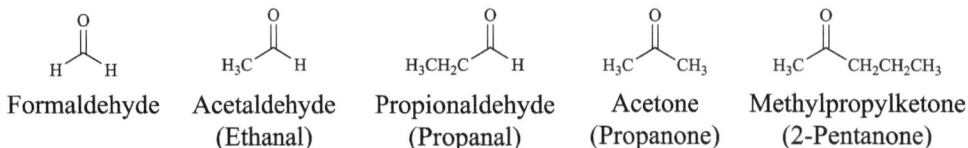

| Formaldehyde | Acetaldehyde (Ethanal) | Propionaldehyde (Propanal) | Acetone (Propanone) | Methylpropylketone (2-Pentanone) |

The common names of aldehydes have similarity with the corresponding carboxylic acids, differing only in suffix. For example, the one carbon aldehyde is form*aldehyde*, whereas the acid is form*ic acid*; with two carbons, these are ace*taldehyde vs* acet*ic acid* and with three carbons, propion*aldehyde vs* propion*ic acid*, and so forth. Indeed, aldehydes are the former oxidation state of carboxylic acids, and hence the names originated from the acid's names. However, in the IUPAC system, the last *e* of alkane is replaced with *al* to name the corresponding aldehyde, *e.g.,* ethanal (ethane), and propanal (propane).

The common name of ketones comes from the constituent radical names suffixed with *ketone* (*e.g.,* methyl-propyl ketone). In the IUPAC system, the longest continuous carbon chain containing a ketone is first identified to give the parent alkane name. Then, the ending *e* of alkane is replaced with *one* suffix, and also the chain is numbered in a way so that the ketone group gets the lowest number (*e.g.,* 2-Pentanone).

Physicochemical Properties of Aldehydes and Ketones

The C=O bond of the carbonyl compounds (aldehydes and ketones) is polar, forming a dipole of negative oxygen and positive carbon ($C^{\delta+} = O^{\delta-}$). The opposite ends of the C-O dipole are capable of forming weak electrostatic interactions (intermolecular) that result in an increased boiling point relative to alkanes but less than alcohols. However, the polar carbonyl group can form hydrogen bonds with water (hydrogen bond donor), increasing its water solubility. In general, aldehydes and ketones with less than 5 carbons are water soluble and as the carbon chain increases, the water solubility decreases.

Due to the polarity of the C=O bond as discussed above, the carbonyl carbon acts as an electrophile while the carbonyl oxygen acts as a nucleophile. Through this property, carbonyl compounds undergo variety of nucleophilic addition reactions of synthetic as well as pharmaceutical importance. The important biological and pharmaceutical addition reaction include the hemiacetal and acetal formations from aldehyde and hemiketal and ketal formation from ketone, with the nucleophilic addition reactions with alcohol (Fig. **1**). Such reactions are responsible for the instability of erythromycin and related aminoglycoside antibiotics as well as the ring structure of sugars and aldosterone in the biological systems.

Aldehyde (X = H)   Hemeacetal (X = H) or   Acetal (X = H)
or Ketone (X = R)   Hemeketal (X = R)   or Ketal (X = R)

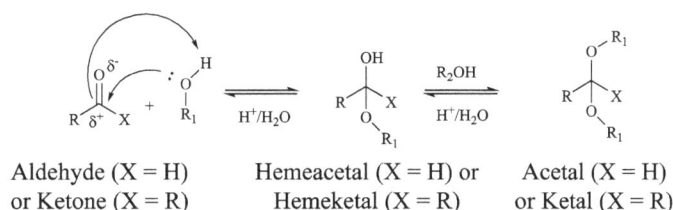

Fig. (1). Hemiacetal and acetal formations from aldehyde and hemiketal and ketal formation from ketone.

Other important reactions that are metabolically significant include oxidation of aldehydes (not ketones) to carboxylic acids by the action of several oxidases, and stereoselective reduction of ketones to respective secondary alcohols.

Amines

Definition and Nomenclature of Amines

A large number of clinically used drugs are amines that are characterized by the presence of an amino (-NH_2) group and with a general formula of RNH_2 (a 1° amine), R_2NH (a 2° amine) or R_3N (a 3° amine). When the lone-pair of electrons of amine is also shared by a fourth alkyl group to give a permanent positive charge on the nitrogen atom it is called quaternary alkylammonium salt or cation (R_4N^+, *e.g.*, choline).

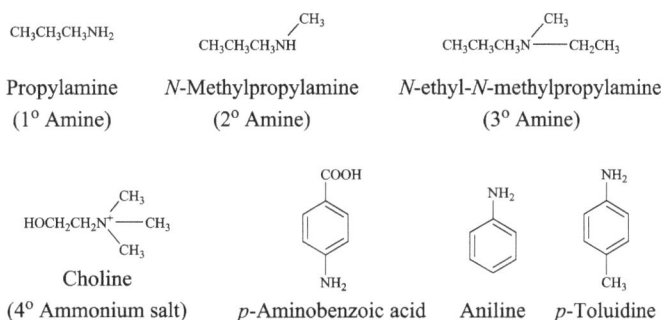

Propylamine   *N*-Methylpropylamine   *N*-ethyl-*N*-methylpropylamine
(1° Amine)   (2° Amine)   (3° Amine)

Choline
(4° Ammonium salt)   *p*-Aminobenzoic acid   Aniline   *p*-Toluidine

The primary amines are named by adding a suffix *amine* to the name of corresponding alkyl group substituted on the amine group (*e.g.*, propylamine). The secondary and tertiary amines are named based on the corresponding primary amine identified by the largest alkyl chain and adding an *N-alkyl* prefix to it (*e.g.*, *N*-methyl-propylamine, *N*-ethyl-*N*-methyl-propylamine). However, when other

functional groups of higher priority are also present in addition to the -NH$_2$ group, then it is considered a substituent and named by adding a prefix *amino* (*e.g., p*-amino benzoic acid). Sometimes amines are known by their trivial names (*e.g.,* aniline, toluidine).

Physicochemical Properties of Amines

Amines are polar compounds, due to N-H and C-N bonds, and form inter- and intra-molecular hydrogen bonding but weaker than alcohols, as nitrogen is less electronegative than oxygen. Thus amines have higher boiling points but still less than alcohols. However, the water solubility is somewhat similar to alcohols with the order: 1° > 2° > 3°. Also, the longer the alkyl chain, the lower the water solubility. An amine group can solubilize up to 5 or 6 carbon alkyl chains. Because nitrogen atoms have a lone-pair of electrons, amines are basic (nucleophilic) in character, a property highly important for drug molecules.

The nucleophilic amine group can attack the carbonyl carbon of acyl chlorides, anhydrides, aldehydes and ketones. Thus, with acyl chlorides and anhydrides it forms amides, while with carbonyl compounds (aldehydes and ketones) it forms imines (also known as *Schiff bases*) with the loss of a water molecule. The amides formed after the reaction with sulfonyl chlorides are known as sulfonamides. When catalyzed by a base, quaternary ammonium salts can produce an alkene and a tertiary amine by a process known as *Hofmann degradation*. Amines (1°, 2°, and 3°) can also react with acids to give salts, *e.g.*, HCl, nitrate and sulfate salts with HCl, HNO$_3$, and H$_2$SO$_4$, respectively. Such reactions are especially important in forming the drug salts for increasing water solubility.

Amine drugs can undergo a variety of metabolic reactions including oxidation and subsequent dealkylation, or deamination, conjugations and *N*-oxidations.

Carboxylic Acids and Derivatives

Definitions and Nomenclatures of Carboxylic Acids and Derivatives

<u>*Carboxylic Acids*</u>

The organic acids with general structure RCO$_2$H are collectively known as carboxylic acids. The R group can be both aliphatic and aromatic forming aliphatic acids and aromatic acids, respectively. The simplest carboxylic acid is formic acid, where R = H. Table **2** lists a few naturally occurring carboxylic acids with their common and IUPAC names.

Table 2. Naturally occurring carboxylic acids with their common and IUPAC names.

No. of Carbon	Common Name	IUPAC Name	Chemical Formula	Occurrence
1	Formic acid	Methanoic acid	HCOOH	Insect stings
4	Butyric acid	Butanoic acid	$CH_3(CH_2)_2COOH$	Rancid butter
6	Caproic acid	Hexanoic acid	$CH_3(CH_2)_4COOH$	Goat fat
8	Caprylic acid	Octanoic acid	$CH_3(CH_2)_6COOH$	Coconuts & breast milk
12	Lauric acid	Dodecanoic acid	$CH_3(CH_2)_{10}COOH$	Coconut oil

The IUPAC name stems from the longest continuous chain hydrocarbon bearing the carboxylic group. The last *e* from the hydrocarbon's name is replaced with *oic acid* to get the corresponding carboxylic acid name (*e.g.,* methan*e* to methan*oic acid*). When occurs in a polyfunctional compound, the carboxylic acid group gets the highest priority. The priority order of the common functional groups is: carboxylic acid > ester > amide > aldehyde > ketone > alcohol > amine > alkene. *p*-Amino benzoic acid and *m*-hydroxy benzoic acid are the examples of aromatic acids.

Esters

Esters are the condensed form of carboxylic acids and alcohols with the release of a water molecule. The -OH (hydroxyl) group of the carboxylic acid is replaced by an alkoxy group (-OR) giving the general formula RCO_2R'. The names of esters originate from the names of the parent acid and alcohol. The alkyl group of the alcohol gives the prefix while the carboxylate group gives the suffix of the ester name (*e.g.,* ethyl acetate (common) or ethylethanoate (IUPAC)). For simple carboxylate esters both common names and IUPAC names are used, but for esters derived from complex carboxylic acids, the IUPAC name is used. When an ester is a part of cyclic ring system, it is known as a lactone (*e.g.,* γ-butyrolactone, or 4-hydroxybutanoic acid lactone).

Ethyl acetate (Common)
Ethylethanoate (IUPAC)

Butyl propionate (Common)
Butylpropanoate (IUPAC)

γ-Butyrolactone (Common)
4-Hydroxybutanoic acid lactone (IUPAC)

Amides

An amide is formed by the condensation of a carboxylic acid and an amine with the release of a water molecule. The -OH group of the carboxylic acid is replaced

by amine (-NR$_2$) group giving the general formula RCONR$_2$, where R can be H or any other alkyl group. Amides derived from ammonia, 1° amine and 2° amine are called 1° amide (RCONH$_2$), 2° amides (RCONHR), and 3° amide (RCONR$_2$), respectively. An amide bond connecting two amino acids is called a **peptide**, while cyclic amides are called **lactams**. Additionally, if the N-atom of the amide bond is attached to a phenyl ring, it is called *anilide*.

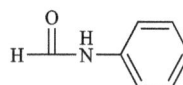

Acetamide (Common) Ethanamide (IUPAC)	Propanamide	*N,N*-dimethylformamide	Formanilide
A 1° Amide	A 2° Amide	A 3° Amide	

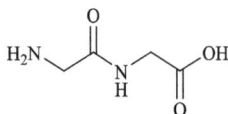

glycylglycine
Glycylglycine

A dipeptide

Pyrrolidin-2-one (Common)
4-Aminobutanoic acid lactam (IUPAC)

In naming the amides, the term *amide* is added, dropping the suffix *ic acid* or *oic acid* of the stem of the parent acid's name. For 2° and 3° amides, the substituents on nitrogen are indicated first in the name (*e.g., N,N*-dimethylformamide). Lactams are named by adding the term *lactam* at the end of the parent carboxylic acid name.

Carbonates

Esters of carbonic acid are called carbonate, also known as organocarbonates or carbonate esters, and have the general formula ROCOOR′, or RR′CO$_3$. In carbonate, a carbon atom is bound to three oxygen atoms, one of which is double bonded. Some examples include dimethyl carbonate, ethyl methyl carbonate and ethyl propyl carbonate. Carbonates are named by adding the suffix *carbonate* at the end of the names of two constituting alcohol portions.

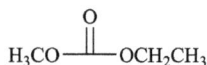
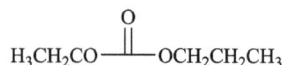

Dimethyl carbonate   Ethyl methyl carbonate   Ethyl propyl carbonate

Carbamates

Carbamates, also called **urethanes,** are esters of carbamic acid and have the general structure NR_2COOR' (where R can be H or any alkyl group; *e.g.,* carbaryl). Carbamates are named similar to carbonates, by its alcoholic group with the addition of the suffix *carbamate* (*e.g.,* naphthalene-1-yl-methylcarbamate). Physostigmine and related drugs are a group of carbamates classified as acetylcholine esterase inhibitors.

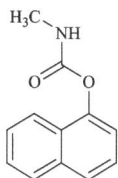

Carbaryl (Common)
Naphthalen-1-yl
methylcarbamate (IUPAC)

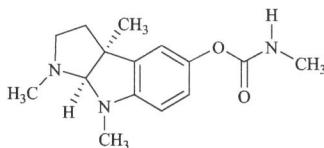

Physostigmine (Common)
$(3\alpha S,8\alpha R)$-1,3α,8-trimethyl-1,2,3,3α,8,8α-
hexahydropyrrolo[2,3-*b*]indol-5-yl
methylcarbamate (IUPAC)

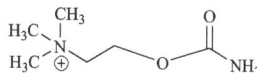

Carbachol

Ureas

Urea (also called carbamide) is the diamide derivative of carbonic acid that has two amide ($-NH_2$) groups joined by a carbonyl (C=O) functional group giving a general formula R_2NCONR_2. The simplest of this family with the structure H_2NCONH_2 is known as urea, was discovered in urine in 1773 by the French chemist Hilaire Rouelle and was first synthesized in 1828 by the German chemist Friedrich Wöhler. The nomenclature is similar to carbonates and carbamates with the exception that *urea* is added as suffix.

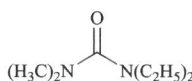

Urea   *N,N*-Diethyl-*N′,N′*-diemethyl urea

Physicochemical Properties of Carboxylic Acids and Derivatives

Carboxylic Acids

The presence of both hydrogen-bond acceptors (the carbonyl) and hydrogen-bond donors (the hydroxyl) make the carboxylic acids polar (Fig. **2**). Hydrogen bonds can form within two molecules of the acid, thus existing as dimeric pairs in nonpolar media, and with water molecules resulting in solubilization. Carboxylic acids with 1 to 5 carbons are normally soluble in water. As the carbon chain is

increased, the water solubility decreases. Because of dimerization, due to hydrogen bonding, they also have high boiling points (more than water).

One of the most important characteristics of carboxylic acids is their ability to dissociate, thus releasing a proton and carboxylate anion, making these weak acids. The carboxylate anion is stabilized through resonance structure (Fig. **2**). The acidity can increase or decrease with the substitution by electron withdrawing or donating groups, respectively. With bases, *e.g.*, NaOH, carboxylic acids can react to form salts (conjugate bases) that increase the water solubility tremendously.

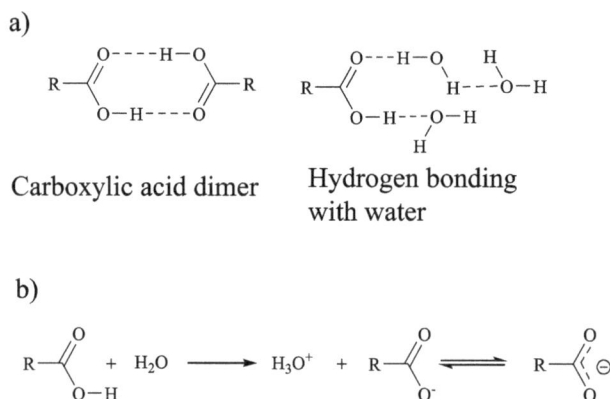

a)

Carboxylic acid dimer   Hydrogen bonding with water

b)

Fig. (2). Hydrogen bonding (**a**) and resonance structure (**b**) of carboxylic acid.

Carboxylic acid groups are most common in drugs, especially the NSAIDs (nonsteroidal anti-inflammatory drugs) and thus their metabolism is relatively straightforward. It can form ester glucuronides and amino acid conjugates, and also undergo β-oxidation to shorten the chain by two carbons, and releasing acetic acid.

Esters

Esters contain a hydrogen bond acceptor but not a hydrogen bond donor, which makes these compounds somewhat polar, more polar than ethers but less polar than carboxylic acids and alcohols. Esters are also nonionizable. All these characteristics make esters slightly water-soluble, and more volatile (lower boiling point) than carboxylic acids of similar molecular weight. Esters are also commonly present in drug molecules, especially the prodrugs of alcohols and carboxylic acids. They are easily hydrolyzed in the body either in alkaline media of the intestine or blood plasma or by the actions of a variety of esterase enzymes to yield the free carboxylic acid or alcoholic active drugs.

Amides

Amides exhibit no clear acid-base property and are considered neutral. Although the C=O bond is shown conventionally in representing the amide bond (-CONH-), it actually stays in two resonance structures A and B, where A represents the C=O and B represents the C=N (Fig. **3**). It implies that the lone-pair of electrons on nitrogen atom is not free to donate but is a part of a resonance structure and therefore, neutral. The C=O (or C-O) dipole is stronger than C=N (or C-N) dipole and acts as a hydrogen bond acceptor. The $1°$ and $2°$ amides can also act as a hydrogen bond donor due to the presence of NH_2 and NH bonds, respectively. Thus, they can participate in hydrogen bonding with water enabling some water solubility that is comparable to esters.

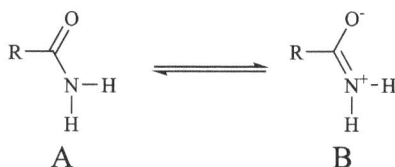

Fig. (3). Resonance structures of amides; A represents the C=O and B represents the C=N.

Metabolically, amides are more stable than esters towards hydrolytic degradations and are more commonly used as active drugs rather than prodrugs.

Carbonates, Carbamates and Ureas

The physicochemical properties of carbonates are similar to those of esters, and those of ureas are similar to amides. However, the urea is sp^2 hybridized and the C-N bonds exhibit significantly more double bond character than amides. Ureas also have higher water solubility than amides. Carbamates containing both ester and amide bonds have similar properties to both amides and esters. All these functional groups have similar metabolic pathways as esters and amides but with different rates due to different nucleophilicity and bond stability.

Few Pharmaceutically Important Heterocyclic Ring Systems

As we proceed through the various drug classes and their activities, we will find that heterocyclic compounds are common in organic, medicinal and bioorganic chemistry. For example, the building blocks of nucleic acids, the nucleotides, are built upon the heterocyclic purine and pyrimidine bases, condensed with a sugar and an inorganic phosphate. To understand the drug classes and their mechanism of action, the knowledge of the basic heterocyclic structures is essential.

Five-Membered Unsaturated Heterocycles

Pyrrole, furan, and thiophene are the most common five membered unsaturated heterocycles, each containing two double bonds and one hetero atom (N, O, or S). Another important five membered aromatic ring system is imidazole which is present in the amino acid histidine and the neurotransmitter histamine. This imidazole contains two nitrogen atoms in the ring. Thiazole is also a five membered ring with sulfur and nitrogen heteroatoms in the ring.

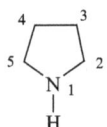 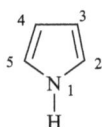 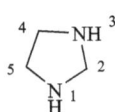 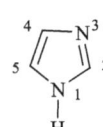 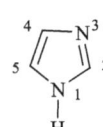

Pyrrolidine   Pyrrole   Imidazolidine   Imidazoline   Imidazole
(pK$_a$: 11.27)   (pK$_a$: 0.4)

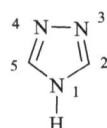 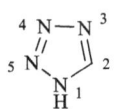

1,2,4-triazole   Tetrazole   Isoxazole   Oxazole   Thiophene   Thiazole

The N, O, and S atoms of pyrrole, furan and thiophene, respectively, are very weak nucleophiles as the nonbonding electrons are less available for bonding to electrophiles. In case of pyrrole, for example, the *N* atom uses all five valence electrons in bonding where the nitrogen lone-pair is a part of the aromatic sextet, and therefore, is less basic than the aliphatic amines (pKa of pyrrolinium ion = ~0).

Six-Membered Unsaturated Heterocycles

Pyridine is a flat aromatic molecule occurring in many biological and drug molecules, which contains five carbon atoms and a nitrogen atom as a member of aromatic sextet. Unlike the pyrrole, the lone-pair of electrons on the nitrogen of pyridine is not involved in bonding and thus is a stronger base (pKa of pyridinium ion = 5.25). However, pyridine is a weaker base than the alkyl amines *e.g.*, pyrrolidine and piperidine.

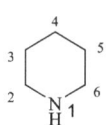

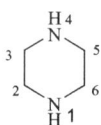

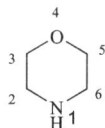

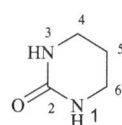

Piperidine   Piperazine   Morpholine   tetrahydro-2-
(pK$_a$: 11.2)   pyrimidinone

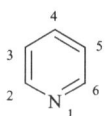

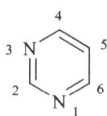

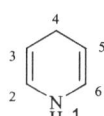

Pyridine   Pyrimidine   1,4-dihydropyridine
(pK$_a$: 5.2)   (pK$_a$: 1.3)

In imidazole, the pyrrole-like nitrogen has a similar basicity to pyrrole and the pyridine-like nitrogen has a similar basicity to pyridine. Another important six-membered aromatic ring is pyrimidine which contains two nitrogens. Pyrimidine is an essential component of nucleic acids. Both the nitrogens of pyrimidine have similar basicity as the pyridine nitrogen.

Fused-ring Heterocycles

Quinoline, isoquinoline, indole, and purine are important fused ring heterocycles occurring in biological and drug molecules. Quinoline and isoquinoline, which are isomers, contain fused rings of benzene and pyridine and have similar properties like pyridine. Indole contains fused benzene and pyrrole rings while purine contains fused pyrimidine and imidazole rings. All of them contain pyrrole-like or pyridine-like N atoms and share similar basic properties.

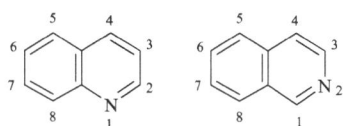

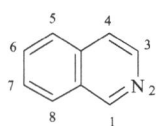

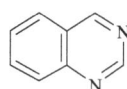

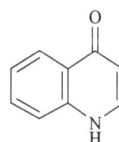

Quinoline   Isoquinoline   Tetrahydroisoquinoline   Quinazoline   4-quinolinone

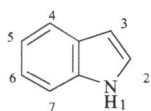

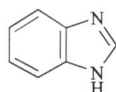

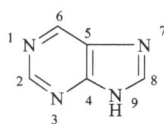

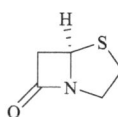

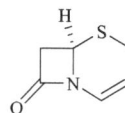

Indole   Benzimidazole   Purine   Penam   Cephem

Quinoline and isoquinoline are present in variety of drug molecules. Indole is an important heterocycle, present in the neurotransmitter serotonin, and in the amino acid tryptophan and in numerous drug molecules. Like pyrimidine, purine is another essential component for the nucleic acids. Few other fused heterocycles commonly occurring in drug molecules are also shown here.

Tricyclic, Tetracyclic and Bridged Ring Systems

The tricyclic ring phenothiazine contains both nitrogen and sulfur heteroatoms in its ring and is common to the classical antipsychotic drugs. The numbering system is shown below. An important property of this ring system is that the two aromatic rings are not coplanar and are aligned by an angle of α, which is important for its dopamine antagonist activity. Its isostere dibenzepin, devoid of any heteroatom, has both bending and twisting with one additional angle β. The other isostere, dibenzazepine, has two additional angles, β and δ. Due to these twists in dibenzepin and dibenzazepine, these are not antidopaminergic rather antidepressants. These compounds are essentially neutral or very weak bases and are hydrophobic in nature. These are metabolized by aromatic hydroxylation.

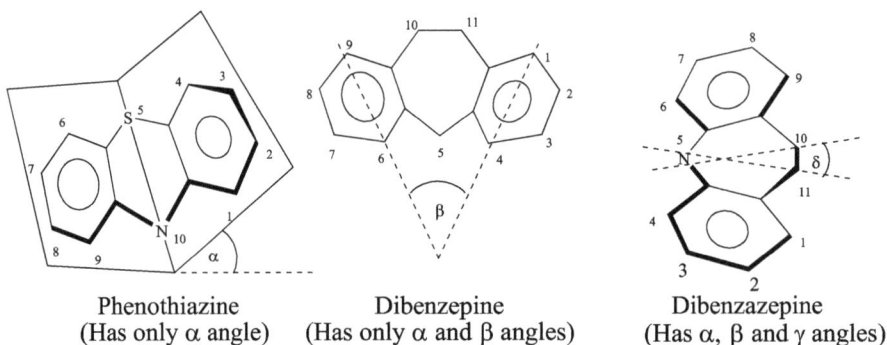

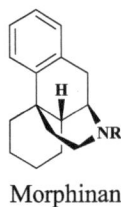

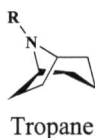

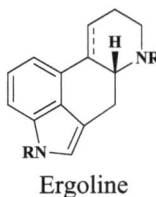

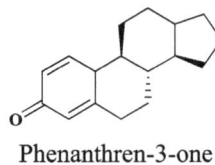

Phenothiazine   Dibenzepine   Dibenzazepine
(Has only α angle) (Has only α and β angles) (Has α, β and γ angles)

Morphinan   Tropane   Ergoline   Phenanthren-3-one

Morphinan and tropane are bridged ring systems while ergoline and phenanthrene are tetracyclic ring systems that occur commonly in drug molecules.

THE BIOMOLECULES

The biomolecules are chemical substances found in living organisms, composed of the same functional groups found in organic compounds. Both inorganic and organic compounds exist in biological systems and the mass composition data for the human body in terms of major types of biomolecules are: water (~70%), inorganic salts (~5%), proteins (~15%), lipids (~8%), carbohydrates (~2%), and nucleic acids (~2%). The major classes of bioorganic substances (carbohydrates, lipids, proteins, and nucleic acids) will be considered.

Carbohydrates

Derived from the abandoned idea of "hydrate of carbon, $(C_n(H_2O)_n$", the term carbohydrate is used to refer a broad class polyhydroxylated aldehydes and ketones commonly called **sugars**. Although their abundance is relatively low in humans, carbohydrates are most abundant in the plant kingdom. Average human diet ideally contains about two-third carbohydrates by mass. Key biological functions of carbohydrates are listed below.

1. Provide energy on biological oxidation and short term energy reserve as stored glycogen
2. Supply carbon atoms for the synthesis of other biomolecules
3. Important structural component in DNA and RNA
4. When linked to lipids (glycolipids), are the components of cell membranes
5. When linked to proteins (glycoproteins), function in a variety of cell-cell and cell-molecule recognition processes

Classification of Carbohydrates

Based on the molecular size, carbohydrates can be classed into monosaccharides, oligosaccharides, and polysaccharides.

Monosaccharides

Monosaccharides, also known as simple sugars, are either polyhydroxy aldehydes (**aldoses**) or polyhydroxy ketones (**ketoses**), which are not converted to any simpler aldoses or ketoses under aqueous hydrolysis conditions. Glucose $(C_6H_{12}O_6)$, the first simple carbohydrate to be obtained pure, is a polyhydroxy aldehyde, while fructose is a polyhydroxy ketone. A monosaccharide with six carbon atoms and an aldehyde function is aldohexose; with five carbon atoms and a keto group is ketopentose and so on. Although the monosaccharides are known by their trivial names, systemic names are in use. The systemic names of straight chain monosaccharides are based on a stem name indicating the number of carbon

atoms, a prefix indicating the configuration D or L, and either the suffix **-ose** (aldoses), or **-ulose** (ketoses), except for trioses which are known as glyceraldehydes (aldose) or dihydroxyacetone (ketose).

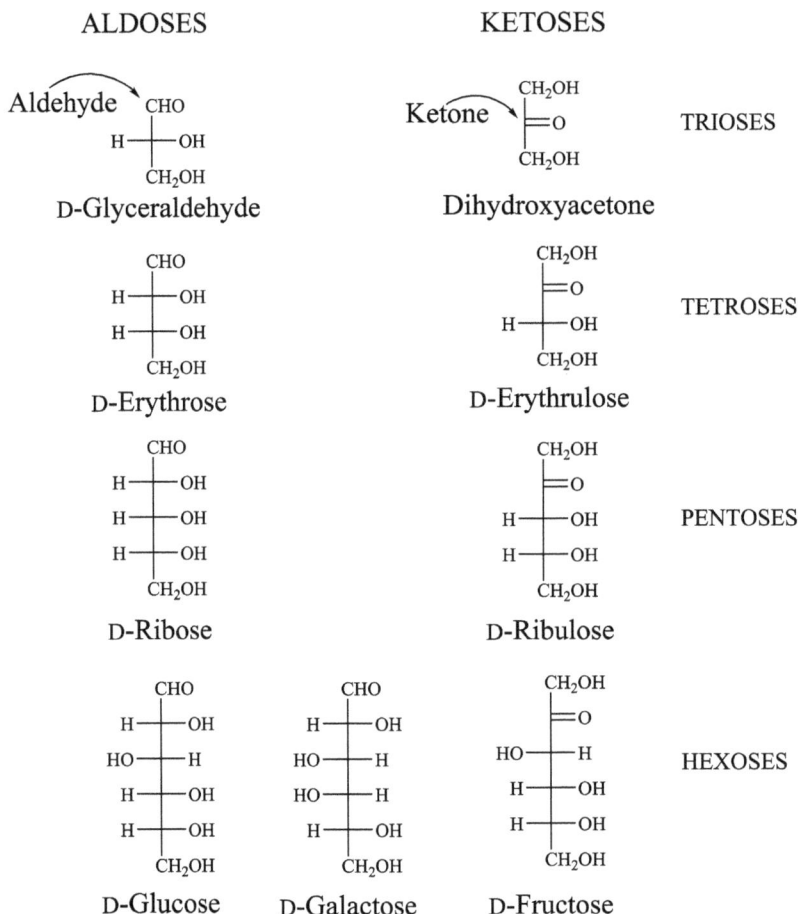

ALDOSES	KETOSES	

Aldehyde ⟶ CHO
H——OH
CH₂OH
D-Glyceraldehyde

Ketone ⟶ CH₂OH / =O / CH₂OH
Dihydroxyacetone

TRIOSES

CHO
H——OH
H——OH
CH₂OH
D-Erythrose

CH₂OH
=O
H——OH
CH₂OH
D-Erythrulose

TETROSES

CHO
H——OH
H——OH
H——OH
CH₂OH
D-Ribose

CH₂OH
=O
H——OH
H——OH
CH₂OH
D-Ribulose

PENTOSES

CHO
H——OH
HO——H
H——OH
H——OH
CH₂OH
D-Glucose

CHO
H——OH
HO——H
HO——H
H——OH
CH₂OH
D-Galactose

CH₂OH
=O
HO——H
H——OH
H——OH
CH₂OH
D-Fructose

HEXOSES

Six of the biochemically most important monosaccharides are: D-glyceraldehyde, dihydroxyacetone, and D forms of glucose, galactose, fructose and ribose. Glyceraldehyde and dihydroxyacetone are important intermediates of glycolytic process. D-Glucose, the primary energy source of cells, is the most important monosaccharide, which is also known as dextrose or **blood sugar**. The normal blood concentration of glucose is 70-100 mg per 100 mL of blood. D-Galactose, an **epimer** of glucose, differs only in the configuration at one chiral center. It is synthesized from D-glucose in the mammary glands for use in lactose (**milk sugar**). It is also known as **brain sugar** since it is a component of glycoproteins found in brain and nerve tissues. Also known as **fruit sugar**, D-fructose is the

sweetest-tasting of all sugars. D-Ribose is a pentose sugar and is an important component of a variety of complex molecules including RNAs and ATP, and 2-deoxy-D-ribose is a component of DNAs.

D and L Configurations

The monosaccharides have several stereogenic centers which can be indicated by the *R/S* system. However, they are indicated historically by the D/L system according to the configuration of the CHOH of the furthest chiral center from the carbonyl group (aldehyde or ketone). In the D form, this OH **projects on the right** of the carbon chain towards the observer whilst in the L form it **projects on the left** of the carbon chain, which are represented by modified **Fischer projections**. In fact, this is based on the structures of the reference compounds D- and L-glyceraldehydes (Fig. **4**). If the D and L forms are mirror images of each other, they are termed **enantiomers**. Monosaccharides with identical configuration except for one carbon atom are termed **epimers**, *e.g.*, α-D-glucose and α-D-galactose. Monosaccharides which are not mirror images and differ in more than one chiral centers are **diastereomers** (also called diastereoisomers).

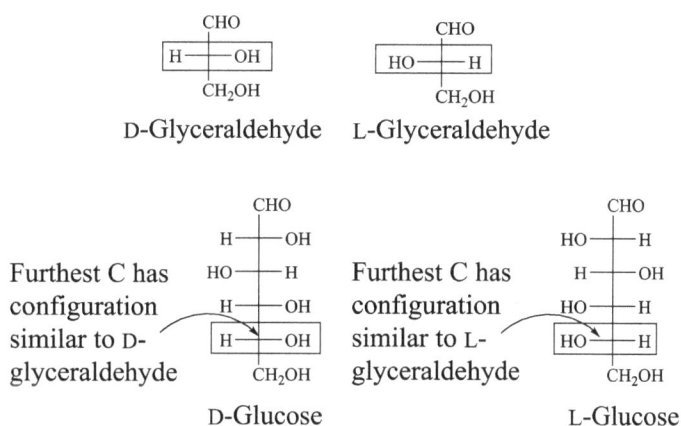

Fig. (**4**). Illustration of D and L configurations of sugars.

Cyclic Structures

Monosaccharides exist as straight chain or cyclic structure – those with five or more carbon chains usually assume five (**furanose**) or six (**pyranose**) membered ring structures – formed by an internal nucleophilic addition between a suitably positioned hydroxyl group and the carbonyl group in the molecule resulting in the formation of the corresponding cyclic hemiacetal or hemiketal (Fig. **5**). The six

membered rings usually occur as chair conformations while five membered rings exist as envelope conformations. This results in conversion of the carbonyl center of the molecule into a new chiral center known as the **anomeric carbon** and the resulting two isomers are known as **anomers**. The isomer with the new -OH on the opposite side of the ring from the -CH$_2$OH group (attached to carbon 5 of glucose for example) is known as **α anomer**, and the one in which both groups are on the same side of the ring is the **β anomer**. In aqueous solution of D-glucose at equilibrium, 63% of the molecules are in β-D-glucose, 37% are in α-D-glucose and <0.01% are in the open chain form.

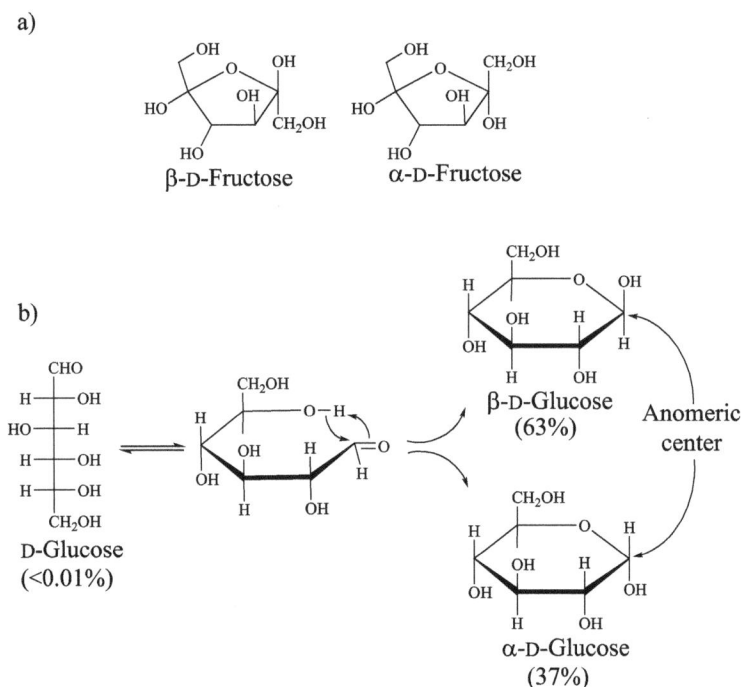

Fig. (5). The furanose (**a**) and pyranose (**b**) structures of fructose and glucose.

Reducing Sugars

Aldoses reduce the weak oxidizing agents, *Tollens' reagent* (Ag$^+$ in aqueous NH$_3$), *Fehling's reagent* (Cu^{2+} in aqueous sodium tartrate), and *Benedict's reagent* (Cu^{2+} in aqueous sodium citrate), to yield the oxidized sugar and a reduced metallic species, and so are called **reducing sugars**. Thus, if Tollens' reagent is used, metallic silver, and if Fehling's or Benedict's reagent is used, a reddish precipitate of Cu$_2$O signals the positive test for reducing sugars (Fig. **6**).

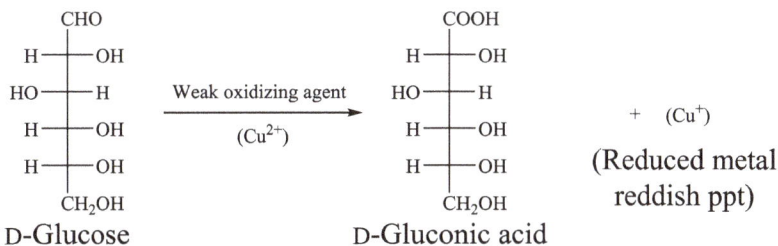

Fig. (6). Oxidation of glucose in presence of weak oxidizing agents.

Oligosaccharides

Oligosaccharides are those which yield two to ten monosaccharide units on hydrolysis. Small oligosaccharides are usually classified according to the number of their monosaccharide residues; *e.g.*, **disaccharide** and **trisaccharide** contains two and three monosaccharide residues, respectively. Disaccharides are most common type of oligosaccharides, which are crystalline water-soluble substances, *e.g.*, sucrose (table sugar) and lactose (milk sugar). Within the body, simple oligosaccharides except disaccharide, are seldom encountered and are usually associated with lipids (glycolipids) and proteins (glycoproteins) in complexes.

Glycosides

In complex carbohydrates (oligosaccharides or polysaccharides), the monosaccharide units are covalently linked together through glycosidic bonds. Compounds with a carbohydrate residue (**glycone**) are also bonded to a non-sugar residue (**aglycone**) by glycosidic linkage to the anomeric carbon of the glycone. Chemically, a glycoside is an acetal formed from a cyclic monosaccharide by the replacement of the hemiacetal carbon –OH group with an –OR group (Fig. **7**). Both α- and β-glycosidic links are known. The most common glycosidic link is an ether link (*O glycosidic link*), but amino (*N glycosidic link*), sulfide (*S glycosidic link*) and carbon-carbon (*C glycosidic link*) are also known. Glycoside formed from glucose is called glucoside; that from galactose is called galactoside, and so on.

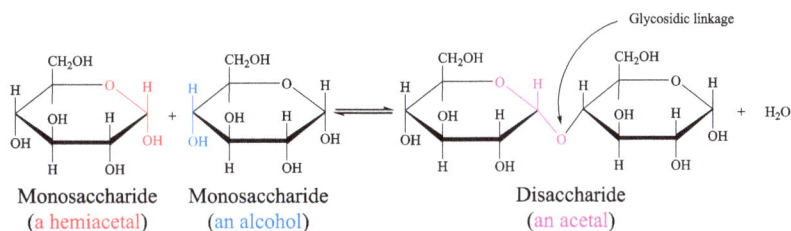

Fig. (7). The glycosidic linkage.

Maltose, lactose and sucrose are three important disaccharides. **Maltose**, often called **malt sugar**, is composed of α(1→4) linkage between two glucose units that form an acetal by carbon 1 of the first glucose (in α configuration) and carbon 4 of the second. It is a reducing sugar because the second glucose unit (on the right side) has a hemiacetal carbon atom, which can be in both α- and β- forms and thus both α- and β-maltose are possible. The milk sugar, **lactose**, is made up of a β--galactose unit and a D-glucose unit joined by a β(1→4) glycosidic linkage, and like maltose, both α and β forms are possible; α-lactose being sweeter and more soluble in water. It is also a reducing sugar since it possesses a hemiacetal carbon. **Sucrose**, the common **table sugar**, is the most abundant of all disaccharides, and is made up of an α-D-glucose and a β-D-fructose by a α,β(1→2) glycosidic linkage. Since the –OH groups of both hemiacetal carbons (of glucose and fructose) are involved in formation of the glycosidic linkage, no free hemiacetal is available, thus sucrose is a **non-reducing sugar** and has only one form – there are no α and β isomers (Fig. **8**).

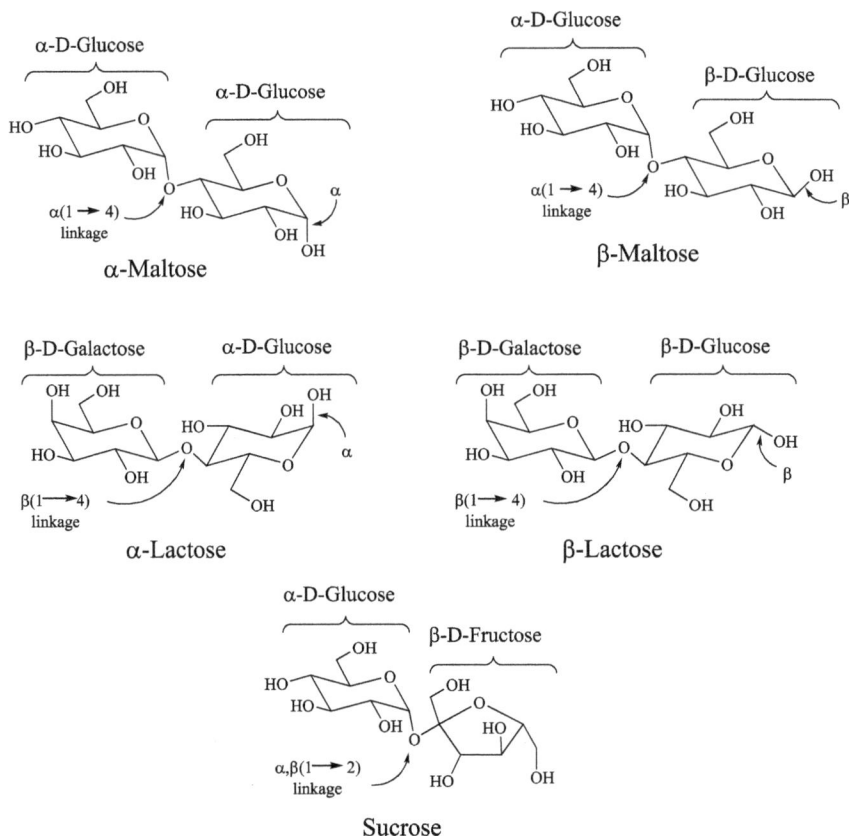

Fig. (8). Structures of a few common disaccharides showing critical configurations.

Polysaccharides

Polysaccharides, also known as *glycans* include monosaccharides joined together by O glycosidic links, yield larger number (up to several thousands) of monosaccharide residues, on hydrolysis. Cellulose and starch are the most commonly encountered polysaccharides. **Cellulose**, primarily a structural material, consists of several thousands of D-glucose units linked by $\beta(1\rightarrow4)$-glycoside bonds. The cellulose molecules tend to have linear (straight-chain) structures. Several linear molecules are aligned side by side, which are held together by inter-chain hydrogen bonding to yield water-insoluble fibers. Human digestive enzymes contain α-glycosidase that catalyzes the hydrolysis of only those polysaccharides that are formed by α-glycosidic links and are thus unable to digest cellulose.

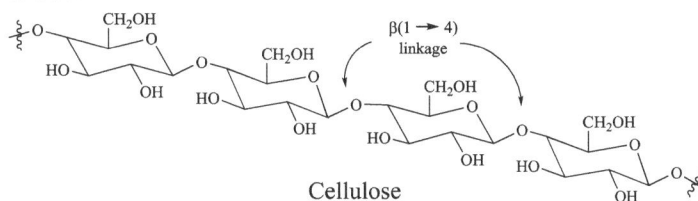

Cellulose

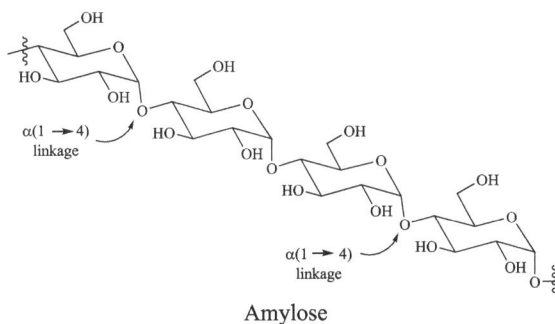

Amylose

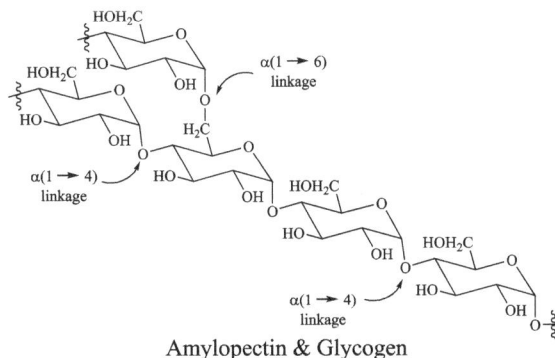

Amylopectin & Glycogen

Starch, on the other hand, is a polymer of glucose which is linked together by $\alpha(1\rightarrow4)$-glycoside bonds. Two fractions of starch are: **amylose** that is insoluble in cold water, and **amylopectin** that is soluble in cold water. Amylose is a linear polymer like starch while amylopectin contains $\alpha(1\rightarrow6)$-glycoside branches approximately every 25 glucose units, thus has an exceedingly complex three dimensional structure. Starch consists of unbranched amylose chains (10-20%) and amylopectins (~80%). The energy storage glucose polymer, **glycogen**, is similar to amylopectin, except that the chains are branched every 8-10 glucose residues through $\alpha(1\rightarrow6)$-glycosidic links.

Other Important Carbohydrates and Derivatives (Fig. 9)

2-Deoxyribose

This is the most common **deoxy sugar** (sugar that is missing an oxygen atom) found in DNA and possesses a furanose structure.

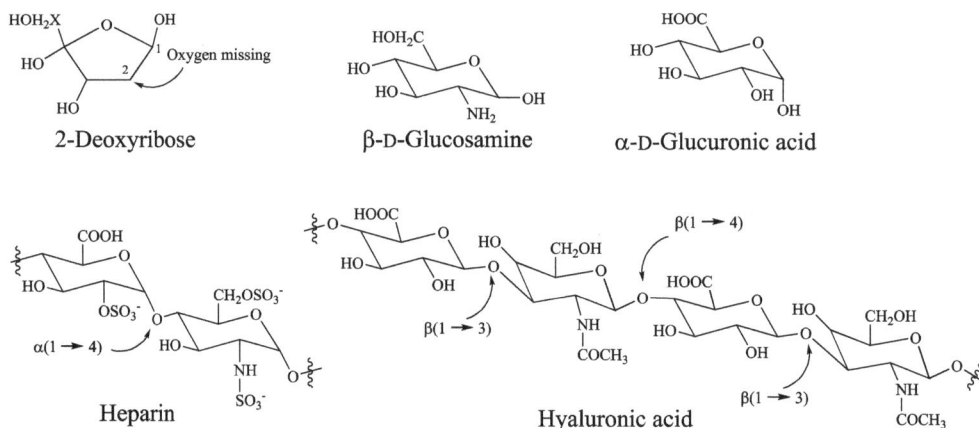

Fig. (9). Structures of other important carbohydrates and derivatives.

Amino Sugars

D-Glucosamine has one of its –OH groups replaced with an –NH_2. Its *N*-acetyl amide form occurs in the polysaccharide **chitin**, a hard crust that protects insects and shellfish. The amino sugars also occur in variety of antibiotics such as streptomycin and gentamycin (see later).

Acidic Carbohydrates

The enzymatic oxidation of the primary alcohol side of glucose in biochemical systems produces **D-glucuronic acid**, which is an important component for conjugation (glucuronidation) of xenobiotics, including drugs. **Hyaluronic acid** contains alternating residues of *N*-acetyl-β-D-glucosamine and D-glucuronic acid, with an alternating pattern of glycosidic bond types, β(1→3) and β(1→4), which is a lubricant in joint-fluids. The anticoagulant **heparin** is made up of a disaccharide repeating unit of D-glucuronate-2-sulfate and *N*-sulfo-D-glucosamine-6-sulfate.

Antibiotics and Cardiac Glycoside

The antibiotics streptomycin and gentamycin are aminoglycosides, while digoxin is a cardiac glycoside.

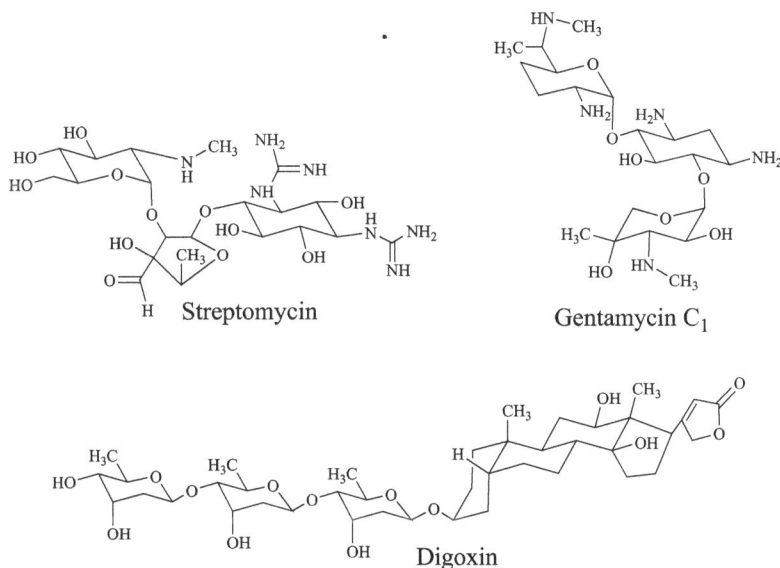

Streptomycin

Gentamycin C$_1$

Digoxin

Lipids

A wide variety of structural features of lipids make it difficult to define them chemically, but are broadly defined as naturally occurring organic compounds that are soluble in nonpolar organic solvents and insoluble or sparingly soluble in water. Fatty acids, glycerides, eicosanoids, steroids, terpenes, phospholipids and glycolipids are examples of the chemical substances that are often classified as lipids. Lipids perform a variety of functions in biological system such as:

1. Providing the major way of storing chemical energy (triacylglycerols)
2. Insulating vital body organs, protecting from mechanical shock, and preventing excessive loss of heat energy (biological waxes, particularly in plants)
3. Serving as basic components of cell membranes (phospholipids, glycolipids and cholesterol)
4. Acting as hormones and messengers (*e.g.*, steroids and eicosanoids) and
5. Emulsification by phospholipids (*e.g.*, dipalmitoylphosphatidyl choline is a lung surfactant)

The chemistry of the most important lipids has been outlined here.

Fatty Acids

Saturated or unsaturated (mono- and poly-unsaturated) straight chain monocarboxylic acids with an even number of carbon atoms (C_4 to C_{22}) occur as isolated molecules and are often found as residues in other lipid structures. In the case of unsaturated fatty acids, both *cis* and *trans* isomers are known, but *cis* isomers are more common.

Omega-3 (ω-3) and Omega-6 (ω-6) Fatty Acids

An unsaturated fatty acid with its terminal double bond three carbon atoms away from its methyl end is a ω-3 fatty acid and the one with six carbon atoms away is a ω-6 fatty acid. Thus, linolenic acid is a ω-3 fatty acid, while linoleic and arachidonic acids are ω-6 fatty acids. Cold-water fish, also called fatty fish, such as albacore tuna, salmon, and mackerel contain more ω-3 fatty acids than leaner warm water fish and are healthier for the heart when compared with ω-6 fatty acids (Table **3**).

Table 3. A few common fatty acids with their structure and names.

Trivial Name	Systemic Name	Structure
Saturated:		
Palmitic acid	Hexadecanoic acid	$CH_3(CH_2)_{14}COOH$
Stearic acid	Octadecanoic acid	$CH_3(CH_2)_{16}COOH$
Monounsaturated:		
Oleic acid	*cis*-9-Octadecenoic acid	$CH_3(CH_2)_7CH=CH(CH_2)_7COOH$
Polyunsaturated:		
[e]Linoleic acid	*cis*-9-*cis*-12-Octadecadienoic acid	$CH_3(CH_2)_3(CH_2CH=CH)_2(CH_2)_7COOH$
[e]Linolenic acid	*cis*-9-*cis*-12-*cis*-15-Octadecatrienoic acid	$CH_3(CH_2CH=CH)_3(CH_2)_7COOH$
Ricinoleic acid	12-hydroxy-*cis*-9-Octadecenoic acid	$CH_3(CH_2)_5CHOHCH_2CH=CH(CH_2)_7$ COOH
Arachidonic acid	*cis,cis,cis,cis*-5,8,11,14-Eicosatetraenoic acid	$CH_3(CH_2)_3(CH_2CH=CH)_4(CH_2)_3COOH$

[e]Essential fatty acids, which are needed in the human body and must be obtained from dietary sources because they cannot be synthesized within the body in adequate amounts.

Glycerides (Acylglycerols)

These are mono-, di-, and tri-esters of glycerol and fatty acids – triesters (triacylglycerols or triglycerides) are the most common naturally occurring among the three. A simple triacylglycerol is a triester formed from the esterification of the glycerol with three identical fatty acid residues, while a mixed triacylglycerol is formed by esterification with more than one fatty acid moiety (Fig. **10**). Fats and oils are the complex mixtures of triacylglycerols and the "liquidity" of oils at room temperature is attributed to their having a higher proportion of unsaturated fatty acids in their structure.

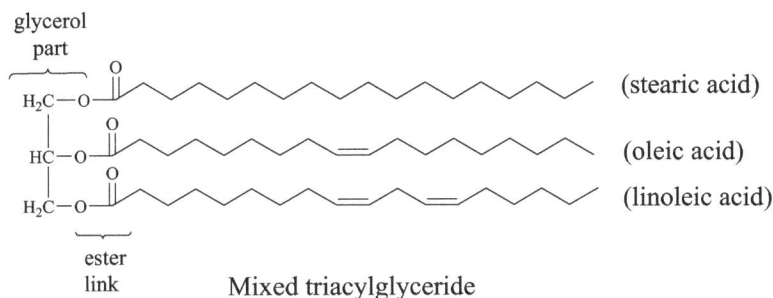

Fig. (**10**). Structure of a mixed triglyceride showing the important components.

Phospholipids

Phospholipids, are a class of lipids and a major component of all biological membranes built upon a nitrogen-containing alcohol such as ethanolamine or an organic compound such as choline. **Phosphoglycerides** and **sphingolipids** are the two main kinds of phospholipids. In cell membranes, phosphoglycerides are more common and are composed of one glycerol bonded to two fatty acids and one phosphate group by an ester linkage. Common phospholipids in cells include phosphatidylcholine, dipalmitoylphosphatidyl choline, phosphatidylethanolamine, phosphatidylinositol and phosphatidylglycerol (Fig. **11**). Mitochondria contain a special phospholipid called cardiolipin, which may be considered to be a product of condensation of phospholipids and contains four acyl chains.

The backbone of **sphingolipid** is **sphingosine**, an 18-carbon monounsaturated aminodialcohol. The long-chain fatty acid is attached to the amino terminal through an amide linkage to yield ceramide of which the terminal hydroxyl is attached to phosphatidylcholine (phosphate ester of choline) through ester linkage to form **sphingomyelin**. Sphingomyelin is present in cell membranes, especially the plasma membrane, and as a major component of myelin, particularly concentrated in the nervous system (Fig. **11**).

Phospholipids are **amphipathic** in character, meaning that they contain both a hydrophobic and a hydrophilic group within the same molecule (Fig. **12**). In phospholipids, the hydrophilic phosphate group comprises the **polar ionic head** and the lipophilic hydrocarbon chain comprises the **nonpolar tail**. In biological systems, phospholipids commonly assemble as bilayers, in which the nonpolar tails line up against one another, forming a membrane with polar heads on both sides facing the water. This membrane is semi-permeable, has fluid properties, in which are embedded proteins, receptors, or ion channels. Another common assembly of phospholipids is in the form of micelles, which are spherical structures formed when amphipathic molecules are added to an aqueous medium, with the hydrophobic tails pointing inwards and the hydrophilic heads facing the aqueous phase. Micelles are pharmaceutically important since they can be used for solubilizing drugs and in human physiology since they enable the digestion of fats.

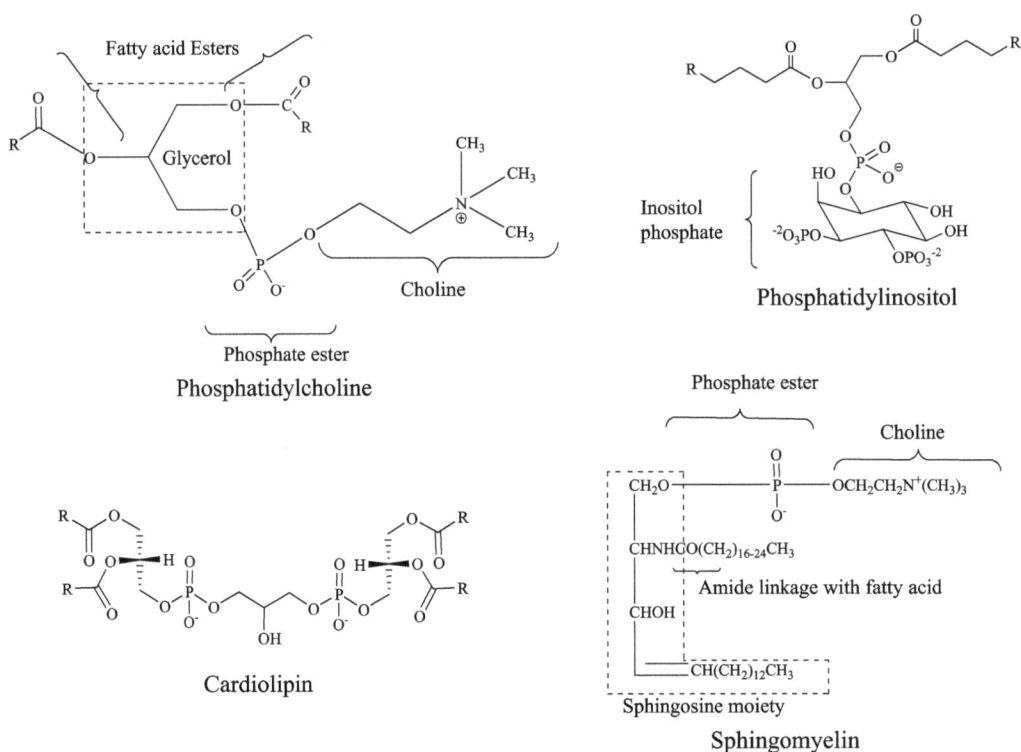

Fig. (11). Structures of representative phospholipids.

Polar ionic (phosphate) head

Nonpolar hydrocarbon tails

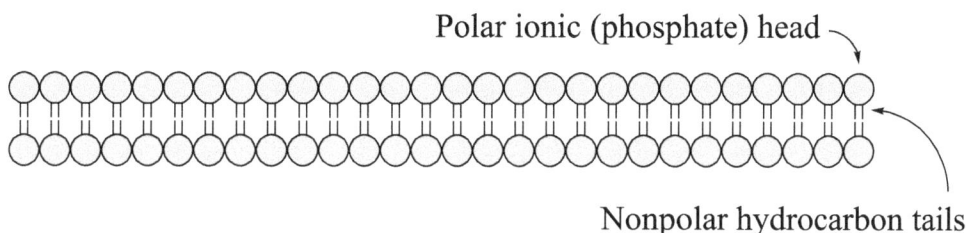

Fig. (12). Amphipathic character of phospholipids.

Eicosanoids

Derived from 20-Carbon (Greek word *eikos* means twenty) fatty acids by oxygenation eicosanoids are messenger lipids, produced in almost all cells except red blood cells. Arachidonic acid is the most common precursor of eicosanoids, synthesized by the action of membrane-bound phospholipase A_2 on complex higher order membrane phospholipids. Unlike hormones, which are transported through blood to sites of action away from the producing cells, eicosanoids are short lived, usually few seconds, locally active substances and are thus termed **autacoids** or local **hormone like mediators** rather than true hormones. The most important functions performed by eicosanoids include:

1. Mediation of inflammatory responses
2. Production of pain and fever
3. Regulation of blood pressure
4. Induction of blood clotting
5. Induction of labor
6. Regulation of sleep/wake cycle and
7. Homeostasis

Three principal **types** of eicosanoids are: prostaglandins, thromboxanes, and leukotrienes (Fig. **13**). **Prostaglandins**, named after prostate gland, have a characteristic oxygenated cyclopentane ring formed by a connection between the C_8 and C_{12} of an arachidonic acid precursor. Usually, C_9 and/or C_{11} is/are oxygenated *via* a keto or hydroxyl functions and C_{15} is hydroxylated in different prostaglandins. **Thromboxanes** contain cyclic ether ring(s) and are named after thrombocytes, more commonly known as platelets. The first **leukotriene** was isolated from leukocytes and contains a conjugated triene system in its structure, hence the name leukotriene.

Prostaglandins

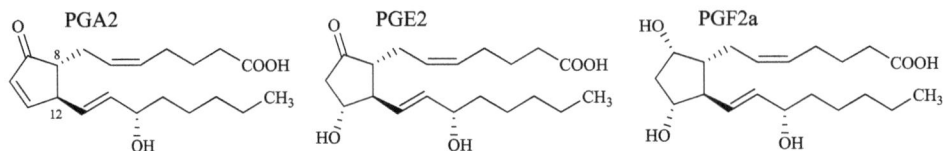

Thromboxanes

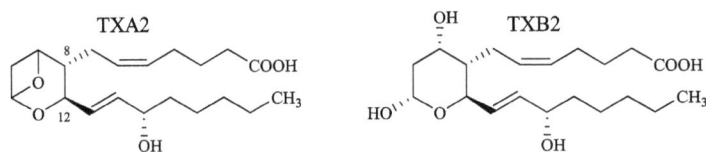

Leukotrienes

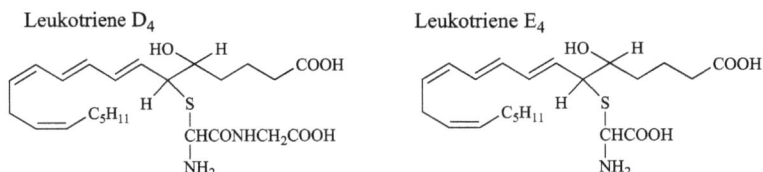

Fig. (13). Structures of eicosanoids. PG, prostaglandin; TX, thromboxane.

Steroids

Chemically, a steroid is any molecule that contains tetracyclic **cyclopentano perhydrophenanatherene** ("perhydro-" means fully saturated; Fig. **14**). The four rings are assigned letters, from A to D (from left to right), with the cyclopentyl ring being D. The carbon atoms comprising the rings are numbered 1 through 17 starting with ring A, as illustrated. The C atoms of the methyl groups at the C/D and A/B fusion are numbered C18 and C19, respectively. The hydrocarbon chain at C17 starts with C20 and continues to C27. Carbon atoms 28 to 30 are found only in precursor molecules. Higher carbon amounts occur in plant and fungal steroids. The structure on the right illustrates the three dimensional aspects of steroids. The top of the molecule is referred to as the *beta face* and the bottom as the *alpha face*. If a carbon atom is saturated, it has two substituents, one axial and one equatorial. The methyl groups at C18 and C19 are oriented in the *beta* configuration, sticking upwards and are called **angular methyls** and define the β face.

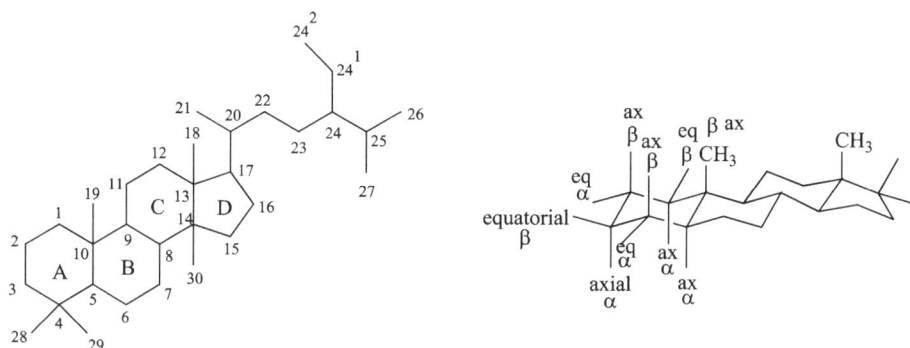

Fig. (14). Structures of steroid showing different conformers position numbering.

The steroids are assigned a **primary steroid name** based on the total number of carbons in the molecule. Lanostanes have 30 carbon atoms, while gonanes have only 17. The important steroid hormones belong to the pregnane, androstane, and estrane families. Cholesterol is an important steroid from which the steroid hormones are biosynthesized.

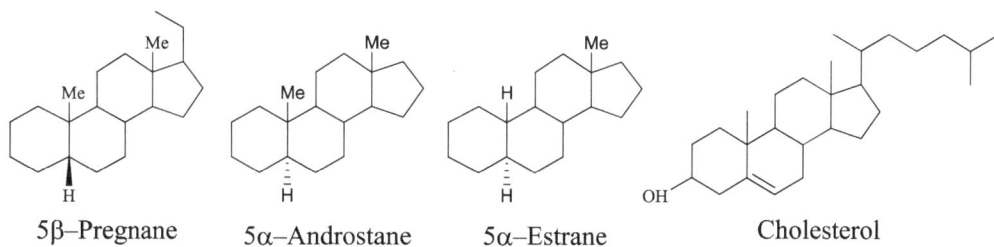

5β–Pregnane   5α–Androstane   5α–Estrane   Cholesterol

The fusion of the rings in steroids is extremely important because it determines the three dimensional shape of the molecule. If the A/B and C/D fusions are arranged in the *trans* configuration, the molecule is essentially **flat**. These are the 5α–steroids. If the A/B fusion is *cis*, the molecule is **L shaped**. These are the 5β–steroids. Bile acids have this type of shape and thus lack hormonal activity. If both fusions are *cis*, the molecule is **C shaped**. The cardiac glycosides (such as digoxin) have this shape. All the clinically used steroids have some form of unsaturation around C5, and thus typically do not contain a *cis* or *trans* A/B fusion, but are still flat. They are known as being in the ***quasi trans* configuration**.

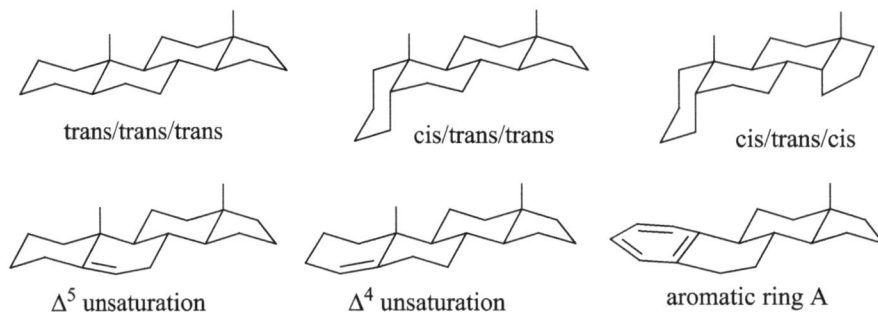

trans/trans/trans cis/trans/trans cis/trans/cis

Δ^5 unsaturation Δ^4 unsaturation aromatic ring A

Unsaturation in ring A is important because it determines not only the shape of the ring but also the orientation of the oxygen atom at C3. Unsaturation between C4 and C5 has little effect on both; however, di-unsaturation flattens the ring and orients the oxygen away from the *alpha face*. Unsaturation can be indicated by various means. If there are various other substituents, as there usually are, the number is placed just before *ene*. A delta sign is used to indicate a *pi* bond and the Carbon number is added as a superscript. If only one number is used, it implies that the *pi* bond is between the atom with that number and the atom with the next higher number. If the *pi* bond is not between consecutively numbered atoms, then both atom numbers must be indicated (Fig. **15**).

The systematic name of steroids is complicated but it does describe the entire structure. Since these are too complex, generic names are used. However, generic names give no indication of the structure. Example of the trivial names and systemic names of glucocorticoid hydrocortisone, female sex hormone estradiol, and male sex hormone testosterone are as follows.

Proteins

Proteins are naturally-occurring polymers built upon monomer amino acid units joined together by amide bonds between $-NH_2$ and $-COOH$ called **peptide** linkages (-CONH-). A large variety of different proteins with different functions exist in biological system including enzymes, certain hormones and some blood components. After a brief introduction to the structure and chemistry of amino acids, peptides and proteins will be considered.

Androst–5–ene   Androst–8–ene   Androst–8(14)–ene

3-one steroid   4-ene-3-one steroid   1,4-diene-3-one steroid

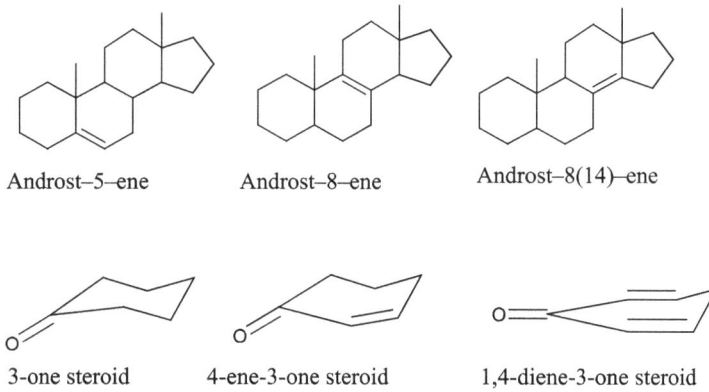

Fig. (15). Nomenclature of steroid structures.

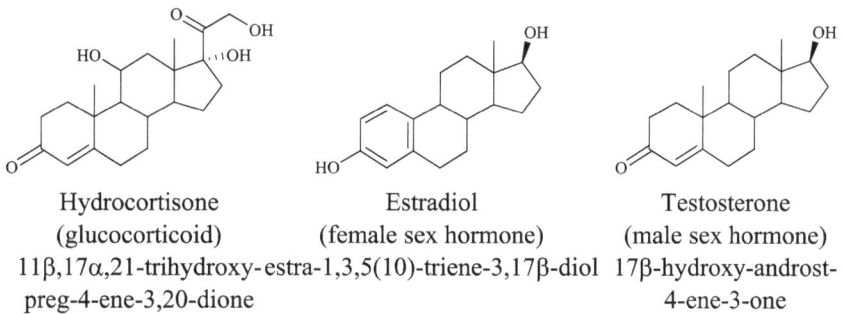

Hydrocortisone
(glucocorticoid)
11β,17α,21-trihydroxy-
preg-4-ene-3,20-dione

Estradiol
(female sex hormone)
estra-1,3,5(10)-triene-3,17β-diol

Testosterone
(male sex hormone)
17β-hydroxy-androst-
4-ene-3-one

Amino Acids

Chemical substances containing both an amino group and a carboxylic acid group are often termed amino acids; α-amino acids are the most common building blocks of proteins where the amino group and carboxyl group are attached to the α-carbon atom (Fig. **16**).

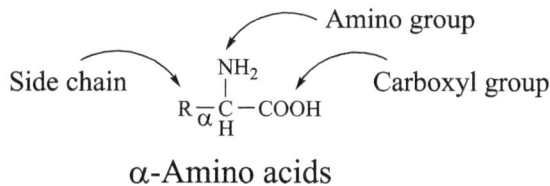

Amino group

Side chain   NH$_2$ Carboxyl group

$$R-\underset{\underset{H}{\alpha}}{C}-COOH$$

α-Amino acids

Fig. (16). General structure of amino acid.

a) Nonessential amino acids

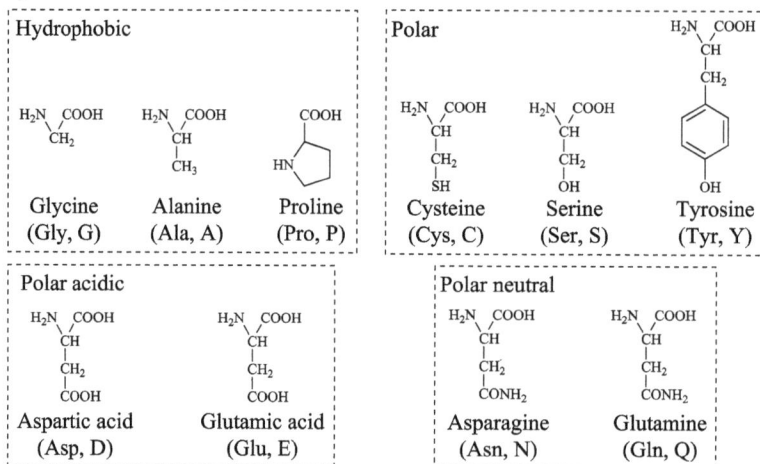

b) Essential amino acids

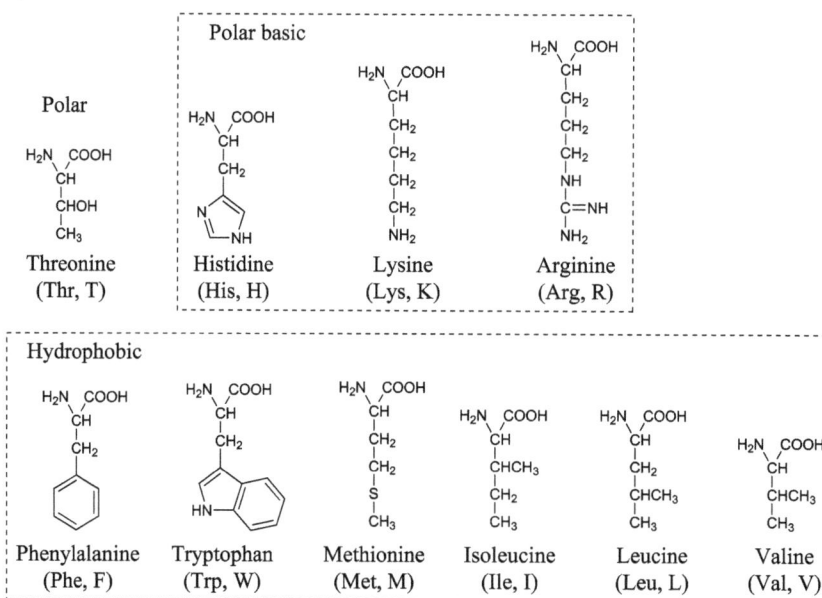

Fig. (17). Structures of amino acids.

The structures of the 20 naturally occurring α-amino acids found in proteins are shown in Fig. **17**. Each of the amino acids is referred to by a three letter code as well as a one letter code, both of which are shown in parentheses. The simplest of all amino acids is glycine (Gly, G). Nine **nonpolar amino acids** that contain hydrophobic or nonpolar side chains are: Gly, Ala, Val, Leu, Ile, Pro, Phe, Met,

and Trp. There are six **polar amino acids** that contain hydrophilic R groups, which are: Ser, Cys, Thr, Asn, Gln, and Tyr. Asp and Glu contain carboxyl function in the side chain and are known as **polar acidic amino acids**. On the other hand, His, Lys, and Arg contain polar basic amino groups on the side chain and are called **polar basic amino acids**. All the 20 amino acids are necessary for protein synthesis, but humans cannot synthesize 10 of them and are essential in diet, so are called **essential amino acids** which are: Phe, Val, Thr, Trp, Ile, Met, His, Arg, Leu, Lys (PVT TIM HALL) [1 - 3]. Tyrosine, arginine, cysteine, glycine, glutamine, and proline are known as **conditionally essential** amino acids because some pathophysiologic conditions lead to an inadequate synthesis of these.

Isoelectric Point of Amino Acids

In aqueous solution, amino acids exist as dipolar ions known as *zwitterions*, their exact structures depending on the pH of the solution. At acidic pH the amino group takes up a proton and develops a net positive charge while at basic pH the carboxyl group donates a proton and develops a net negative charge. However, at a certain pH known as the **isoelectric point (pI)**, the aqueous solution of an amino acid becomes electrically neutral, as the proton donated by the carboxyl group is taken up by the intramolecular amino group forming zwitterions (Fig. **18**).

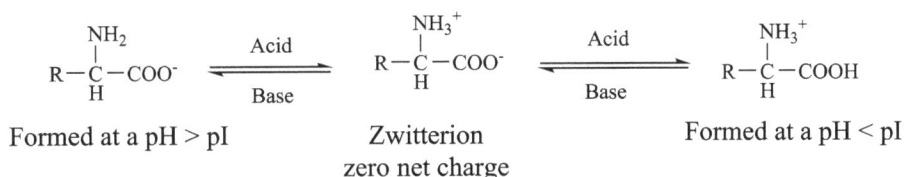

Fig. (**18**). Isoelectric point of amino acids.

Chirality of Amino Acids

Most amino acids (except glycine) are optically active and are indicated by D/L system (not *R/L* system; Fig. **19**). According to the **Fisher projection**, the –COOH group is placed at the top and the R group at the bottom. If the horizontal –NH$_2$ group positioned on the right denotes the D isomer and if it is on the left, denotes the L isomer. Most naturally occurring amino acids are L-amino acids. However, some bacteria also possess D-amino acids, which are the basis for the development of some antibacterial agents.

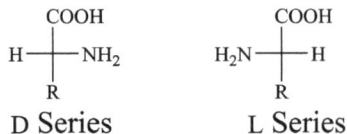

Fig. (19). Chirality of amino acids.

Sulfhydryl-Disulfide Redox Pair

The polar amino acid **cysteine** is a sulfhydryl containing amino acid which can be oxidized to form a dimer known as cystine that contains a disulfide bond. This is a unique property that contributes significantly in protein structure and shape, and also in the function of many redox active enzymes and their substrates, *e.g.*, glutathione (GSH)-glutathione reductase (GR) system (Fig. **20**).

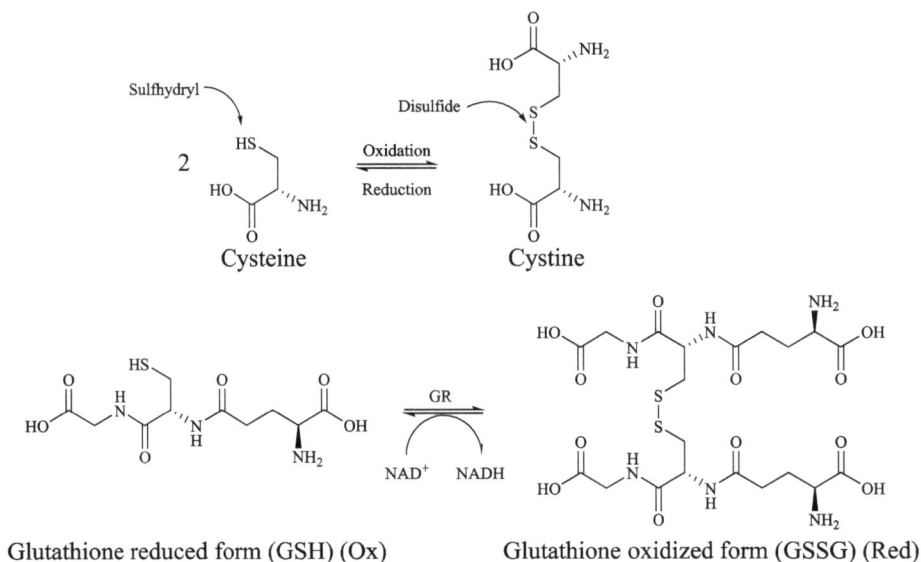

Fig. (20). Oxidized and reduced forms of cysteine and glutathione.

Peptides and Proteins

Peptides and proteins consist of amino acid residues linked together by covalent **amide bonds** known as **peptide bonds** (Fig. **21**). The carboxyl group of one amino acid condenses with the amino group of the other amino acid with the removal of a water molecule to produce the peptide bond – this is known as dehydration synthesis. In a peptide bond the lone pair of electrons on nitrogen is able to interact with the π electrons of the carbonyl group and is usually shown by a resonance structure.

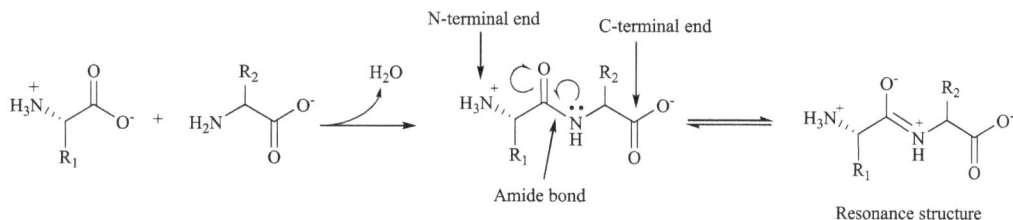

Fig. (21). Formation of peptide bonds.

Compounds with a small number of amino acid residues are termed **peptides**, larger amino acid residues with molecular mass 500 - 2000 are **polypetides** and those with molecular mass >2000 are termed **proteins**. Glutathione is a tripeptide as it contains three amino acid residues (Glu-Cys-Gly). In polypeptide or protein chains, one end has a free amino acid termed the **N-terminal end** and the other end has a free carboxylic acid termed the **C-terminal end**. By convention, the direction of the peptide chain is always N-terminal end → C-terminal end; that is the N-terminal end is always on the left and the C-terminal end is always on the right. Proteins are relatively more complex and are termed **simple proteins** if composed of only amino acid residues or **conjugated protein** if they contain other residues in addition to the amino acids in their structures. Angiotensin II is an example of a simple octapeptide that functions as a blood-pressure-regulating chemical, while hemoglobin is a conjugated protein (hemoprotein) because it contains a non-amino acid heme residue. The non-amino acid residue of a conjugated protein, involved in its biological function, is known as a **prosthetic group**, *e.g.*, heme that is involved in O_2-career function. Glycoproteins contain carbohydrates in their structure while lipoproteins are biochemically assembled lipids and proteins.

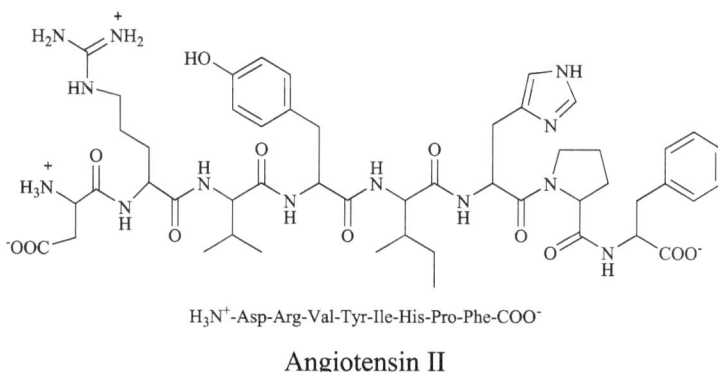

H_3N^+-Asp-Arg-Val-Tyr-Ile-His-Pro-Phe-COO$^-$

Angiotensin II

Primary, Secondary, Tertiary, and Quaternary Protein Structures (Fig. 22)

• The primary (1°) protein structure is the sequence of amino acid residues in

peptides and proteins, that is the order in which the amino acids are connected to each other.

- The secondary (2°) protein structures are the local regular and non-random conformations assumed by sections of peptide chain found in structure of peptide and protein, which are mainly α-helix, and β-plated sheet, and are believed to be formed due to intramolecular hydrogen bonding between different sections of the peptide chain.
- Tertiary (3°) protein structure is the overall shape of the molecule, often formed by folding of the chain stabilized by S-S (disulfide) bridges, H-bonding, salt bridges and van der Waals' forces within and around the peptide chain. Hydrophobic interactions also play an important role in the overall shape of the peptide molecules.
- Quaternary (4°) structures are the 3-D structures formed by the noncovalent association of a number of individual peptide and polypeptide molecules known as subunits. Hemoglobin has four subunits, two α- and two β-units held together by H-bond and salt bridges.

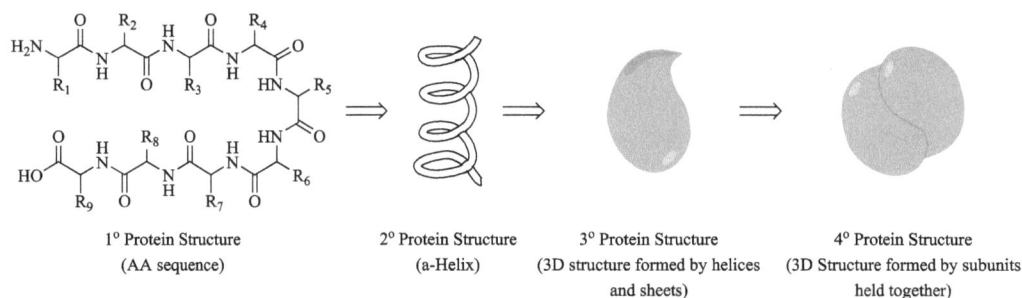

1° Protein Structure	2° Protein Structure	3° Protein Structure	4° Protein Structure
(AA sequence)	(a-Helix)	(3D structure formed by helices and sheets)	(3D Structure formed by subunits held together)

Fig. (22). Primary, secondary, tertiary, and quaternary protein structures.

Based on their conformations, that is the secondary, tertiary and quaternary structural features, proteins can be **globular** (usually water-soluble) or **fibrous** (usually not water-soluble). In a globular protein, the peptide chains are folded into roughly spherical or globular shapes, while in fibrous proteins, peptide chains are arranged in long strands or sheets. Hemoglobin and γ immunoglobulins (antibodies) are examples of globular proteins, while collagen is an example of a fibrous protein. **The water solubility of proteins is at minimum at its p*I*. The charge on a protein may affect its ease of transport across the plasma membrane and thus its biological availability and activity.**

Nucleic Acids

Two general types – the **deoxyribonucleic acid (DNA)** that contains β-D-deoxyribose sugar and **ribonucleic acid (RNA)** that contains β-D-ribose sugar –

both consist of long polymer chains based on repeating **nucleotide** units. Each nucleotide consists of a purine or pyrimidine base bonded to the 1' Carbon atom of a sugar residue by a β-N-glycosidic linkage (the **nucleosides**), which are linked through the 3' and 5' carbons of their sugar residues by phosphate units to form the nucleic acid polymer chain.

The Nucleotides and Nucleosides

Nucleotides can exist as individual molecules with one or more phosphate or polyphosphate groups attached to the nucleoside residue. When sugar is attached to the specific purine or pyrimidine bases through a glycosidic linkage, it is termed as the nucleoside. The ribose **nucleosides** (RNA) are named after their bases but either with suffix **-osine** or **-idine** and those for DNA nucleosides are prefixed by **deoxy-**. The base rings are numbered conventionally while primes (') are used for the sugar residue numbers. Numbers are not included in the name if the phosphate unit is at position 5' but are indicated by appropriate location for phosphates attached to any other position. The bases and sugars occur in DNA and RNA nucleotides and nucleosides are shown below. The purine bases adenine (A) and guanine (G) and pyrimidine base cytosine (C) occur in both DNA and RNA. However, the pyrimidine base thymine (T) occurs in DNA but its replacement uracil (U) occurs in RNA (Fig. **23** and Table **4** & **5**).

Table 4. The Nucleosides (Base + Sugar).

Base	Ribonucleoside	Deoxyribonucleoside
Adenine	Adenosine	Deoxyadenosine
Guanine	Guanosine	Deoxyguanosine
Cytosine	Cytidine	Deoxycytidine
Uracil	Uridine	-
Thymine	-	Deoxythymidine

Table 5. The Ribonucleotides and Deoxyribonucleotides (Base + Sugar + Phosphate).

Base	Monophosphate	Diphosphate	Triphosphate
Ribonucleotides			
Adenine	Adenosine monophosphate (Adenylic acid, AMP)	ADP	ATP
Guanine	Guanosine monophosphate (Guanylic acid, GMP)	GDP	GTP
Cytosine	Cytidine monophosphate (Cytidylic acid, CMP)	CDP	CTP
Uracil	Uridine monophosphate (Uridylic acid, UMP)	UDP	UTP
Deoxyribonucleotides			
Adenine	d-Adenosine monophosphate (Adenylic acid, d-AMP)	d-ADP	d-ATP
Guanine	d-Guanosine monophosphate (Guanylic acid, d-GMP)	d-GDP	d-GTP
Cytosine	d-Cytidine monophosphate (Cytidylic acid, d-CMP)	d-CDP	d-CTP
Thymine	d-Thymidine monophosphate (Thymidylic acid, d-TMP)	d-TDP	d-TTP

a) Pyrimidine bases

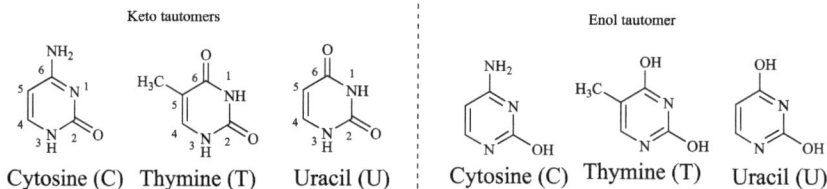

b) Purine bases

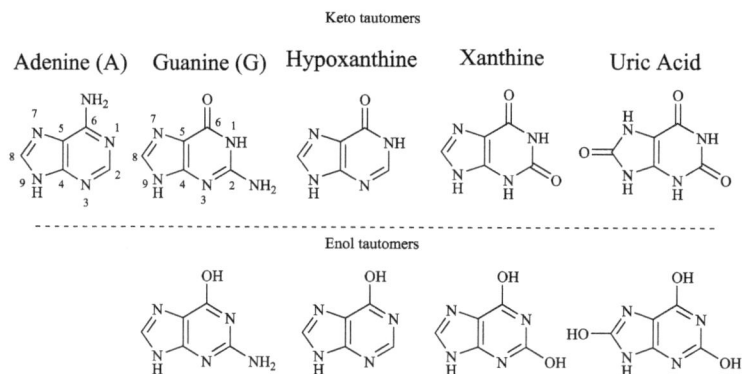

c) Sugars

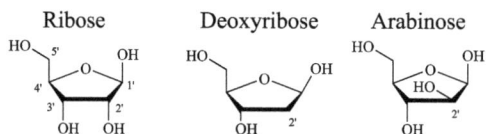

Fig. (23). The structures of bases and sugars present in nucleic acids.

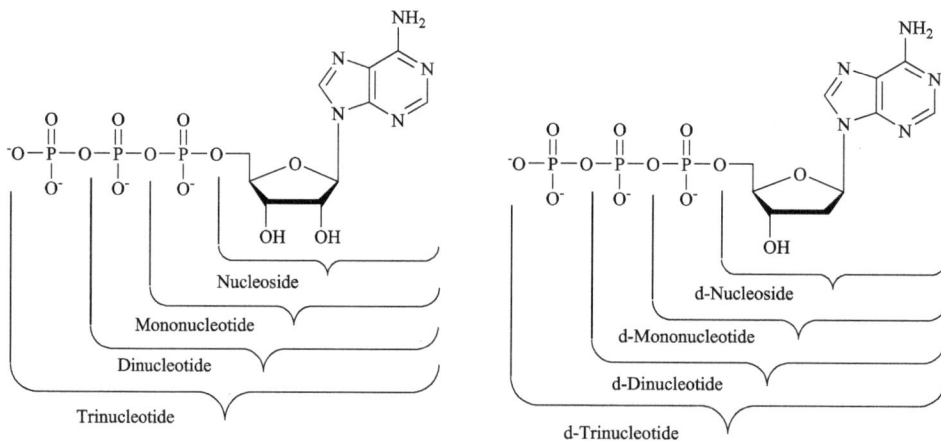

The adenine nucleotides for both DNA and RNA with different components are shown below as an example.

a) From Nucleotide to DNA

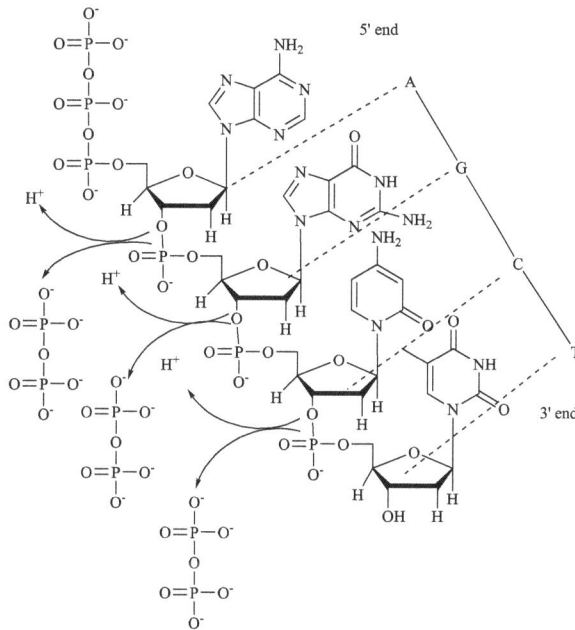

b) Base pairing

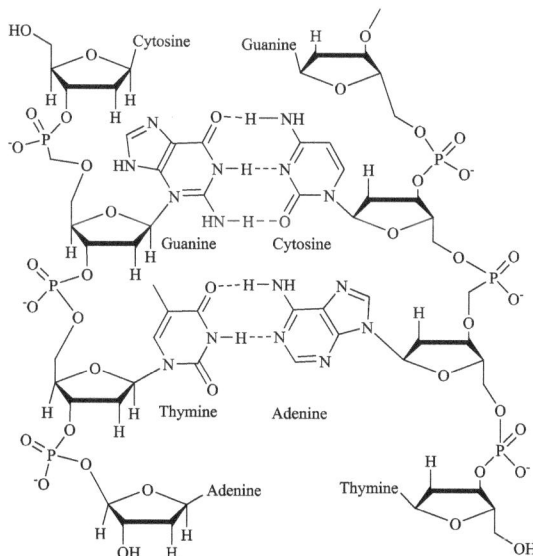

Fig. (24). DNA structure and base pairing.

DNA Structure

DNA molecules, formed by the condensation of nucleotides by the action of DNA polymerase, are large with a molecular mass of up to one trillion (10^{12}) consisting of two strands held together through hydrogen bonding between complementary base-pairs (A-T and G-C) to form a double helix (Fig. **24**). The **complementary base pairs** that form the internal structure of the helix are hydrogen bonded in such a way that their flat structures lie parallel to one another across the inside of the helix. The base pair A-T (or A-U in RNA) form two hydrogen bonds while the base pair C-G forms three hydrogen bonds. Interestingly, the amount of the complimentary bases in a particular organism is always equal, for example, human DNA contains 30% adenine, 30% thymine (%A = %T), 20% guanine and 20% cytosine (%C = %G).

The two polymer chains are aligned in opposite direction. At the terminal, one chain has a free 3'-OH while the other has 5'-OH group. The double helical chain of DNA is folded, twisted and coiled into quite compact shapes. Some DNAs are cyclic, which are also coiled and twisted into specific shapes, referred to as supercoils, supertwists or superhelices. The natural DNA has about 10 bases per turn of the helix and the outer surface has two grooves – the minor and major grooves – which act as binding sites for many drug molecules (Fig. **25**).

1 turn = 10 base pairs = 3.4 nm

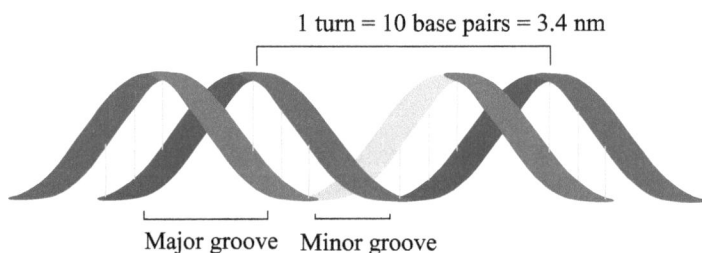

Major groove   Minor groove

Fig. (25). DNA helix showing major and minor grooves.

A segment of a DNA strand, about 1000 – 3500 nucleotide units long in humans, that contains the base sequence for the production of a specific heterogeneous nuclear RNA (hnRNA)/mRNA molecules is known as **gene**. The gene segment that conveys the genetic information is termed an **exon** and the gene segment that does not convey the genetic information is termed an **intron** (Fig. **26**). All the genetic information contained in the chromosomes of an organism is known as **genome**.

Fig. (26). Illustration of a gene structure showing exon and introns.

DNA Replication

DNA molecules are able to reproduce exact copies of themselves in a process called **replication** (Fig. **27**), which occurs during the synthesis or S-phase of the cell cycle. DNA replication begins with the unwinding of the DNA double helix starting at either end of the molecule or more commonly in the central section guided by a sequence of nucleotide bases in the DNA known as the origin of replication or ORI. Bacterial chromosomes generally have only one ORI, while human chromosomes may have multiple ORIs to aid in speeding up replication. The unwinding of the double helices occurs under the influence of **DNA helicase** – unwinding is important for the creation of a physical niche enabling the entry of enzymes that will continue the process of DNA replication. The separated strands act as templates for the formation of a new **daughter** strand, which is an exact copy of the original. **New** individual nucleotides bind these separated strands by H-bonding to the complementary parent nucleotides. As the nucleotides H-bond to the parent strand, they are linked to the adjacent nucleotide, which is already H-bonded to the parent strand, by the action of **DNA polymerases**. As the daughter strands grow, the DNA helix continues to unwind and both daughter strands are formed simultaneously in the 5' to 3' direction. This means that the growth of the daughter strand that starts at the 3' end of the parent strand can continue smoothly as the DNA helix continues to unwind. This strand is known as **leading strand**. However, this smooth growth is not possible for the daughter strand that started from the 5' end of the parent strand. This strand is known as lagging strand and is formed in a series of sections, each of which also grows from 5' to 3' direction. Each section is known as an **Okazaki fragment**, all of which are ultimately joined together by **DNA ligase** to form the second daughter strand.

RNA Structure and Transcription

RNA is synthesized using DNA as a template in a cellular process called **transcription** which occurs in the nucleus. RNA is found in three forms: (1) ribosomal or rRNA; (2) messenger or mRNA and (3) transfer or tRNA. rRNA and tRNA are located in the cytoplasm, while mRNA is synthesized in the nucleus and is subsequently transported to the cytoplasm to be used as a template for

protein synthesis. Unlike DNA, RNA is single stranded, consisting of the same base as DNA, except for thymine, which is replaced by uracil in RNA. Also unlike DNA, RNA does not contain equal amounts of complimentary bases. RNA is formed in the nucleus from unwound DNA in the 5' to 3' direction, which proceeds smoothly from the 3' end towards 5' end of the parent DNA strand. The reaction is commenced and catalyzed by the enzyme **RNA polymerase (RNA Pol)**. Since only complementary base paring is possible between the DNA template and the nascent RNA strand, the sequence in the parent DNA template determines the sequence of the new RNA strand, thus transcribing the genetic information. The **start** and **stop signals** in DNA control the size of the RNA molecule. RNA thus formed is much smaller than DNA, ranging from 75 nucleotides to a few thousand, and is known as *heterogeneous nuclear RNA (hnRNA), premessenger RNA (pre-mRNA) or primary transcript RNA (ptRNA)* and in humans is subjected to a process known as post-translational modification where chemical changes are made to the RNA that make it less prone to lysis.

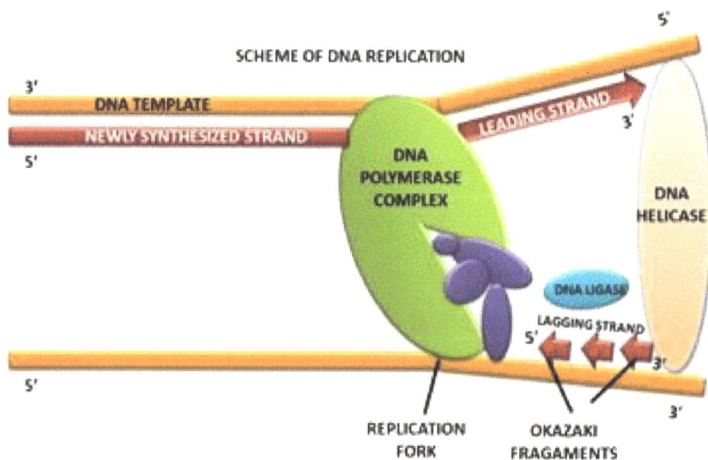

Fig. (27). Illustration of DNA replication process.

Messenger RNA (mRNA)

mRNA is synthesized (Fig. **28**) as required and is degraded once the corresponding protein is synthesized. However, the half-life of tRNAs varies immensely, based on the type of protein and its cellular function. In eukaryotes, intervening sequences called **introns** are removed and **exons** or expressed sequences are spliced together to form a continuous mRNA sequence. This then leaves the nucleus through a nuclear pore and carries the message to the ribosome in the cytoplasm, where it dictates the amino acid sequence of the particular

protein it synthesizes. The information is carried in the form of a trinucleotide code (**codon**), which is indicated by a sequence of letters corresponding to the 5' to 3' order of bases in the trinucleotide. Some amino acids are coded for by more than one codon, and the complete code made by all mRNA's codons is called the **genetic code**. Since three-base sequences (codons) are formed out of the 4 naturally occurring bases (A, C, G and U), there would be $4^3 = 64$ possible codons (Table **6**). It is now known that 61 out of 64 codons are related to specific amino acids while the other three are termed **"stop" codons** that terminate protein synthesis. When it occurs as the first codon in an amino acid sequence, the codon AUG functions as an initiator of protein synthesis (**start codon**) and codes for the amino acid methionine. Interestingly, in bacteria AUG codes for a chemically modified form of methionine called formyl-methionine (fMET). Thus, all bacterial polypeptides start with fMET. Since fMET is different from MET, it serves as an important therapeutic target for chemotherapeutic drugs.

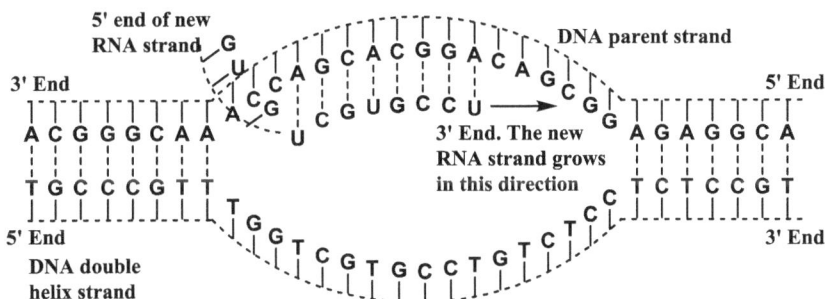

Fig. (28). Synthesis of mRNA from DNA.

Transfer RNA (tRNA)

Each amino acid has its own tRNA looks like a hairpin in its conformation (Fig. **29**) to be transported as an amino acid-tRNA complex. The addition of an amino acid to its specific tRNA is said to "activate" or charge the tRNA. Amino acids are transported from the cell's amino acid pool to the mRNA bound to the ribosome, the site of protein synthesis, as charged amino acid-tRNA complexes. The **anticodon** on the loop of the tRNA recognizes the complementary codon on the mRNA, where it binds and delivers the specific amino acid at a specific initiation point. Translation or protein synthesis begins with the formation of the peptide bond (-CONH-) between the incoming amino acid and that already present at the ribosomal site attached through the codon-anticodon pairing. By repeating this process, a polypeptide chain is formed as directed by the genetic code on the mRNA. The growth of the polypeptide chain occurs from the N-terminal end by addition of amino acids.

Table 6. The universal genetic code. (Start codon is always AUG; some codons work as start and stop signals; codons are always written left to right, 5' to 3').

First Position (5' end)	Second Position U C A G				Third Position (3' end)
U	Phe	Ser	Tyr	Cys	U
	Phe	Ser	Tyr	Cys	C
	Leu	Ser	Stop	Stop	A
	Leu	Ser	Stop	Trp	G
C	Leu	Pro	His	Arg	U
	Leu	Pro	His	Arg	C
	Leu	Pro	Gln	Arg	A
	Leu	Pro	Gln	Arg	G
A	Ile	Thr	Asn	Ser	U
	Ile	Thr	Asn	Ser	C
	Ile	Thr	Lys	Arg	A
	Met	Thr	Lys	Arg	G
G	Val	Ala	Asp	Gly	U
	Val	Ala	Asp	Gly	C
	Val	Ala	Glu	Gly	A
	Val	Ala	Glu	Gly	G

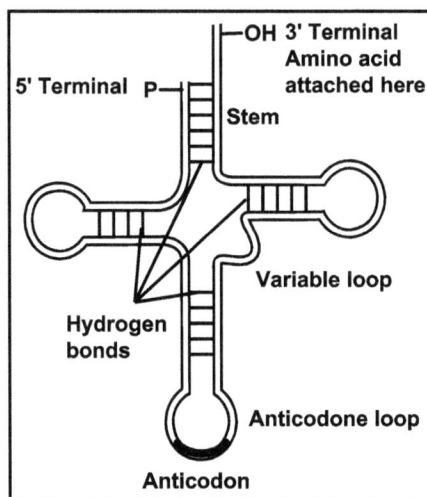

Fig. (29). The hairpin structure of tRNA.

Ribosomal RNA (rRNA)

rRNA is found in ribosomes (about 35% protein and 65% rRNA) and its sequence varies considerably from species to species. The strand is folded and twisted to form a series of single stranded loops separated by sections of loops formed by H-

bonding between complementary base pairs of the long stretch of the RNA. The significance of this is the ultimate globular shape of the ribosome which is achieved as a consequence of this packaging. The general pattern of loops and helices is very similar between species even though the sequences of nucleotides are different. rRNA, aside from being an important structural component of ribosomes, serves as sites of attachment for activated tRNAs during translation.

STUDENT SELF-STUDY GUIDE

1. Identify each functional groups from given structures.
2. Understand the IUPAC nomenclatures of the given organic functional groups.
3. What is the Hückel rule of aromaticity? Apply this rule to determine if a given structure is aromatic or not.
4. What are polar, nonpolar, ionizable and hydrogen bonding functional groups? Further, which functional groups are water soluble and which ones are soluble in organic solvents? Why?
5. What is the difference between ester, amide, carbonate, carbamate and urea? Amide *vs* lactam *vs* peptide *vs* anilide. Ester *vs* lactone.
6. Structural aspects of carbohydrates: straight chain or cyclic structure, furanose or pyranose structure and their formation, hemiacetal and hemiketal.
7. The anomeric carbon and anomers – assigning a anomer and b anomer
8. Assigning D and L configurations of carbohydrates. What is enantiomers, diastereomers and epimers?
9. What is a reducing sugar? How do you determine from structure if it is a reducing sugar?
10. What is a glycosidic link? How to assign α- and β- glycosidic link?
11. Few important disaccharides with structures (maltose, sucrose, lactose) with their α- and β-assignments
12. Difference between cellulose, amylose, amylopectin and glycogen
13. Structure and function of few important carbohydrates: Deoxyribose, glucosamine, glucuronic acid, hyaluronic acid
14. Define lipids with examples
15. What are ω-3 and ω-6 Fatty Acids?
16. Difference between simple lipid, sphingolipid, and phospholipid with examples
17. Steroid structure and nomenclature
18. Difference between prostaglandins, thromboxanes, and leukotrienes
19. Amino acid structures and assign: polar or nonpolar, acidic or basic, essential or nonessential
20. Zwitterion, positively or negatively charged amino acids, influence of pH
21. Difference between peptides and proteins
22. Difference between cofactor, coenzyme, prosthetic group

23. Primary, secondary, tertiary and quaternary structure of proteins
24. Structures of few heterocycles as shown in your handout
25. What are important bases, sugars and other components in a nucleotide?
26. Difference between nucleotide and nucleoside
27. Formation of DNA from nucleotides. No of nucleotide per complete DNA turn. The DNA minor and major grooves
28. What are exon and intron in a gene?
29. Definitions of DNA polymerase, DNA ligase, lagging strand, leading strand, and Okazaki fragments.
30. What is a hairpin turn?
31. What is a codon and anticodon? What are start codon and stop codons?

STUDENTS SELF-ASSESSMENT

Part I: Directions: Each of the numbered items or incomplete statements in this section is followed by answers or by completions of the statements. Select the **one** lettered answer or completion that is **best** in each case.

1. Which one is not an essential amino acid?
   A. Phenylalanine
   B. Tryptophan
   C. Histidine
   D. Glutamic acid
   E. Lysine

2. All of the following carbohydrates are considered to be polysaccharides except:
   A. Murein
   B. Amylose
   C. Amylopectin
   D. Lactose
   E. Cellulose

3. Which of the following compounds is considered the building block of DNA?
   A. AMP
   B. Adenosine
   C. UMP
   D. Purines
   E. Amino acids

4. The following is the structure of:

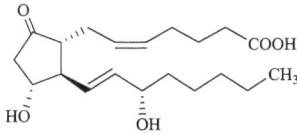

A. Prostaglandin E_2
B. Thromboxane A_2
C. Linolenic acid
D. Leukotriene D_4
E. Arachidonic acid

5. The systemic name of hydrocortisone (structure shown) is:

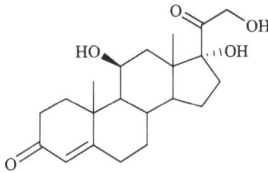

A. glucocorticoid
B. 11β,17α,21-trihydroxy-preg-4-ene-3,20-dione
C. 11α,17α,21-trihydroxy-preg-4-ene-3,20-dione
D. 11α,17β,21-trihydroxy-preg-4-ene-3,20-dione
E. 11β,17α,21-trihydroxy-estr-4-ene-3,20-dione

6. Phenyl alanine (structure shown) is not:

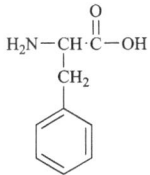

A. hydrophobic amino acid
B. nonpolar amino acid
C. aromatic amino acid
D. essential amino acid
E. polar acidic amino acid

7. The heterocyclic ring present in cimetidine is:

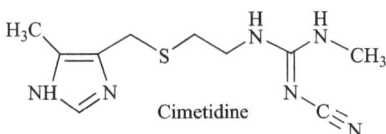

Cimetidine

A. pyrrole
B. furan

C. thiophene

D. imidazole

E. guanidine

8. See the following structure and find the <u>INCORRECT</u> statement from the keys A – E afterwards:

A. The fusion type between ring A and ring B is a *trans* fusion

B. Steroid hormones have unsaturation in ring A and thus they possess *trans* type of A/B ring fusion

C. Saturation of steroid hormones produces metabolite with a *cis* fusion which is active

D. Bile acids have *cis* A/B fusion

E. Cardiac glycosides have *cis* A/B and C/D fusions

9. If a protein chain is 28 amino acids long, at least how many complete rotations are represented in its original DNA sequence assuming no introns?

A. 2

B. 3

C. 9

D. 10

E. 11

10. A reducing sugar by definition:

A. contains an exposed hemiacetal

B. can be hydrolyzed into its base sugars

C. will not branch

D. contains an alpha or beta 1,4-linkage

11. <u>Statement 1</u>: Humans normally digest polysaccharides with alpha glycosidic linkages; and can also digest lactose that contains beta glycosidic linkage but needs special enzyme lactase.

<u>Statement 2</u>: Essential amino acids include: methionine, lysine, leucine and alanine.

A. Statement 1 is true but statement 2 is false

B. Statement 2 is true but statement 1 is false

C. Both statements are true

D. Both statements are false

12. <u>Statement 1</u>: Primary protein structure is the sequence of amino acid residues in peptides and proteins.

<u>Statement 2</u>: The DNA helicase join together the Okazaki fragments in the lagging strand.

    A. Statement 1 is true but statement 2 is false

    B. Statement 2 is true but statement 1 is false

    C. Both statements are true

    D. Both statements are false

1. Which one of the following polysaccharides has an 1,6-linked glycosidic branches every 8-10 glucose units?

    A. Cellulose

    B. Glycogen

    C. Amylose

    D. Amylopectin

2. If a protein chain is 28 amino acids long, at least how many complete rotations are represented in its original DNA sequence assuming no introns?

    A. 2

    B. 3

    C. 9

    D. 10

    E. 11

3. A reducing sugar by definition:

    A. contains an exposed hemiacetal

    B. can be hydrolyzed into its base sugars

    C. will not branch

    D. contains an alpha or beta 1,4-linkage

Part II: Directions: Each item below contains three suggested answers of which **one or more** is correct. Chose the answer

    A. if **I only** is correct

    B. if **III only** is correct

    C. if **I and II** are correct

    D. if **II and III** are correct

    E. if **I, II, and III** are correct

1. The name(s) of the following structure is(are):

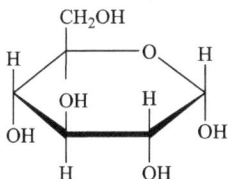

    I. β-D-Glucose.

    II. α-D-Glucose.

    III. Glucopyranose

2. **Linolenic acid (structure shown) is:** $CH_3(CH_2CH=CH)_3(CH_2)_7COOH$
    I. essential fatty acid
    II. omega-3 fatty acid
    III. octadecatrienoic acid

3. When certain functional groups are introduced into a benzene nucleus, they tend to <u>decreaselipophilicity</u> of benzene. These groups include
    I. a hydroxyl group
    II. a carboxylic acid group
    III. a chlorine group

4. Functional groups present in the molecule shown include:

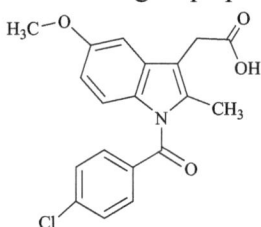

    I. an ester
    II. a tertiary amine
    III. a carboxylic acid

5. The functional groups and/or ring systems present in Penicillin G, as shown, include:

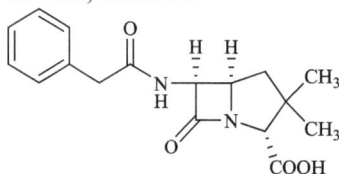

    I. Carboxylic acid
    II. β-Lactam
    III. Amide

6. Diazinon has:

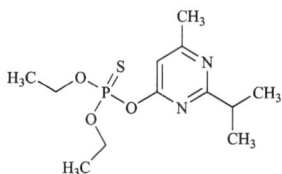

    I. Isopropyl group
    II. Thiophosphate ester,
    III. Pyridine ring system

Part III: Directions: The group of items in this section consists of lettered options followed by a set of numbered items. For each item, select the one lettered option that is most closely associated with it. Each lettered option may be selected once, more than once, or not at all.

1. Match the statements/name (A-E) with the appropriate amino acid structures I-V.

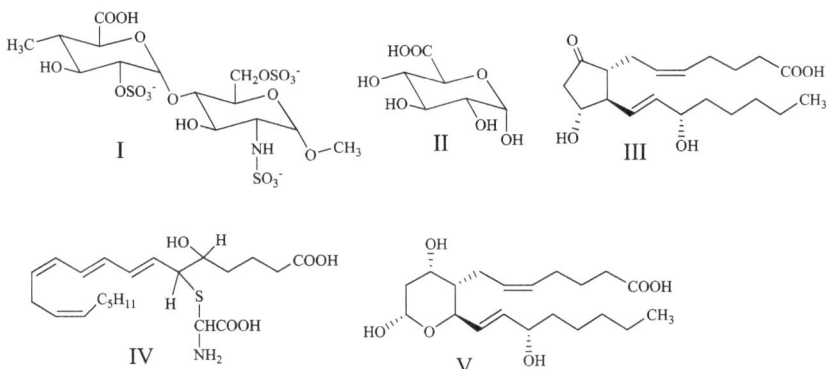

A. Prostaglandin E_2
B. Thromboxane B_2
C. Leukotriene D_4
D. Sugar component important in glucuronidation
E. Sugar repeating unit in heparin

2. Match the appropriate statement (A-E) with the amino acid structures I-V.
   A. An essential, hydrophobic amino acid capable of charge transfer interaction
   B. An essential polar amino acid capable of hydrogen bonding but no ionic interactions
   C. An essential amino acid capable of ionic interaction with acidic drugs
   D. A nonessential amino acid capable of ionic interaction with amine drugs
   E. A nonessential hydrophobic amino acid important for protein folding

3.

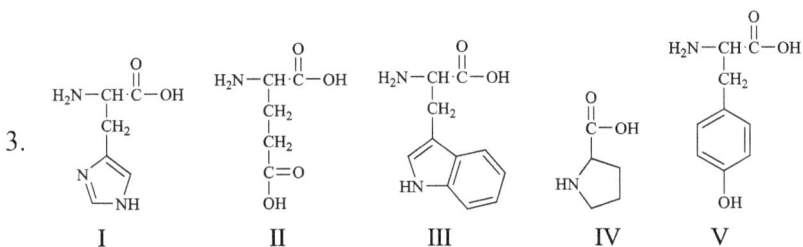

4. Match the functional group/ ring system names A-E with the circled and/or colored structural components (I-V) of eszopiclone as shown.

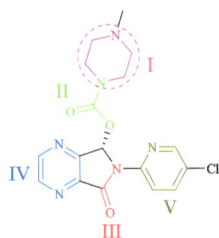

 A. Carbamate
 B. Lactam
 C. Piperazine
 D. Pyrazine
 E. Pyridine

5. Match the name of the fused ring systems A-E present in drug structures I-V.

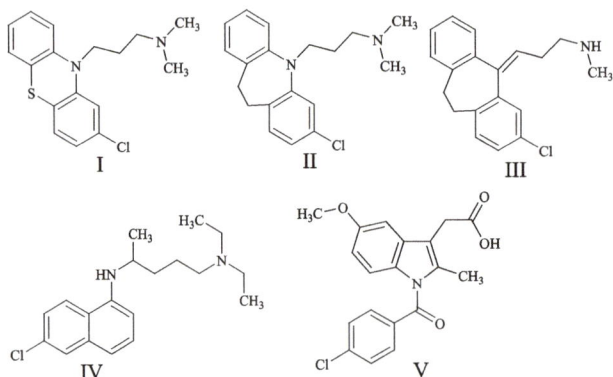

 A. Phenothiazine
 B. Indole
 C. Quinoline
 D. Dibenzazepine
 E. Dibenzepin

6. Match the name of the mono-heterocyclic rings A-E that are present in drugs structures I - V as shown below.

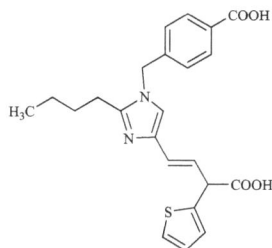

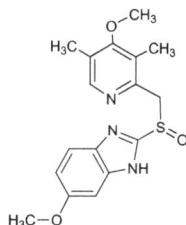

I. Eprosartan   II. Omeprazole   III. Famotidine

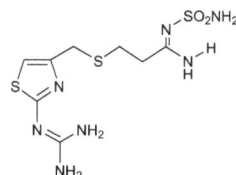

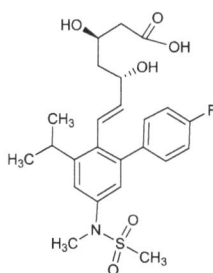

IV. Rosuvastatin   V. Enalapril

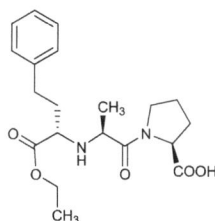

A. Thiazole
B. Pyrrolidine
C. Pyrimidine
D. Imidazole
E. Pyridine

CONSENT FOR PUBLICATION

Not applicable.

CONFLICT OF INTEREST

The authors declare no conflict of interest, financial or otherwise.

ACKNOWLEDGEMENT

Declared none.

REFERENCES

[1]   Reeds PJ. Dispensable and indispensable amino acids for humans. J Nutr 2000; 130(7): 1835S-40S. [http://dx.doi.org/10.1093/jn/130.7.1835S] [PMID: 10867060]

[2]   Gutiérrez-Preciado A, Romero H, Peimbert M. An evolutionary perspective on amino acids. Nature Education 2010; 3(9): 29.

[3]   Kopple JD, Swendseid ME. Evidence that histidine is an essential amino acid in normal and chronically uremic man. J Clin Invest 1975; 55(5): 881-91.
[http://dx.doi.org/10.1172/JCI108016] [PMID: 1123426]

Further Reading

1. Rakoff H, Rose NC. Organic Chemistry, New York, NY: The Macmillan Company; 1966.

2. Lemke TL. Review of Organic Functional Groups: Introduction to Medicinal Organic Chemistry. Philadelphia, PA: Lippincott Williams & Wilkins; 1992.

3. Glossary of class names of organic compounds and reactive intermediates based on structure, International Union of Pure and Applied Chemistry, Organic Chemistry Division. http://www.chem.qmul.ac.uk/iupac/class/

4. McMurry J. Organic Chemistry: A Biological Approach. Belmont, CA: Thompson Brooks/Cole; 2011.

5. Wermuth CG, Ganellin CR, Lindberg P, Mitscher LA. Mitscher. Glossary of terms used in medicinal chemistry (IUPAC Recommendation 1998). http://www.chem.gmul.ac.uk/iupac/medchem/

CHAPTER 3

Acid-base Chemistry and Salt Formation

Hardeep Singh Saluja[1] and **M. O. Faruk Khan**[2,*]

[1] *Department of Pharmaceutical Sciences, College of Pharmacy, Southwestern Oklahoma State University, Weatherford, Oklahoma, USA*

[2] *Department of Pharmaceutical Sciences and Research, Marshall University School of Pharmacy, Huntington, West Virginia, USA*

Abstract: This chapter provides a brief review of the concepts of acid-base chemistry with its applications on drug molecules. It also highlights the significance of salt formations in pharmaceutical products. After studying this chapter, students will be able to:
• Comprehend the Arrhenius, Brønstead-Lowry, and Lewis concepts of acids and bases.
• Comprehend the ionization, factors controlling ionization, and the acids-base strengths.
• Identify the acids and bases occurring in drug molecules and their significance.
• Apply the concept of acid-conjugate base and base-conjugate acid in pharmacy.
• Apply Henderson-Hasselbalch equation and in pharmacy and drug action and bioavailability.
• Interpret pH partition theory and its significance in drug pharmacokinetics.
• Illustrate the purpose of salt formation with drug molecules and the acidity or basicity of the salts.

Keywords: Acids, Bases, Henderson-Hasselbalch equation, Ionization, Partition coefficient, pH partition theory, Salt formation, Solubility.

INTRODUCTION

The chemical structure of a drug determines its interactions with the body, and is therefore, critical to its clinical utility. Physicochemical properties of a drug determined by its chemistry include melting & boiling points, crystal structure, ionization and solubility. It is generally accepted that if a drug is ionized, it will have greater water solubility than lipid solubility. The opposite is true if it is unionized. The solubility of a drug in turn influences its pharmacokinetic parameters such as absorption, distribution, metabolism and excretion (ADME),

* **Corresponding author M.O. Faruk Khan**: Department of Pharmaceutical Sciences and Research, Marshall University School of Pharmacy, Huntington, WV, USA; Tel: 304-696-3094; Fax: 304-696-7309; E-mail: khanmo@marshall.edu

M.O. Faruk Khan & Ashok Philip (Eds.)

which in turn influence the clinical parameters, such as onset, duration, peak time, and half-life. Almost 75% of all drugs are bases, 20% are acids, and 5% are either non-ionizable or amphoteric in nature [1]. Since amphoteric drugs possess both acidic and basic functional groups, >95% of all drugs can be categorized as acids or bases.

ACID–BASE THEORIES

Theories put forth to define the acid-base characteristics of a drug are discussed below.

Arrhenius Concept

According to Arrhenius concept, an acid is a substance that donates hydrogen ion in aqueous medium and thus increases hydrogen ion concentration *(H^+, also represents as hydronium ions H_3O^+).* A base on the other hand, donates hydroxide ion (represented as OH^-) in aqueous medium and thus increases hydroxide concentration. *The following example illustrates the aforementioned concept.*

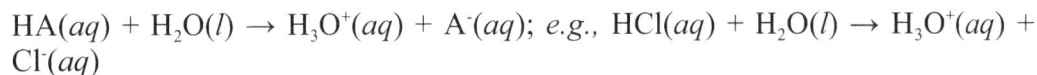

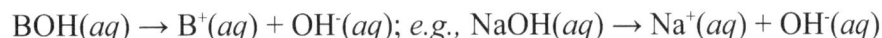

$HA(aq) + H_2O(l) \rightarrow H_3O^+(aq) + A^-(aq)$; *e.g.,* $HCl(aq) + H_2O(l) \rightarrow H_3O^+(aq) + Cl^-(aq)$

$BOH(aq) \rightarrow B^+(aq) + OH^-(aq)$; *e.g.,* $NaOH(aq) \rightarrow Na^+(aq) + OH^-(aq)$

An aqueous solution of acid is denoted by $HA(aq)$ and its ionized form is denoted by $A^-(aq)$ while an aqueous solution of base is denoted by $BOH(aq)$ (or simply B) and its ionized form by $B^+(aq)$. A strong acid, for example hydrochloric acid and a strong base, for example sodium hydroxide are completely ionized in aqueous solution producing the hydronium ion and hydroxide ion, respectively. Table **1** shows the list of strong acids and bases. The definition of Arrhenius is applicable only to an aqueous solution and is a major limitation of the concept. H^+ ion is the source of acid character, and the OH^- ion is the source of base character. It should be noted that the concepts of pH and pK_a were developed from Arrhenius theory.

Table 1. Common strong acids and bases.

Strong Acids	Strong Bases
Perchloric acid ($HClO_4$)	Lithium hydroxide (LiOH)
Sulfuric acid (H_2SO_4)	Sodium hydroxide (NaOH)
Hydroiodic acid (HI)	Potassium hydroxide (KOH)
Hydrobromic acid (HBr)	Calcium hydroxide ($Ca(OH)_2$)
Hydrochloric acid (HCl)	Strontium hydroxide ($Sr(OH)_2$)
Nitric acid (HNO_3)	Barium hydroxide ($Ba(OH)_2$)

Brønsted-Lowry Concept

This is the most applicable concept to drug molecules in general, which defines an *acid as a proton (H⁺) donor and a base as a H⁺ acceptor* (solvent does not have to be water in any case) in a proton transfer reaction. The general equations for Brønstead-Lowry acids and bases are (Fig. **1**):

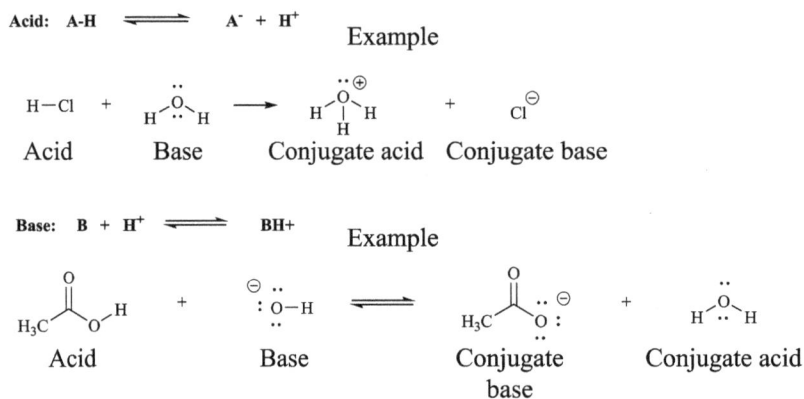

Fig. (**1**). Brønsted-Lowry acids and bases.

Note that the above two reactions are reversible. A⁻ is a proton acceptor and according to the above definition, it makes it a "base". When gaseous hydrogen chloride dissolves in water, for example, a polar HCl molecule acts as an acid and donates a proton, while a water molecule acts as a base and accepts the proton, yielding hydronium ion (H_3O^+) and chloride ion (Cl^-). Similarly, in the second reaction, BH^+ is a proton donor which makes it an "acid" as per the above definition. In this case, hydroxide (OH^-), accepts a proton from the Brønstead-Lowry acid, acetic acid. In both cases, the corresponding conjugates of the reaction are shown on the right (Fig. **1**). To summarize this concept, unionized parent acids are termed as **HA acids** while the bases that become acids after accepting a proton are termed as **BH⁺ acids.**

Lewis Concept

The realization that certain acid-base reactions do not involve the proton transfer required a more generalized way to represent acid-base reactions. According to this concept, a **Lewis acid** is a species that can form a covalent bond by "*accepting an electron pair*" from another species and a **Lewis base** is a species that can form a covalent bond by "*donating an electron pair*" to another species. Thus, a *Lewis acid is an electrophile* while a *Lewis base is a nucleophile*. This definition includes all kinds of acids and bases and thus provides a more versatile definition of acids and bases.

Remember,   Lewis acid = Electron acceptor;   Lewis base = Electron donor

Fig. **2** shows that ammonia donates an electron-pair to proton to form the N-H bond. Thus, according to the Lewis concept, ammonia is a base and proton is an acid. Again, ammonia is accepting a proton and the proton is donating itself. Thus, ammonia is a base, and proton is an acid according to Brønsted-Lowry concept. Lewis theory therefore gives a more generalized definition of acids and bases and helps in understanding the strength of acids and bases in terms of the nucleophilicity or electrophilicity of certain groups.

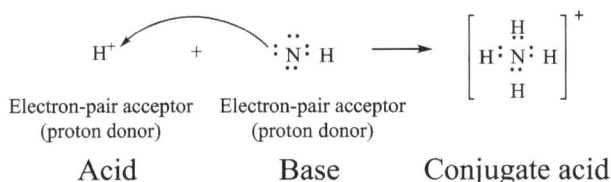

Electron-pair acceptor (proton donor)	Electron-pair acceptor (proton donor)	
Acid	Base	Conjugate acid

Fig. (2). Illustration of Lewis acid concept using ammonia.

Application of Lewis and Brønsted-Lowry Definitions to Imidazole

An imidazole is a five-membered heterocycle found in more than 20 FDA approved drugs. The N3 nitrogen of imidazole is basic, because the lone-pair of electrons acts as a **Lewis base** and donates an electron pair to a **Lewis acid** (HCl, hydrochloric acid).

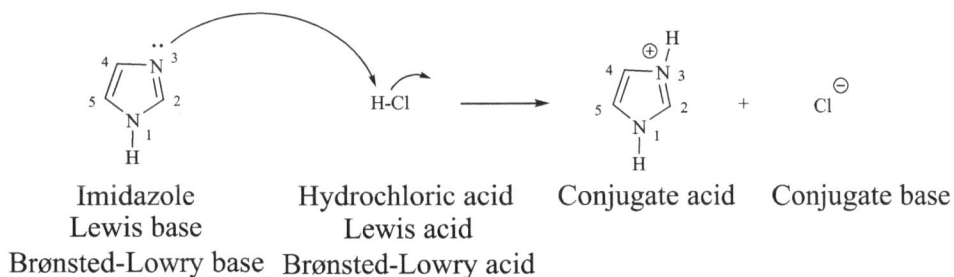

Imidazole	Hydrochloric acid	Conjugate acid	Conjugate base
Lewis base	Lewis acid		
Brønsted-Lowry base	Brønsted-Lowry acid		

From understanding that the *lone-pair of electrons on the imidazole N3 nitrogen drives the reaction*, one can easily surmise that the N3 nitrogen can also be thought of as a **Brønsted-Lowry base**, accepting a proton from HCl due to the availability of the N3 lone pair. This then makes HCl a **Brønsted-Lowry acid**, as it donates a proton (H^+) in this reaction. A special consideration in this reaction is that only the N3 nitrogen is basic, while the N1 nitrogen is not. This is due to the location of the N3 lone pair, in an sp^2 orbital, which is perpendicular to the aromatic π-electrons located in the p-orbitals. The imidazole N1 nitrogen is not basic, because the lone pair of electrons is located in a p-orbital, and therefore delocalized with the other aromatic π-electrons occupying p-orbitals in this heterocyclic system. Similar to the imidazole N1 nitrogen, the N1 nitrogen of the heterocycle pyrrole is not basic, as its lone pair of electrons is also located in a p-orbital and delocalized throughout the π cloud of the ring. All pyrrole-like nitrogens in any heterocycle behave in similar way.

ACIDIC AND BASIC FUNCTIONAL GROUPS OCCUR IN DRUG MOLECULES

Oxy Acids

The most common acid groups that occur in drugs are called oxy acids because the acidic proton that is donated is attached to an oxygen atom. The strength of the acid is based on the strength of the O–H bond and the stability of the anionic product. The strength of a few oxy acids in descending order there are:

Sulfonic acids > phosphoric acids > carboxylic acids > enols > arenols (aromatic hydroxyls) > oximes > alcohols (alkyl hydroxyls).

The O–H bond is polarized because the oxygen atom is more electronegative than the hydrogen. However, the polarization may be increased if the oxygen is attached to an electron withdrawing group. For example, the very strong sulfonyl group attached to hydroxyl group polarizes the O–H bond to such an extent that the sulfonic acids have negative pKa values. On the other hand, the phosphonyl group has relatively less electron withdrawing capability compared to sulfonyl group, thus phosphate esters are weaker acids with pKa values around 2. In carboxylic acids, the electron withdrawing group is the carbonyl, which is even weaker than the phosphonyl. Therefore, carboxylic acids have pKa values in the range of 3 to 5. Fig. **3** shows a few examples of drugs containing carboxylic and sulfonic acid functions.

Since unsaturation is electron withdrawing, enols, arenols and oximes are acidic. The olefinic bond produces enols which have a pKa range of 3 to 7. Aromatic rings, such as the phenyl ring, are mild electron withdrawers, thus arenols are weak acids with a pKa range of 9 to 10. Oximes are weakly acidic due to the unsaturation of the imine. Even though oximes are isosteres of enols, they are much weaker acids than enols, with pKa values above 10. The examples of enolic and phenolic drugs are shown in Fig. **4**.

Acidity of sulfuric, phosphoric and carboxylic acids:

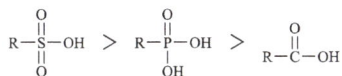

Drug examplres:

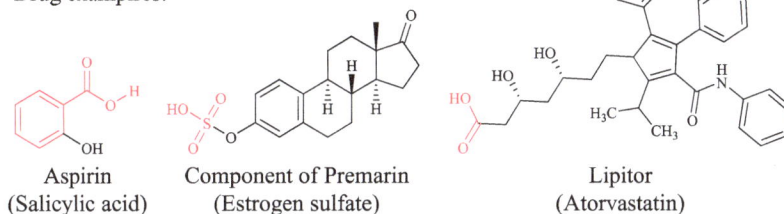

| Aspirin (Salicylic acid) | Component of Premarin (Estrogen sulfate) | Lipitor (Atorvastatin) |

Fig. (3). Sulfuric, phosphoric, and carboxylic acids with drug examples.

Acidity of enol, phelol and oxime:

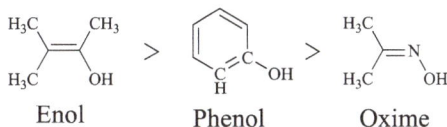

Enol   Phenol   Oxime

Drug examples:

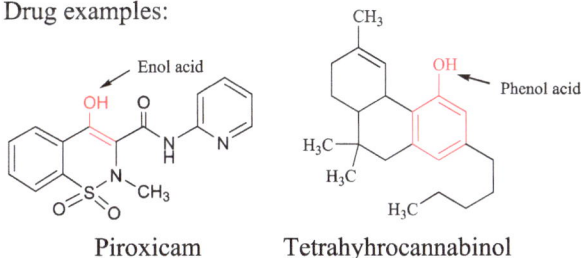

Piroxicam   Tetrahyhrocannabinol

Fig. (4). Enolic and phenolic acids with drug examples.

Alkyl groups are electron donating; therefore, they decrease the polarizability of the O–H bond and in turn strengthen the O–H bond. This makes alkyl alcohols extremely weak acids, weaker than water. Their pKa values are above 15 and therefore, alcohols are considered neutral. Furthermore, the stability of the product is also affecting the acid strength by affecting the ionization equilibrium. If the product is stable, there is a low tendency for the reverse reaction. The greater tendency for the forward reaction results in a greater degree of proton donation making it a stronger acid. However, if the product is unstable, the reverse reaction predominates which resists the proton donation making it a weak acid. The stability of the product depends on the ability to delocalize the extra electron(s). The greater the resonance stabilization, the lower is the tendency for the reverse reaction, and thus stronger the acid.

In the above example, sulfonic acids have three oxygen atoms to help delocalize the charge. Phosphate ester and carboxylic acids (as carboxylates) have two oxygen atoms to delocalize the charge. The π system of the enols, aromatic rings, and oxime delocalize the charge by moving the electrons through a π bond system. In the case of the aliphatic alcohols, the ability of the oxygen to handle the charge is decreased by the electron donating effect of the alkyl group, so the reverse reaction prevails, producing very weak acids. Indeed, the alcohols have pK_a values above 14 and are essentially neutral.

Thio Acids

An isostere of the -OH group is the -SH group, the thiol group. The thiol is also known as the sulfhydryl, or mercapto group. Ethanol has a pKa of 15.9, too weak to be considered acidic. This is not surprising since the ethyl groups are electron donors. Ethylthiol, on the other hand, has a pKa of 10.5 making it a weak acid. The reason that the S–H bond is weaker than the O–H bond can be attributed to the size of the sulfur atom. The valence electrons in sulfur are the 3s and 3p. These orbitals are larger than oxygen's 2s and 2p. Therefore, the overlap between the sulfur's large orbitals and hydrogen's small 1s orbital is very poor, resulting in a weak S–H bond. The anionic form, S^- is also more stable than the O^-. The ion is more stable because the larger 3s and 3p orbitals provide more volume to disperse the charge; the charge is not as concentrated. Thus, thio acids are stronger acids than the corresponding oxy acids, whether it is attached to an electron donor or an electron withdrawing group. An extremely important result of the poor bonding between sulfur and hydrogen is also manifested in poor hydrogen bonding. This results in sulfur containing compounds having lower water solubility than their oxygen analogs. This difference in solubility plays a major role in biological activity.

Amine Bases

Organic chemistry tells us that the classical organic basic group is the amine because they can accept a proton. All amines are basic because they contain a nitrogen atom. The nitrogen atom in amine actually accepts protons or donates electrons, the process that makes the group basic. Recall that the nitrogen has 5 electrons in the outer most shells. Three of the five electrons are used to form covalent bonds with other atoms. The remaining two electrons are called lone pair, non-bonding or n-electrons because they are not ordinarily used to form a bond. The geometry of the amine is similar to that of a carbon atom in that it forms a tetrahedron with one of the apexes occupied by the n-electrons. It is this pair of electrons that are donated to the proton; therefore, the availability of the electron pair is important. The availability of these electrons determines the basic strength of the amine. The availability of the n-electrons depends on the substituents on the nitrogen. Electron donor groups increase the availability of the electrons (strong base), while electron withdrawing groups decrease availability of the electrons (weak base).

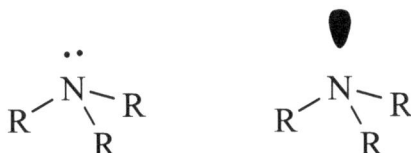

The most common electron donors are alkyl groups producing alkyl amines. The most common electron withdrawers are aromatic rings, producing aromatic amines. Thus aliphatic amines are much stronger bases than aromatic amines. Ammonia, with no donors or withdrawers, has a pKa of 9.26. Electron donors increase the pKa as illustrated by: methylammonia (10.64), dimethylammonia (10.71), trimethylammonia (9.72). The general pattern is ammonia < primary amines < tertiary amines < secondary amines. Tertiary amines are weaker bases than secondary amines because of the steric effect. The pK_a values of different substituted aliphatic amines are shown below [2].

MeNH$_2$ (10.64)	EtNH$_2$ (10.63)	PrNH$_2$ (10.58)	i-BuNH$_2$ (10.61)
Me$_2$NH (10.71)	Et$_2$NH (10.93)	Pr$_2$NH (10.98)	i-Bu$_2$NH (11.00)
Me$_3$N (9.72)	Et$_3$N (10.87)	Pr$_3$N (10.74)	i-Bu$_3$N (10.65)

Aromatic Amines

The phenyl ring is a mild electron withdrawer. But has drastic effects on the availability of the n-electrons. Consider ammonia (pK$_a$ 9.26), and aniline with one phenyl has a pKa of 4.63, that is aniline is 46,300 times weaker as a base than ammonia. Further additions of electron withdrawers produce even weaker bases. Diphenyl amine has a pK$_a$ of 0.8, triphenylamine has a pK$_a$ of -5, too weak to be considered basic. The general pattern for aromatic amines is: ammonia > primary amines > secondary amines > tertiary amines.

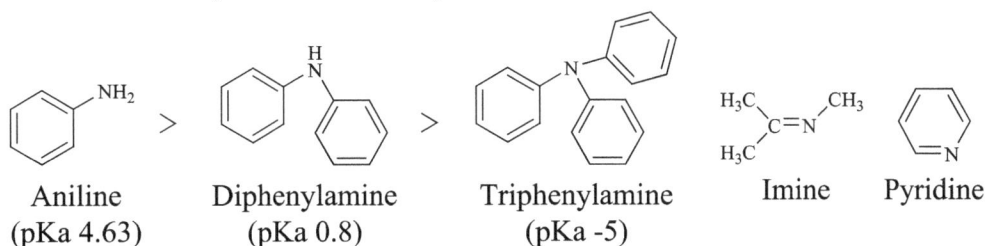

| Aniline (pKa 4.63) | Diphenylamine (pKa 0.8) | Triphenylamine (pKa -5) | Imine | Pyridine |

Pyridine and imine are other nitrogen containing groups, which are not amines but are basic. The imine is a carbon-nitrogen double bond, which is a weak base because of the unsaturation that functions as an electron withdrawing group. The pyridine is an aromatic ring containing nitrogen. The pyridine nitrogen is weakly basic because the aromatic ring functions as an electron withdrawing group, making the n-electrons less available. Since the pyridine ring is an isostere of the phenyl group, it is encountered often in drug structures. The pK_as of these groups are usually less than 4.

Generally speaking, all amines are basic but differ in ability to accept a proton. Since the effects can be so drastic, amines are usually grouped into: aliphatic amines (relatively strong bases), example is ephedrine (pKa 9.58), aromatic amines (very weak bases), example is dapsone (pK_a 1.0) (Fig. **5**).

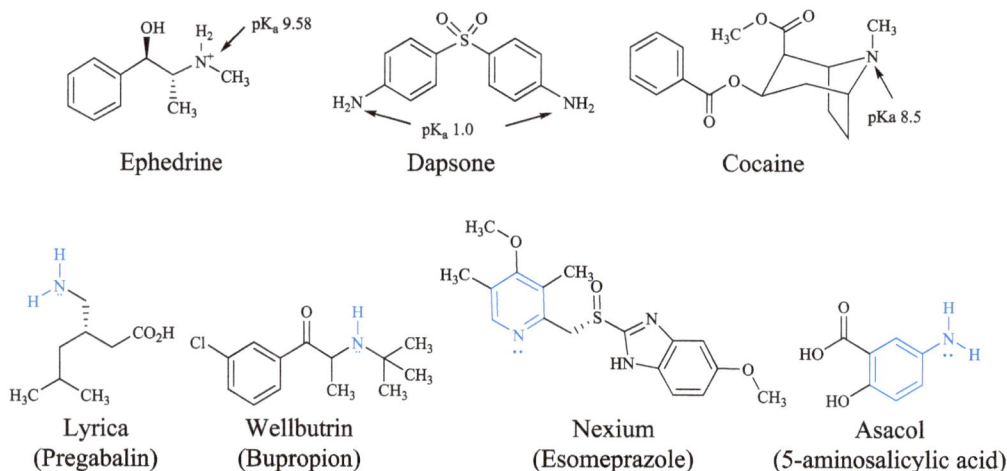

Ephedrine   Dapsone   Cocaine

Lyrica   Wellbutrin   Nexium   Asacol
(Pregabalin)   (Bupropion)   (Esomeprazole) (5-aminosalicylic acid)

Fig. (5). Drugs with aliphatic and aromatic amine functional groups.

There remain three similar groups which are strongly basic, more basic than aliphatic amines, but are less common in drugs (Fig. **6**). These are the amidine, guanidine, and biguanidine groups. The amidine is an isostere of the amide where the carbonyl oxygen is replaced by nitrogen.

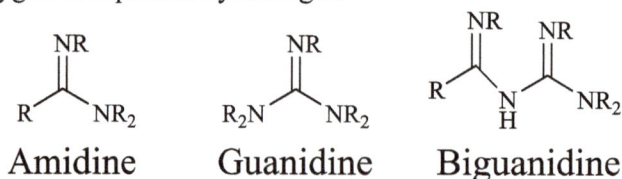

Amidine   Guanidine   Biguanidine

The C=N functional group is called an imine and, as discussed above, the nitrogen atom is basic. However, unlike the aforementioned imine, this nitrogen in amidine, guanidine and biguanidine, is very basic because of two reasons. The NR_2 functions as a very strong electron donor. In the amide, the carbonyl, C=O, oxygen functions as an electron withdrawer, making the amine NR_2 nitrogen's n-electrons unavailable. The NR_2 is neutral, thus the amide is neutral. The imine C=N functions as the carbonyl C=O, making the NR nitrogen's n-electrons again are unavailable, and thus neutral. However, the NR_2 functions as an electron donor to the imine C=N nitrogen making it more basic than a simple imine. Contributing to the basic nature is stability of the cationic product. The charge can be delocalized over two nitrogens. In aliphatic amines the charge remains only on one nitrogen atom. Thus, while aliphatic amines have pK_a values in the range of 9-10, the pK_a values of amidines are >10. The guanidine group is more basic than the amidine for the same two reasons. The imine C=N nitrogen's n-electrons are even more available because this group has two NR_2 groups functioning as electron donors. Also the protonated product is more stable because the charge can be delocalized over three nitrogens. The pKa range is 10-13. The biguanidine groups functions as two guanidines sharing a common NR atom. Thus, it is a dibasic group, accepting two protons, one at a time. Biguanidines are just as basic as guanidines.

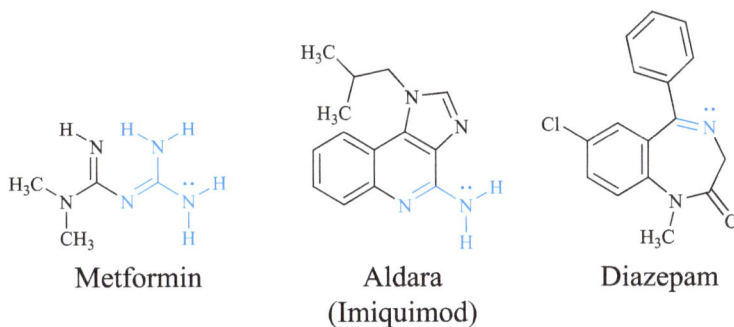

Metformin   Aldara (Imiquimod)   Diazepam

Fig. (6). Drugs containing imine, amidine and guanidine.

Acidic Amides (Azo Acids)

Adding a stronger electron withdrawer to the nitrogen has significant effects on the availability of the n-electrons. The carbonyl is a strong electron withdrawer. Substituting a carbonyl on the nitrogen produces an amide where the inductive effect is so strong the n-electrons of nitrogen are not available. Amides cannot accept a proton, thus are not basic. The phosphonyl is a stronger electron withdrawer than the carbonyl; substituting a phosphonyl on nitrogen produces a phosphamyl. The inductive effect is so strong that not only are the n-electrons not available but the N–H bond is polarized, weakening the bond, resulting in a weak acid. Since the acidic proton is on a nitrogen. These are called **azo acids** (Fig. 7).

Substituting the stronger sulfonyl on the nitrogen produces sulfamyl, slightly stronger azo acids. This group is found on many diuretics and antibacterial sulfonamides. Substituting two carbonyls on the nitrogen produces an imide. The inductive effect of two carbonyls produces acids generally stronger than sulfamyls. The imide is found in barbiturates and several antiepileptic drugs (Fig. **7**). Substituting one carbonyl and one sulfonyl produces the sulfamidocarbonyl. The combined effect produces acids nearly as strong as carboxylic acids. This group is found in the sulfonylurea oral hypoglycemics, sulfa antibacterials and barbiturates (Fig. **7**).

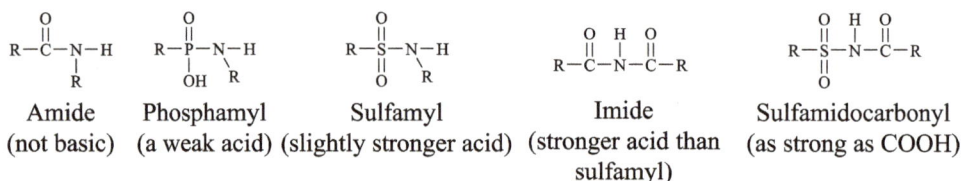

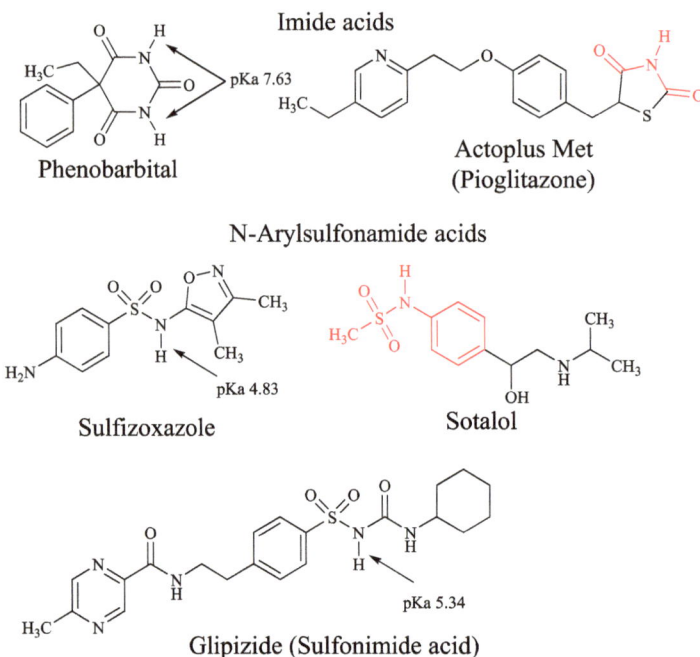

Amide (not basic) Phosphamyl (a weak acid) Sulfamyl (slightly stronger acid) Imide (stronger acid than sulfamyl) Sulfamidocarbonyl (as strong as COOH)

Imide acids

Phenobarbital Actoplus Met (Pioglitazone)

N-Arylsulfonamide acids

Sulfizoxazole Sotalol

Glipizide (Sulfonimide acid)

Fig. (7). Azo acids and their drug examples.

ß-Diketone

This last functional group is rare in clinically used drug molecules. Some consider it a carbon acid because the acidic proton is on a carbon surrounded by two carbonyls. It can be envisioned that the electron-pull from the two carbonyls polarizes the C—H bond (weakens it) as we saw in the case of the imide.

However, it can also be envisioned that one of the carbonyls tautomerizes to an enol, which is acidic. Most ketones have little tendency to enolize, but the tendency to tautomerize is increased by the presence of the second carbonyl. The product is stabilized by delocalizing the charge over two oxygens, just as in the carboxylate anion. Thus, diketones can be as acidic as enols or carboxylic acids (Fig. **8**).

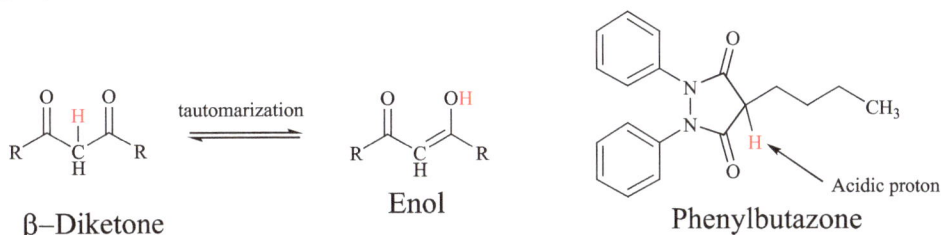

Fig. (8). β-Diketone: tautomerization and drug example.

ACID-BASE STRENGTHS

Acetic acid releases a proton to form acetate ion (conjugate base) with an ionization constant of K_a, and acetate accept the proton to form the acetic acid with an ionization constant of K_b (shown later). Due to the fact that acids and bases are produced after ionization, the strength of the acid or base depends on their ionization potential. The primary measure of ionization potential used in pharmacy, as elsewhere, is the pK_a, which is defined as the negative log of the ionization constant of a weak acid. By definition, weak bases have a pK_b and also a pK_a. We can measure the strength of a base with its pK_a, which is equal to $14-pK_b$. Thus, the ionization for both acids and bases is expressed with pK_a. For example, methylamine accepts a proton to form its conjugate acid methyl ammonium ion with an ionization constant of K_b, which releases a proton with an ionization constant of K_a.

The pK_a of acetic acid is 4.76, which means that its acid species can release the proton to produce the acetate conjugate base with a pK_a value of 4.76. On the other hand, the pK_a of methylamine 10.64 means its conjugate acid methylammonium releases proton with a pK_a value of 10.64 while methylamine has a pK_b value of (14-10.64) = 3.36. If we compare the acidity of the two, acetic acid with lower pK_a (thus higher ionization constant K_a) is stronger acid than methylammonium.

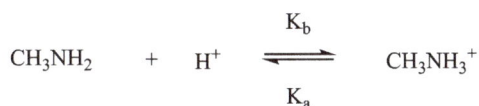

$$CH_3NH_2 \quad + \quad H^+ \quad \underset{K_a}{\overset{K_b}{\rightleftharpoons}} \quad CH_3NH_3{}^+$$

In general, the lower the pK_a the higher is the acid strength. *However, a pK_a value by itself is not useful unless one knows whether the value refers to an acid or a base.* Generally, an acidic drug with $pK_a < 2$ is considered strong acid that yields after removal of proton a conjugate base with no basic properties in H_2O. Acidic drugs with pK_a 4-6 are weak acids that yield weak conjugate bases, those with pK_a 8-10 are very weak acids yielding stronger conjugate bases; and those with $pK_a > 12$ are essentially neutral in water and yield even stronger conjugate bases.

Acid-Conjugate Base and Base-Conjugate Acid

An acid after releasing H^+ yields a conjugate base, which may be too weak to accept an electron or be very strong and excellent H^+ acceptor. Usually a strong acid yields weak conjugate base and *vice versa*. The product arising from addition of H^+ to an acid is conjugate base. Therefore, using the concept of conjugate acids and bases the general equation for both Brønsted-Lowry acids and bases are shown in Fig. **9**.

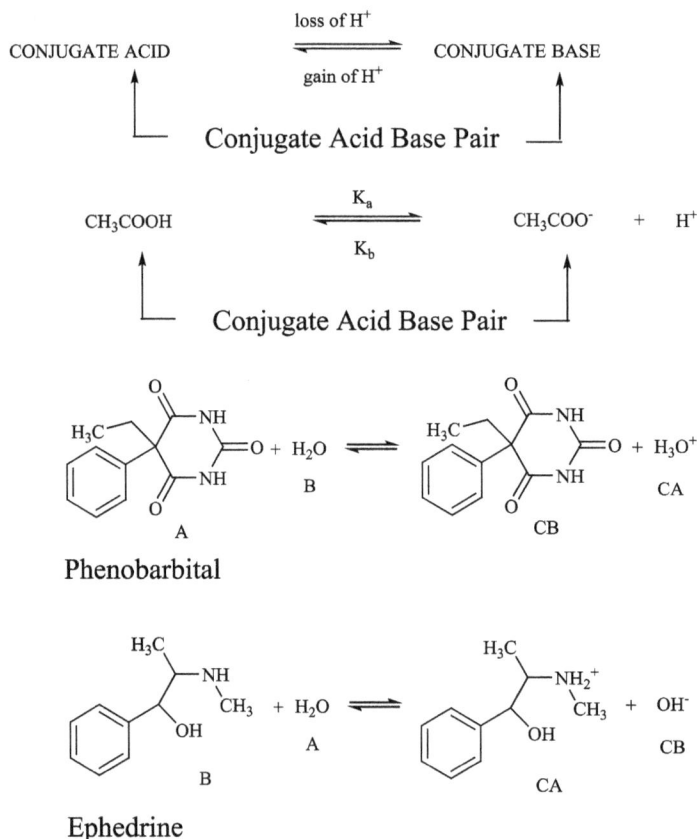

Fig. (9). Conjugates acids and bases with drug examples.

This leads to a very important concept of the Brønsted-Lowry theory, the concept of conjugate acid-base pairs. Acetic acid and acetate form the conjugate acid-base pair with the release and acceptance of one proton with ionization constants K_a and K_b, respectively. This concept states that *an acid will donate a proton and form a substance capable of accepting a proton – a conjugate base of the original acid*. Further, *a base will accept a proton and form a substance capable of donating a proton – a conjugate acid of the original base*. Conjugate acids differ from their conjugate bases in the following manner:

Structurally the conjugate acid has one more H^+ atom and one more positive charge (or one less negative charge) than its conjugate base. If the conjugate acid is strong then its conjugate base is weak. Likewise, the conjugate base has one less H^+ atom and one more negative charge (or one less positive charge) than its conjugate acid. If the conjugate base is strong then its conjugate acid is weak.

With this simple basic knowledge, it is possible to draw the conjugate of any substance. However, before this can be done, one must know whether the starting compound is an acid or a base. Pharmaceutical terminology uses the word conjugate differently. Drugs are referred to as the parent and the conjugate; the conjugate is the charged species and thus the parent is uncharged (Fig. **9**). This is because the primary interest with the drug is its solubility and it is the charged species (conjugate) that is water soluble, no matter whether it is an acid or a base.

Most of the drug molecules are weak acids or weak bases which are partially ionized in the biological systems yielding corresponding conjugate bases or conjugate acids, respectively. Thus, drugs in the biological system, or in the solution dosage forms, can be considered as the buffer solutions and the Henderson-Hasselbach (HH) equation can be used to determine the fraction ionized in biological pH as long as the pK_a values are known. As discussed later, the knowledge of the fraction ionized of a drug molecule in the biological system or in the solutions of a liquid dosage forms are very important for drug's bioavailability and thus potency and sometimes toxicity and duration of action. Use of the HH equation provides the following generalities:

HH equation for acids:

$$pH = pKa + \log_{10}\left(\frac{[\textit{Ionized}]}{[\textit{Unionized}]}\right)$$

HH equation for bases:

$$pH = pKa + \log_{10}\left(\frac{[Unionized]}{[Ionized]}\right)$$

Employing the concept of conjugate acid-base pairs the equation gets this general form:

$$pH = pKa + \log_{10}\left(\frac{[Base]}{[Acid]}\right)$$

For drugs which are acidic in their parent unionized forms (known as the HA acids), the unionized form predominates at pH < pK_a, and for those which are bases in their parent unionized forms, the unionized form predominates at the pH > pK_a. When the pH = pK_a, the ionized fraction is equal to the unionized fraction for all acidic or basic drugs.

Percent Ionization

Stronger acids are always more ionized than weaker acids and stronger bases are always more ionized than weaker basses, regardless of pH of the solution. The percent ionization of an acid or base is an important property since it dictates the amount of absorption of drugs through the lipid barriers. Only the unionized forms can diffuse through the lipid membranes. The % ionization can be calculated by using modified form of HH equations for acids and bases.

For HA acids: %Ionization = $100/(1 + 10^{(pKa - pH)})$
For BH^+ acids: %Ionization = $100/(1 + 10^{(pH - pKa)})$

Let us now examine a few drug examples (Fig. **10**). **Aspirin** is an HA acid, which yields acetate ion as its conjugate base at higher pH than its pK_a (3.48) after ionization, but will be practically unionized at lower pH. Ketoconazole is an interesting example where four nitrogens are present. The amide nitrogen at the right side of the structure on its piperazine ring and the pyrrole-like nitrogen on the imidazole ring are practically neutral. Only the aniline nitrogen fused with the piperazine ring and the sp^2 nitrogen on the imidazole ring are basic, sp^2 imidazole nitrogen being more basic and will produce the BH^+ conjugate acid with a pK_a of 6.54. Since it is a BH^+ acid, it will be ionized at lower pH while be unionized at higher pH than its pKa value. Ampicillin is a bifunctional drug containing both HA acid and the BH^+ acid groups with pK_as 2.61 and 6.79, respectively. At physiologic pH (7.4) its HA acid will be completely ionized to the carboxylate function but the amino function will be partly ionized (about 20%) furnishing overall negatively charged drug molecules with some zwitterions.

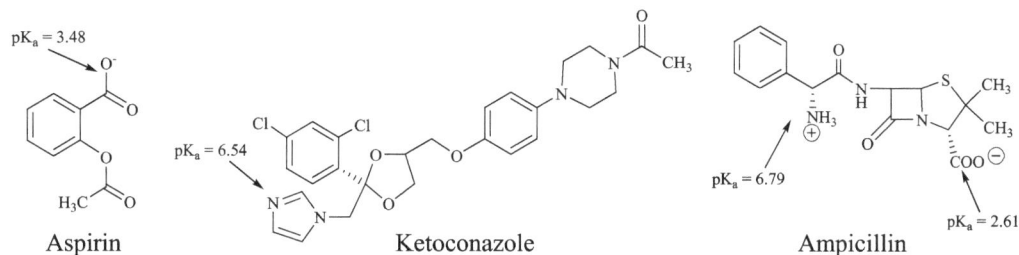

Fig. (10). Drugs with the pKa value of their acidic and basic functional groups.

The percent ionizations of all these drugs or their acidic or basic functional groups can be calculated using the appropriate equation for % ionization as given above. However, the approximate rule of thumb for percent ionization of the weak acids or bases can be summarized as follows:

For Weak Acids (HA Acids)

At pH = pKa   *drug is ~50% ionized*

At pH = pKa + 1   *drug is ~90% ionized*

At pH = pKa + 2   *drug is ~99% ionized*

At pH = pKa + 3   *drug is ~99.9% ionized*

At pH = pKa + 4   *drug is ~99.99% ionized*

For Weak Bases (BH+ Acids)

At pH = pKa   *drug is ~50% ionized*

At pH = pKa - 1   *drug is ~90% ionized*

At pH = pKa - 2   *drug is ~99% ionized*

At pH = pKa - 3   *drug is ~99.9% ionized*

At pH = pKa - 4   *drug is ~99.99% ionized*

Factors Controlling Ionization

Three important factors controlling the ionization of an acid or base are: electronegativity, bonding electron distance, and electronic or inductive effects.

Electronegativity

Linus Pauling who introduced the concept first defined electronegativity as the *power of an atom in a molecule to attract electrons towards it.* All atoms including hydrogen have an electronegativity value that increases as one goes from left to right and from bottom to top of the periodic table. On the Pauling scale, the most electronegative element is fluorine with a value of 3.98 while the least electronegative element is francium with a value of 0.7. Types of bond formed between atoms depend upon differences in electronegativity:

1. Large differences result in the more electronegative atom abstracting the electrons from the weaker atom forming an ionic bond (electronegativity difference >1.7)
2. Equal electronegativity results in equal sharing of electrons resulting in a covalent bond (electronegativity difference <0.4)
3. Small differences in electronegativity result in polar covalent bonds (electronegativity difference is >0.4 but <1.7).

Bonding Electron Distance

The further the bonding electrons are from the nucleus, the less energy required to ionize them completely. Nuclear size and thus the bonding electron distance increases from top to bottom of the periodic table.

Electronic or Inductive Effects

Some chemical groups produce an important electronic effect referred to as an inductive effect. Since this alters the physical and chemical properties of a drug, the biological activity may also be affected. Inductive or electrostatic effects result from electronic shifts along sigma bonds caused by the electronic character of certain groups. These inductive effects are the result of electronegativity differences between atoms. Groups which attract electrons more than hydrogen display a negative inductive effect (–I) called electron withdrawing. Groups that attract electrons less than hydrogen show a positive inductive effect (+I) called electron donating. The terms electronegative and electron withdrawing are somewhat related but are not synonymous. Thus all electronegative atoms cannot be considered electron withdrawing. The terms electron withdrawing and electron donating are used when comparing connected groups or atoms. Thus it is possible for a group to be an electron donating group in some situations but an electron withdrawing group in others depending on the relative strength of connected groups.

Electron Withdrawing Effect

Common electron withdrawing groups are shown in Table **2**. Although amines are sometimes electron withdrawing inductively (but donating through resonance), protonated amines and quaternary amines are stronger withdrawers since they have no π electrons to donate. The cyano, sulfonyl, and carbonyl groups are strong withdrawers. Among the halogens, fluorine is the strongest withdrawer. The triflumethyl is considered a halogen–like substituent. The hydroxyl and ether groups plus their isosteres, the thiol and thioethers are also considered as the withdrawing groups. The phenyl group and other unsaturated groups are generally electron withdrawers.

Table 2. Electron donating and withdrawing groups.

Electron Donating Groups	Electron Withdrawing Groups	
–CH$_3$	-NO$_2$, -NH$_3^+$, -NR$_3^+$	O=S=O
–CH$_2$R		-SH, -SR
–CHR$_2$	-COOH, -COOR	-OH, -OR
–CH$_3$	-CHO, R$_2$C=O	-F, -Cl, -Br, -I
–CH$_2$R	-CONH$_2$	-CH=CH$_2$, –C≡CH
–COO$^-$	–F > -Cl > -Br > -I	–C$_6$H$_5$
–O$^-$, –S$^-$		–CR=CR$_2$

Through an inductive effect electron withdrawing substituents increase the acid nature by weakening the O–H bond and by stabilizing the anion. Since the electron withdrawing power of the halogen group is in the order: F > Cl > Br > I, the acidity of the different halogen substituted acids follow the order:

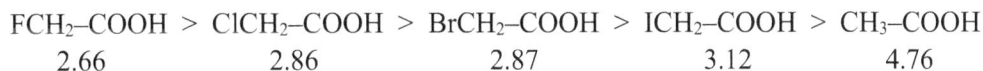

$$FCH_2{-}COOH > ClCH_2{-}COOH > BrCH_2{-}COOH > ICH_2{-}COOH > CH_3{-}COOH$$
$$2.66 \qquad\qquad 2.86 \qquad\qquad 2.87 \qquad\qquad 3.12 \qquad\qquad 4.76$$

The number of electron withdrawing substituents also affects the acidity. The more the number of electron-withdrawing substituents on the carbon adjacent to the acidic function, the more is the acid strength. Thus the acid strength of the following compounds is in the order:

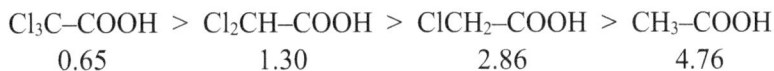

$$Cl_3C{-}COOH > Cl_2CH{-}COOH > ClCH_2{-}COOH > CH_3{-}COOH$$
$$0.65 \qquad\qquad 1.30 \qquad\qquad 2.86 \qquad\qquad 4.76$$

The inductive effect, thus the acid strength, decreases as the distance from the electron withdrawing substituents increases:

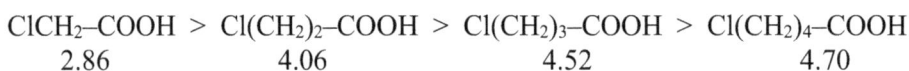

$$ClCH_2-COOH > Cl(CH_2)_2-COOH > Cl(CH_2)_3-COOH > Cl(CH_2)_4-COOH$$
$$2.86 \qquad\qquad 4.06 \qquad\qquad 4.52 \qquad\qquad 4.70$$

Electron Donating Effect

The first column of Table **2** lists groups that commonly exhibit electron donating effects through inductive effects or through resonance effects when found in a drug molecule. Generally, only alkyl and anionic conjugate bases are considered electron donating. Among the n–alkyls the donating effect increases as the chain increases. Branching also increases the donating effect. Electronic effects can contribute to acid strength as well as basic strength. Electron donating groups decrease the acid nature by strengthening the O–H bond and destabilizing the anion. The effects on basic strength are opposite; electron withdrawers decrease basic strength and electron donors increase basic strength.

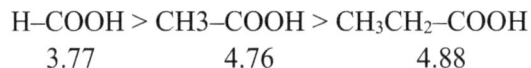

$$H-COOH > CH3-COOH > CH_3CH_2-COOH$$
$$3.77 \qquad\qquad 4.76 \qquad\qquad 4.88$$

PH PARTITION THEORY

The majority of the drugs are either weak organic acids or bases, which exist in ionized or unionized forms depending on the pH of the media. The HH equation is employed to estimate the fraction ionized or unionized. *In general, a weak base will be ionized at pH < pK_a and a weak acid will be ionized at pH > pK_a.* The more the difference between pK_a and pH the more is the fraction ionized. At pH = pK_a, both acidic and basic drugs will be 50% ionized. This phenomenon of drugs is especially important to understand as it has tremendous influence on the rate of drug absorption through the lipid barriers, and thus the bioavailability. Usually the unionized drugs have considerable solubility in the lipid while the ionized forms are more soluble in water and practically insoluble in lipid or oil. Since most of the biological membranes through which drugs are passively diffused into the cell or from gastrointestinal tract (after oral administration) into the systemic circulation are lipid bilayers, it is the unionized lipid soluble fraction of drug that is absorbed in a considerable rate. If we consider the lipid bilayer as a barrier between two compartments, *e.g.*, the extracellular and the intracellular compartments or the intestinal lumen and the systemic circulation, then the absorption or the diffusion is considered as the partitioning of the unionized drug between the two compartments towards the concentration gradients. Drugs moves from higher concentration to lower concentration compartment until the concentration in both compartments is equal. Furthermore, since the fraction unionized is determined by the pH of the compartment and the pK_a of the drug, the partitioning of the drug is governed by these two factors, which is explained by the *pH partition theory*, illustrated in Fig. **11**.

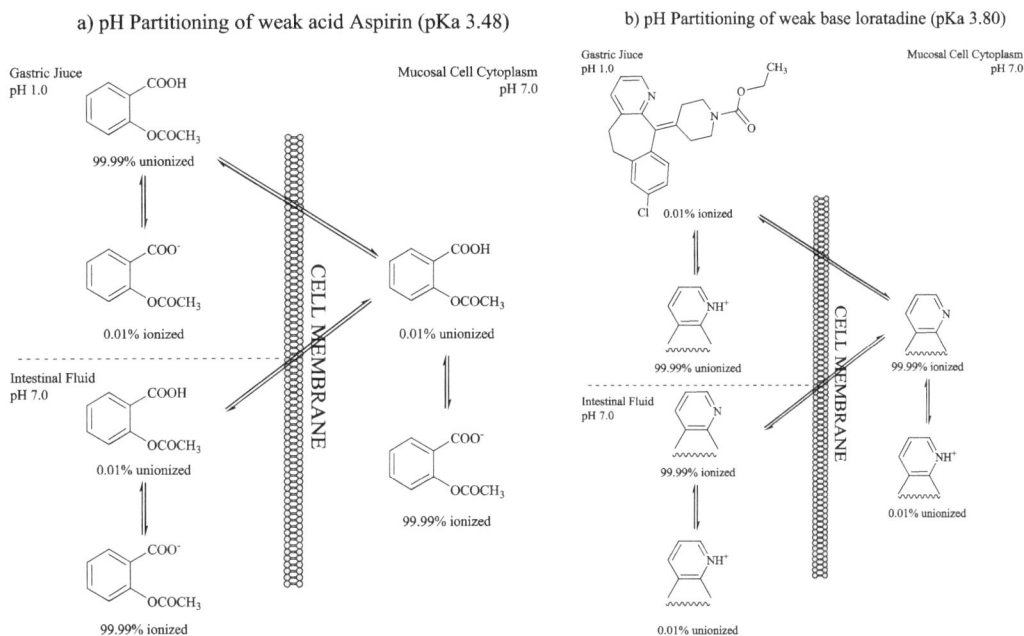

Fig. (11). Illustration of pH partitioning of: (**a**) weak acid aspirin and (**b**) weak base loratadine. It is only the unionized drug that is assumed to partition through the lipid membrane.

As shown in Fig. **11a**, aspirin is a weak acid with pK_a 3.48. Thus it is practically unionized at gastric pH (1.0) as estimated by HH equation. As we know the unionized form is the absorbable form of a drug. According to the pH partition theory, aspirin is absorbed with higher rate from the stomach and duodenum (However, because of the large surface area, it is also absorbed from the intestine in considerable amount). On the other hand, the weak base loratadine with pK_a 3.80 is completely ionized at gastric pH and thus is not absorbed at all from stomach and duodenum, but is absorbed at a high rate from the intestine (pH > 7.0) where it is in completely unionized form, which is the absorbable lipid soluble form (Fig. **11b**).

Although the *pH partition theory* is an oversimplification of more complex process of drug absorptions, it provides a basic framework for understanding the absorption of many drugs. There are several other factors that govern drug absorption and therefore explaining the phenomenon of drug absorption based only on above concept may be inaccurate. Some drugs even though are largely unionized in the intestine may still have low absorption rate if are polar and have low lipid solubility. In such cases the lipid/water partition coefficient plays an important role in guiding their rate of absorption. For example, the weak acid barbital has a pK_a value of 7.9 which is essentially unionized in gastric juice and about 90% unionized at pH 7 has a chloroform/water partition coefficient of 0.7 and absorbed only about 12% from rat colon compared to 50.7% absorption of secobarbital with pK_a of 7.9 (and 12.6), which is unionized in the same extent. The higher absorption of secobarbital than barbital accounts for its higher chloroform/water *partition coefficient* of 50.7. Certain quaternary ammonium drugs and some anions may also be absorbed in the small intestine at a much slower rate, which was not considered in this theory. Other factors that cause deviations from this oversimplification may include the mucosal *membrane thickness*, its *stagnancy*, and the difference between the luminal pH and microclimate or virtual pH at the cell membrane, to mention a few.

SALT FORMATION

Definition of Salts

Compound formed by interaction of an acid and a base is known as salt. Salt formation is an important process for improving the water solubility of otherwise insoluble parent drugs. The weakly acidic or basic drugs are usually converted to their salts by treating with bases or acids respectively. All salts thus formed usually contain two parts, the cation and the anion which are separated apart into water to produce solutions. In pharmaceutical formulations variety of purposes are served through the salt formation. For example, the lipid soluble acidic or basic drugs are converted to their water soluble salts to prepare IV injection formulations. Steroidal drugs are also sometimes modified to produce water soluble prodrugs for preparation of the injectables. Generally, salts are formed by four types of reactions – 1) reaction of organic base with an inorganic acid, 2) reaction of organic acid with an inorganic base, 3) reaction of organic acid with an organic base, and 4) reaction of inorganic acid with inorganic base (less common since drugs are usually organic, an example, though, would be lithium carbonate (Li_2CO_3).

Organic Base Reacting with an Inorganic Acid

The general reaction of the **Type I salt** is shown below:

$$B + HX \rightleftharpoons BH^+ \; X^-$$

The major inorganic acids are hydrochloric, nitric, sulfuric and phosphoric acids. Hydrochloric and nitric acids are monoprotic so there is no doubt what X^- is. There can be a problem with di– and triprotic acids such as sulfuric and phosphoric acids. In chemistry one meaning of "bi" is one less hydrogen (or proton) than possible. Hence, a sulfate salt is composed of the drug plus the SO_4^{2-} anion while a bisulfate salt is the drug plus HSO_4^-. Likewise, phosphate salts can form with $H_2PO_4^-$ (biphosphate), HPO_4^{2-} (phosphate) or PO_4^{3-} (orthophosphate). However, with phosphate salts naming conventions have not been followed with the root word phosphate since a salt labeled "Drug Phosphate" could be any of the three possible anions. Drawing the structure of the organic portion, BH^+, is more interesting.

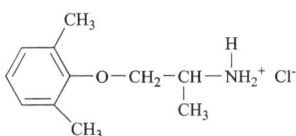

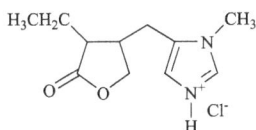

Mexiletine hydrochloride   Pilocarpine hydrochloride   Doxylamine hydrochloride

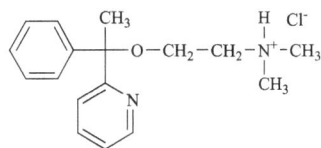

The structure of the parent compound, B, can be obtained from standard reference texts such as the Merck Index, USP/NF, USAN and the USP Dictionary of Drug Names. An excellent source on the Web is the PubChem project by the National Library of Medicine at "http://pubchem.ncbi.nlm.nih.gov".

It should be obvious that the base contains a nitrogen atom which is capable of accepting a proton. Therefore, the proton attaches to the basic nitrogen atom. If there are more than one basic nitrogen atom, the most basic nitrogen atom gets the proton.

The chemical formula of a salt should never be ignored. This can be found in standard references like the Merck Index. The chemical formula indicates not only what the acid is used for but also the ratio of acid molecule to base. The case of mexiletine hydrochloride is the simplest example. Chemical formula of mexiletine from Merck Index is $C_{11}H_{17}NO$ and that of its hydrochloride salt is $C_{11}H_{17}NO \cdot HCl$. Notice how this clearly separates the salt from the parent compound. Chemical formula from other sources: $C_{11}H_{18}ClNO$. This does not clearly separate the salt from the parent drug.

The empirical formula shows type of acid (as does the name of salt) and the ratio of acid to base. The drug to salt ratio for Mexiletine, is 1:1, it has only one nitrogen atom. The hydrochloric acid can donate only one proton. Therefore, HCl (HA acid) donates the proton, leaving Cl– (A– base). The Mexiletine molecule

(B) accepts the proton on the nitrogen to form BH$^+$ acid.

In the case of pilocarpine nitrate the situation is similar except the pilocarpine molecule has two nitrogen atoms. Chemical formula for pilocarpine is $C_{11}H_{16}N_2O_2$ and that of its hydrochloride salt is $C_{11}H_{17}ClN_2O_2$. Since only one of the nitrogen is basic there is only one place for the proton to attach and the ratio of acid to base, or drug to salt, is 1:1, as its name implies.

Doxylamine hydrochloride presents the next level of complexity. It contains two nitrogens and both are basic. Therefore, the hydrochloride salt could form with one, or both, of the basic amines. Its chemical formula is $C_{17}H_{22}N_2O$ and for its hydrochloride salt is $C_{17}H_{23}ClN_2O$. Since the chemical formula shows that only one proton is transferred, the proton goes on the more basic nitrogen as mentioned previously in this chapter. If both nitrogen atoms were protonated, then two molecules of hydrochloride would be required for each drug molecule, so the drug to salt ratio is 1:2 and the salt would likely be referred to as doxylamine dihydrochloride. And the chemical formula would be $C_{17}H_{24}Cl_2N_2O$ or $C_{17}H_{22}N_2O.2HCl$. If the acid is di– or triprotic, the chemical formula and oftentimes the name of the salt are indispensable. Consider the following sulfate salts:

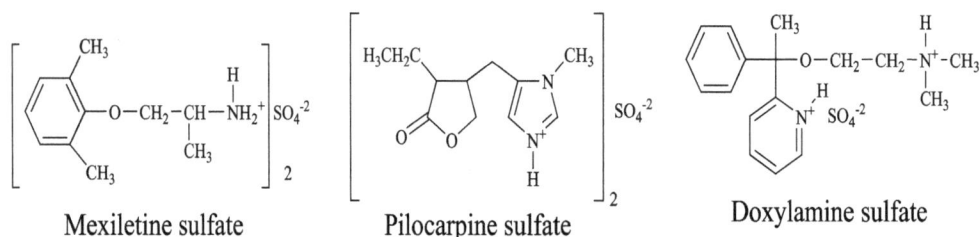

Mexiletine sulfate   Pilocarpine sulfate   Doxylamine sulfate

All three are properly called sulfate salts since the anion formed is SO_4^{2-}. Another possibility would be the bisulfate salt. In the first case two molecules of drug combine with one of sulfate since sulfate can donate two protons and each molecule of drug can only accept one. In the second instance, the drug can accept one or two protons; however, only one is accepted thus the drug to sulfate ratio is again 2:1. In the last instance one drug molecule accepts two protons and so the ratio of drug to sulfate is 1:1.

Organic Acid Reacting with an Inorganic Base

The general reaction of **Type II salt** is shown below:

$$A-H + M(OH)n \rightleftharpoons A^- M^+ + n H_2O$$

The major inorganic bases are sodium, calcium and potassium hydroxides. Silver, magnesium and aluminum salts are also seldom used. In the above reaction the hydroxide ion removes the proton from the acid, leaving the conjugate base (A^-) and the metal ion (M^+). The chemical formula is useful since it shows how many metal ions are present in the salt and also indirectly implies to the number of protons removed. The total positive charge on the metal in the salt is equal to the number of protons removed. The following examples illustrate these points:

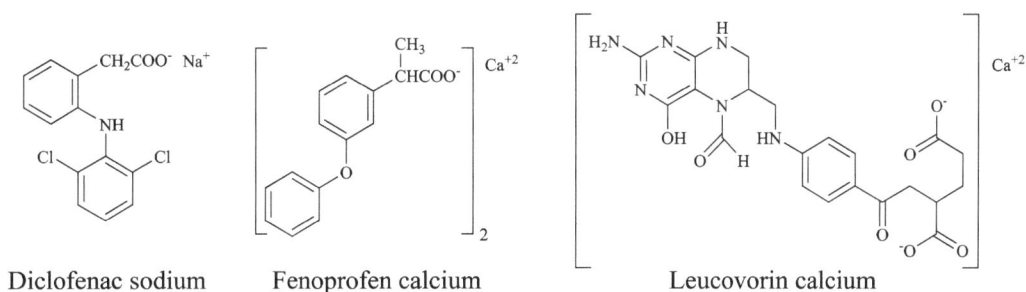

Diclofenac sodium   Fenoprofen calcium   Leucovorin calcium

The first example shows a drug (diclofenac) with a single acidic functional group that interacts with a monovalent metal base (NaOH) to form a salt (diclofenac sodium) with a drug to salt ratio of 1:1. There are no other possibilities. The second case shows a drug (fenoprofen) with a single acidic functional group interacting with a divalent metal base to form a salt (fenoprofen calcium) with a 2:1 molar ratio. The third case shows a drug molecule (leucovorin) with two acidic functional groups interacting with a divalent metal base to form a salt (leucovorin calcium) with a 1:1 molar ratio.

Consider the case of diclofenac sodium. The chemical formula for diclofenac is $C_{14}H_{11}Cl_2NO_2$ for its sodium salt is $C_{14}H_{10}Cl_2NaNO_2$. The metal sodium has a charge of plus one. Assume that the metal came from a metal hydroxide, NaOH. It is capable of removing only one proton. Therefore, diclofenac (AH acid) must have lost one proton and the drug to salt ratio is 1:1. This can be verified from the chemical formula.

The chemical formula for fenoprofen is $C_{15}H_{14}O_3$ and for its calcium salt is $C_{30}H_{26}CaO_6$. The metal calcium is divalent. If we assume it came from $Ca(OH)_2$, it must have reacted with two protons. Since fenoprofen has only one acid proton, it takes two acid molecules to provide the two protons. Thus the ratio of drug to salt is 2:1. The chemical formula verifies this.

In leucovorin calcium, the salt has a divalent metal, Ca^{++}. The chemical formula for leucovorin is $C_{20}H_{23}N_7O_7$ and for its calcium salt is $C_{20}H_{21}CaN_7O_7$. The acidic leucovorin molecule has two acidic protons. Thus the molar ratio of drug to salt can be 1:1 or, if only one proton is removed from each leucovorin molecule, 2:1. Using the chemical formula we can find what the molar ratio is. Thus both are lost in the reaction.

There are two more things that one must be aware of. If there are two acidic protons and only one is lost, the one donated comes from the strongest acid group. One has to know the charge on the important cations.

Organic Acid Reacting with an Organic Base

The general reaction of **Type III salt** is shown below:

$$B + A-H \rightleftharpoons BH^+ + A^-$$

In drawing the structure of salts of this nature calls for applying the techniques used in Type 1 and Type 2 salts. As an example, erythromycin estolate salts. Erythromycin has the chemical formulae $C_{37}H_{67}NO_{13}$ and for its estolate salt $C_{52}H_{97}NO_{18}S$. This would verify that it is a 1:1 drug salt ratio. A more complicated example would be escitalopram oxalate. It is a dicarboxylic acid and, thus, could form a 1:1 or 2:1 ratio of drug to salt. Looking at the chemical formula for escitalopram ($C_{20}H_{21}FN_2O$) and for escitalopram oxalate ($C_{22}H_{23}FN_2O_5$) it is an obvious 1:1 ratio of drug to salt.

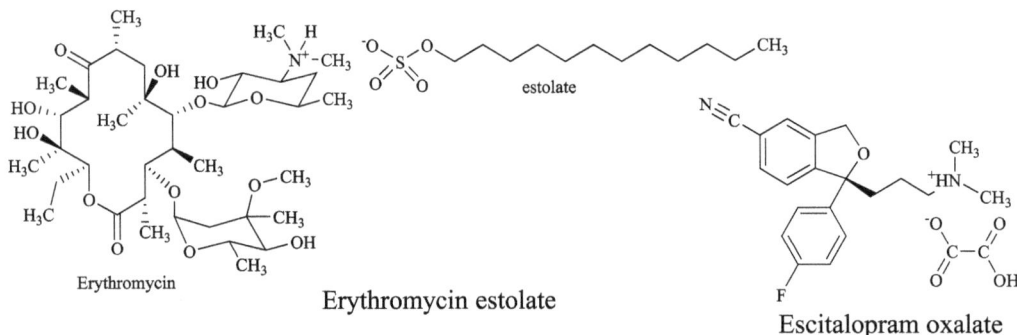

Erythromycin

Erythromycin estolate

Escitalopram oxalate

One complication with the salt of organic acid and organic base is that it may not actually form a salt but rather a covalent bond to become part of the molecule. Take for example erythromycin ethylsuccinate. It is actually esterified to the erythromycin molecule and is not a salt. This is not apparent from the name. Chemical formula for erythromycin ethylsuccinate is $C_{43}H_{75}NO_{16}$. Chemical formulae can also show the level of hydration. Working with a hydrated molecule, the numbers of waters of hydration should be included in order to use the molecular formula for determining the molar ratio of drug to salt. For examples, quinine has a chemical formula of $C_{20}H_{24}N_2O_2$ and that of quinine phosphate is $C_{60}H_{83}N_6O_{14}P$. The chemical formula of quinine phosphate in fact represents three molecules of quinine in one trivalent phosphate anion and four molecules of water of hydration.

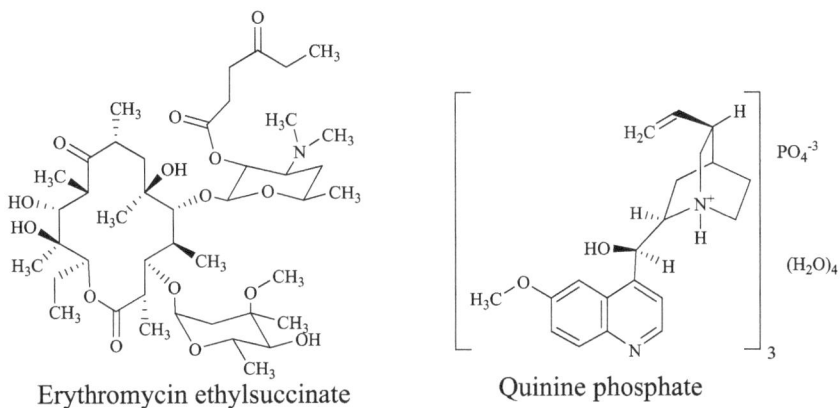

Erythromycin ethylsuccinate   Quinine phosphate

Acidity or Basicity of Salts

Salts are usually completely dissociated in water by a process called *solvation* or *hydrolysis* (as water breaks the anion and cation species). Depending on the strength of acid or base components from which the salt is formed, it can be acidic, basic or neutral thus making the pH of the water solution lower, higher or unchanged respectively.

The *see saw* analogy works well in determining the acid base property of the salts. If the component acids and bases are strong (heavy) both sides balance producing a neutral salt, for example NaCl, which is produced from the strong acid HCl and strong base NaOH. Neither of the ions (Na^+ or Cl^-), produced after dissociation in water, can react with water to produce either H_3O^+ or OH^- in any considerable amount to change the pH of water. Similarly, the salt of a weak (light) acid and a weak (light) base will be neutral, *e.g.*, ammonium acetate ($NH_4^+CH_3COO^-$), where both ions produce equal amount of H_3O^+ and OH^- to neutralize each other.

However, if the salt results from the reaction of a strong acid and a weak base, like see saw balance, the equilibrium goes towards the stronger side giving an acidic salt product. Thus, all the salts of weak bases and strong acids shown in the previous section are weak acids. The formation and dissociation of doxylamine hydrochloride is illustrated in Fig. **12**.

As shown in Fig. **12**, the ammonium cation produced from doxylamine and HCl increases H_3O^+ ions in water solution resulting in decrease of pH and thus an acidic solution. The pH of such solution is calculated by using the following equation (where c is the molar concentration).

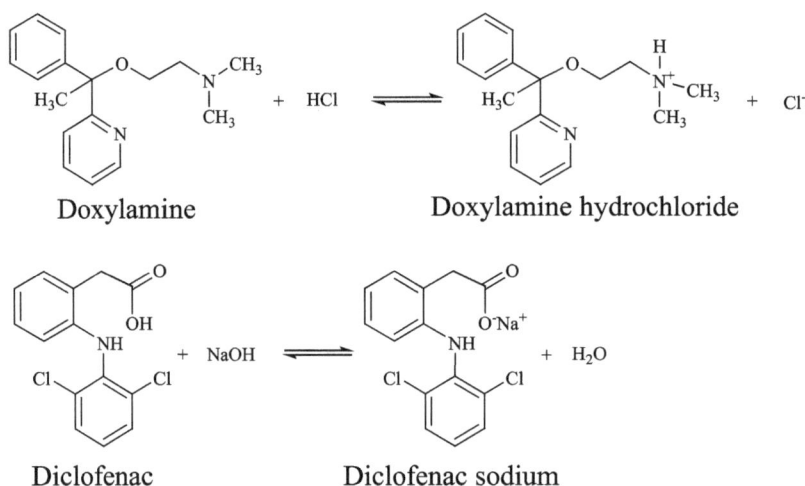

Fig. (12). The formation and dissociation of doxylamine hydrochloride and diclofenac sodium.

$$pH = \frac{1}{2} pK_a - \frac{1}{2} \log c$$

The salts resulting from the reaction of a strong base and a weak acid afford basic solutions in water based on the same principle. Thus diclofenac sodium, fenoprofen calcium and leucovorin calcium, shown in the previous section, are all weak bases and will produce alkaline solutions in water.

The formation of diclofenac sodium salt and its dissociation in water to yield alkaline solution is also illustrated in Fig. **12**. Na^+ does not react in any significant extent, however, the carboxylate anion formed from diclofenac, a conjugate base, react with water to produce OH^- ions (or NaOH) and thus produce the basic solution of higher pH. The pHs of such solutions can be calculated by using the following equation.

$$pH = pK_w - \frac{1}{2} (pK_b - \log c)$$

This is to be noted that the see saw analogy is not a precise measure but an estimate for understanding the acid-base properties of the salts and is summarized as follows:

Strong acid + Strong Base   =   Neutral salt
Strong acid + Weak base   =   Acidic salt
Weak acid + Strong base   =   Basic salt
Weak acid + Weak base   =   Neutral salt

CASE STUDIES

Case 1: OW and YL are having some burning sensations during urination, itching outside the vagina and having abnormal vaginal discharge with unpleasant odor. The pathological tests confirmed that both of them have developed bacterial infection, which are sensitive to miconazole nitrate (pKa of 6.07), a topical antifungal agent. OW is a 60-year-old lady with a vaginal pH of 6.0 and YL is a 35-year-old lady with a vaginal pH of 3.8. Both of these ladies will be treated with miconazole nitrate and the doctor asks your opinion about systemic *vs.* local effects of the drug to determine the dosages. How do you respond? Make sure you respond with acid-base character of the drug and its extent of ionizations in these two ladies.

Solution: Miconazole nitrate is an acidic salt (nitric acid, a strong acid) of weakly basic drug miconazole; it will be available more in unionized form at higher pH than at lower pH, when it will be more ionized. For a basic drug, the higher the pH the lower is the ionization. Therefore, at pH 6.0, about half of the drug will be present as unionized form and chances of its absorption into systemic circulation will be high. On the other hand, at pH of 3.8, the drug will mostly be present in the ionized form. Therefore, the drug will have less systemic absorption and will be available to produce more local effect. The %ionization of the drug in both pH's can be quantified by using HH equation.

Case 2: Ryan has recently passed his NAPLEX and MPJE exams and is currently looking for a full time position in a Hospital. He applied at various hospitals and got an interview call for the ABC hospital. The interview panel consisted of pharmacy director and two staff pharmacist. The director asked Ryan the following questions to check his understanding about the subject.

Q1. If aspirin (pKa 3.48) is administered orally at what pH (1 or 8) do you think will the drug be mostly absorbed systemically?

Answer: Ryan answered that the drug will be mostly absorbed at pH 1 because at this pH most of the drug will be unionized and only unionized drug can cross the G.I membrane. At pH 8 the drug will be mostly ionized that cannot cross G.I. membrane.

Q2. The committee then followed up by asking if the is then absorbed only from stomach or also from intestine?

Answer: Ryan answered the drug will be absorbed easily from stomach as it is mostly unionized at the pH there. However, it can still be absorbed from intestine where pH is basic that can be explained based on the pH-Partition Theory as well as the surface area of the intestine.

The committee was impressed by Ryan's answer and asked him to show the pH partition theory for aspirin diagrammatically, and explain why the drug can still be absorbed considerably from intestine. He gladly did it (see the text for it).

Q3. A list of four drugs was presented to Ryan with their pKa's. Ryan was asked to identify if the drugs are acidic or basic and which drug will be least absorbed at pH 8 and why?

Drug	pKa
Loratadine	3.8
Quinine	8.4
Ephedrine	9.6
Dextromethorphan	9.2

Answer: Ryan explained that it is not possible from the pKa's of the drugs to determine whether these are acidic or basic. He needs the structures of the drugs to determine its acidity or basicity. And only after knowing if these are acidic or basic he can tell the ionizability of the drugs that will determine the cell permeability.

Ryan was given the drug structures as follows:

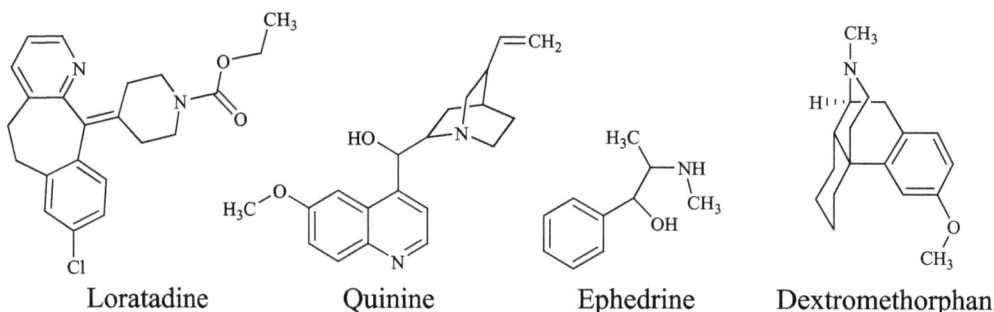

| Loratadine | Quinine | Ephedrine | Dextromethorphan |

Ryan recalled his concepts of acid-base chemistry and determined that all these drugs are basic in character and the order of basicity is: Dextromethorphan > Ephedrine > Quinine > Loratadine. He then recalled his concepts of physical pharmacy and realized that basic drugs will be more ionized at lower pH. The higher the basicity the more ionized the drug will be and less the drug will be permeable through lipid membranes. Therefore, based on the pKa's of the drug, the amount of drug absorption will be as follows: Loratadine > Quinine > Ephedrine > Dextromethorphan

Ryan replied that dextromethorphan will be least absorbed.

The interview board was totally impressed with his understanding of the drugs' physicochemical properties. Now they asked if he can quantify the percentage of ionized and unionized drugs.

STUDENT SELF-STUDY GUIDE

- Definitions of acids and bases
- Why thio acids are stronger than oxy acids?
- Comparative acid-base strength of functional groups shown in the handout
- Identify acid, base or neutral groups in drug molecules
- Compare basicity of amines (both aliphatic and aromatic) and imines
- What is an azo acid? Find it in drug structures.
- What is β-diketone? Is it acid, base or neutral? Drug examples.
- Application of pKa in acidity or basicity of drugs
- What is conjugate acid and conjugate base?
- Application of Henderson-Hasselbalch equation. Determination of percent ionization of drugs.
- The electron withdrawing and donating groups and their effects on acidity and basicity of drugs/organic acid and bases
- Different methods of salt formation. Acidity and basicity of salts with examples. Why salt formation is important?

STUDENT SELF-ASSESSMENT

Part I: Multiple Choice Questions

1. According to the _____ concept of acids and bases, and acid is a proton donor, and a base is a proton acceptor.
   A. Arrhenius
   B. Bronsted-Lowry
   C. Medicinal Chemistry
   D. Lewis
2. An un-ionized weak base will be more than 50% ionized at a pH that is _____ its pKa.
   A. Greater than
   B. Equal to
   C. Less than
   D. Double the value of
3. In regards to greater acidity, of an acidic drug, which pKa denotes the stronger acid?
   A. pKa = 0.1

B. pKa = -0.5
C. pKa = 9.0
D. pKa = 12.5
E. Acidity strength cannot be determined from pKa

4. Which of the following is true in regards to acid strength (strongest to weakest)?
A. FCH_2-COOH > $ClCH_2$-COOH > $BrCH_2$-COOH > Cl_2CH-COOH
B. $Cl(CH_2)_4$-COOH > $Cl(CH_2)_3$-COOH > $Cl(CH_2)_2$-COOH > $ClCH_2$-COOH
C. CH_3-COOH > FCH_2-COOH > ICH_2-COOH > $ClCH_2$-COOH
D. Cl_3C-COOH > FCH_2-COOH > $ClCH_2$-COOH > CH_3-COOH
E. Halogen substituents have no determinacy in acid strength

5. _____ has the acidic H^+ on the N atom.
A. Azo acids
B. Thio acid
C. Oxo acid
D. Phosphor acid
E. None of the above

6. Which of the following is the term used for bases that become acids after accepting a proton?
A. OH^- bases
B. HA bases
C. BH^+ acids
D. BH^+ bases
E. HA acid

7. MC, a 35-year-old male, arrives to the emergency room of the hospital where you work. The patient is given ten times the dose of phenobarbital sodium (pKa = 7.4) ordered due to a transcription error. The patient has a urinary pH of 6.4. What would the percent of ionized Phenobarbital sodium in MC's urine?
A. 1
B. 10
C. 90
D. 99
E. 99.9

8. When compounding an ophthalmic solution of an acidic drug, it is best reacted with _____ to form a water soluble salt.
A. NaOH
B. NaCl
C. HCl
D. a weak acid
E. a quaternary ammonium salt

9. According to, an acid is a substance that releases and thus increases hydrogen ion concentration in water; and a base is a substance that releases and thus increases the hydroxide ion concentration in water.
   A. Arrhenius concept
   B. Bronsted-Lowry concept
   C. Lewis concept
   D. All of the above

10. All of the statements are true about acid - base chemistry <u>except</u>:
    A. Strong Acid + Strong Base = Neutral Salt
    B. Most drugs are either weak acids or weak bases
    C. Ionized species easily cross through the cell membrane
    D. Weaker acid yields a strong conjugate base
    E. Strong acid yields a weak conjugate base

Part II: K-Type Question

Choose the answer:
   A. If **only I** is correct
   B. If **only III** is correct
   C. If **I and II** are correct
   D. If **II and III** are correct
   E. If **I, II, and III** are correct

11. Which of the following statements about salts is/are correct?
    I. Salts can be acidic, basic or neutral
    II. The reaction of a strong base with a strong acid **or** a weak acid with a weak base will yield a neutral salt
    III. When a strong acid reacts with a weak base, or a weak acid reacts with a strong base, the acidity or basicity of the salt is determined by which component was stronger

12. Which of the following statements is/are correct?
    I. Alkyl groups are generally considered electron-donating
    II. Fluorine atoms can withdraw electron density *via* the inductive effect
    III. Acidity decreases as the number of electron-withdrawing groups on the adjacent carbon increases

13. If you are working in chemistry lab and have isolated an amino acid, which of the following would you likely to do for recovering as much raw amino acid as possible from the solution?
    I. Increase the pH to 12 to form a precipitate
    II. Decrease the pH to 2 to form a precipitate
    III. Neutralize the solution to form a precipitate

14. Which of the following is true regarding the Bronsted-Lowry concept?

      I. Acids are proton donors & Bases are proton acceptors

      II. Most applicable concept to drug molecules

      III. Solvent has to be water

15. Which statements are correct?

      I. Thio acids are stronger acids than oxy acids mainly due to the stronger bond between the sulphur and hydrogen atoms.

      II. Compounds containing oxygen produce better hydrogen bonding versus compounds with sulphur, which causes compounds with oxygen to be more water soluble.

      III. Due to its smaller 2s and 2p orbitals compared to sulphur's larger 3s and 3p orbitals, oxygen has less stability as an ion.

Part III: Matching Question.

16. Match the drug structures I-V with the key statements (A-E) from acid-base standpoint

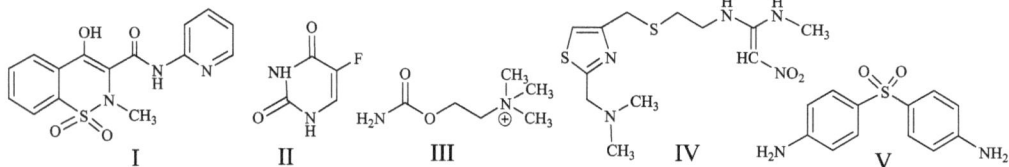

    A. An azo acid (imide)

    B. An enolic acid

    C. Charged, but neutral

    D. Base (imine + amine)

    E. Neutral amine

CONSENT FOR PUBLICATION

Not applicable.

CONFLICT OF INTEREST

The authors declare no conflict of interest, financial or otherwise.

ACKNOWLEDGEMENT

Declared none.

REFERENCES

[1]   Wells JI. Pharmaceutical Preformulation. London, UK: Ellis Horwood Ltd 1988.

[2]   Stenlake JB. Foundations of Molecular Pharmacology. London: The Athlone Press 1979; Vol. 1.

CHAPTER 4

Solubility and Lipid-Water Partition Coefficient

Hardeep Singh Saluja[1] and **M. O. Faruk Khan**[2,*]

[1] *Department of Pharmaceutical Sciences, College of Pharmacy, Southwestern Oklahoma State University, Weatherford, Oklahoma, USA*

[2] *Department of Pharmaceutical Sciences and Research, Marshall University School of Pharmacy, Huntington, West Virginia, USA*

Abstract: This chapter is brief review of solubility and related concepts, lipid-water partition coefficients and their significance in drug bioavailability. After study of this chapter, students will be able to:

• Analyze and predict the solubility of drugs.
• Identify factors affecting the solubility of drugs.
• Project the polarity, hydrophilicity, hydrophobicity, lipophilicity of drugs and their influence on solubility.
• Predict the water or lipid solubility of drugs from their structures.
• Comprehend the concept of partition coefficient.
• Apply the concept of lipid water partition coefficient (LWPC) and its significance in drug bioavailability and action.

Keywords: Hydrophobicity, Hydrophilicity, Lipid-water partition coefficient, Lipophilicity, Solubility, Water solubilizing potential.

SOLUBILITY

A solution is a system in which drugs or other substances (solutes) are dissolved in a vehicle (solvent). The ability for a drug to dissolve in a solvent is referred to as *solubility,* which is measured in terms of the maximum amount of drug dissolved in a solvent at equilibrium, forming a saturated solution under fixed set of conditions like temperature, pressure, pH, to name few. In other words, solubility can be defined as the extent to which molecules from a solid are removed from its surface by a solvent. Solubility ranges widely, from *infinitely soluble* such as ethanol in water (miscible), to *poorly soluble,* such as steroidal drugs in water and can be expressed in one of the following ways [1]:

---

* **Corresponding author M. O. Faruk Khan**: Department of Pharmaceutical Sciences and Research, Marshall University School of Pharmacy, Huntington, WV, USA; Tel: 304-696-3094; Fax: 304-696-7309; E-mail: khanmo@marshall.edu

M.O. Faruk Khan & Ashok Philip (Eds.)

- Very soluble, when less than 1 part solvent needed to dissolve 1 part solute
- Freely soluble, when 1 - 10 parts solvent needed to dissolve 1 part solute
- Soluble, when 10 - 30 parts solvent needed to dissolve 1 part solute
- Sparingly soluble, when 30 - 100 parts solvent needed to dissolve 1 part solute
- Slightly soluble, when 100 - 1000 parts solvent needed to dissolve 1 part solute
- Very slightly soluble, when1000 - 10,000 parts solvent needed to dissolve 1 part solute
- Practically insoluble, when >10,000 parts solvent needed to dissolve 1 part solute, which is a very rare occurrence in practice

The numerical value of solubility is commonly expressed as a concentration, either mass concentration (g of solute per kg of solvent, or g per 100 mL (dL) of solvent) or molarity, or molality, or mole fraction or parts per parts of solvent (parts per million, ppm), or similar. The relatively less precise expressions which are also used include: insoluble, very slightly soluble, slightly soluble, sparingly soluble, soluble, freely soluble, and very soluble [1, 2]. Solubility is of fundamental importance in pharmacy for drug formulations, pharmacokinetics, and pharmacodynamics. A variety of physicochemical parameters concerning the solute-solvent interactions are important subject matters for the drug formulation, however they are beyond the scope of this book. Only the drug-solubility in water (since the biological fluids are mostly water) and in lipids (since the biological membranes are lipids) and thus, the partitioning of drugs in lipid-water system will be focused in this chapter.

Factors Affecting Solubility

The solubility is determined by the balance of inter-molecular forces between the solvent and solute and the entropy change that accompanies the solvation. Although chemical structure and the functional groups present in a drug are of fundamental importance in solubilizing drugs, factors such as surface area of the solute, temperature and pressure can alter this balance, and thus can change the drug solubility. The presence of other species dissolved in the solvent, for example, complex-forming anions (ligands) in liquids, the excess (or deficiency) of a common ion (common-ion effect) in the solution, and to a lesser extent, the ionic strength of liquid solutions also influence solubility. Some other less common factors may affect solubility such as the crystal (or droplet) size of the solute. Typically, solubility will increase with the decreasing crystal size (thus increasing the surface area) for crystals much smaller than 1 μm [3]. For highly defective crystals, solubility may increase with the increasing degree of disorder. Another important factor that can affect a drug's solubility is *polymorphism* and *amorphousism*. A drug can exhibit more than one crystal form, this property is termed as polymorphism and different crystal forms are called polymorphs. In general different polymorphs have different crystal lattice energy (enthalpy of fusion), melting points and solubility. The difference in solubility of one polymorph from another is predominantly due to difference in crystal lattice energy and melting point. Usually, drugs with lower crystal lattice energy and lower melting point are more water soluble. For example, two different polymorphs α and γ of indomethacin, a nonsteroidal anti-inflammatory drug, exhibit different physical properties. Melting temperature and enthalpy of fusing for γ form are 162°C and 102 J/g while for α form being 156°C and 101 J/g, respectively. Thus α form has a higher aqueous solubility than γ form. The solubility ratio of α to γ crystal at 45°C in water was found to be 1.1. On the other hand, amorphous drugs do not have a well-defined crystal lattice and their molecules are disorderly arranged. Amorphous drugs do not exhibit a melting temperature unlike crystalline drugs. Typically, amorphous drugs have higher intrinsic solubility compared to its crystalline form. The solubility ratio of amorphous indomethacin to γ crystal was found to be 2.8 at 45°C in water [4]. Usually, amorphous drugs are thermodynamically unstable and may convert to a crystalline form over a period of time. Organic compounds nearly always become soluble in most solvents as the temperature is raised.

Chemical reactions also play an important role in solubilizing many drug molecules. Weakly acidic drugs salt formation at higher pH than its pKa, and weakly basic drugs salt formation at lower pH than its pKa are the most common strategies employed to increase their solubility in water. The reverse is true for their improved solubility in lipid solvents that is the parent neutral species are usually more soluble in lipid solvents. Biological membranes are considered as lipid membranes, the solubility in which is important for the drug diffusion, and thus absorption.

Solvation

Solvation is an important phenomenon of the solvents through which, they cause dissolution of the solute molecules into solvent and thus forming a solution. In simple terms, *solvation is the process of attraction and association of solvent molecules with solute molecules.* As the solute dissolves in a solvent, the solute molecules disseminate and become surrounded by solvent molecules. When the solvent is water, it is termed as hydration. Water, due to its high polarity, exhibits its solvation action by separation of cationic and anionic species of the solute molecule and subsequent orientation of solvent molecule around the ions. This type of solubilizing action is common with the salts of weakly acidic or basic drugs and also strong acids (*e.g.,* HCl). The miscibility of water with low molecular weight alcohols, *e.g.,* CH_3OH or C_2H_5OH accounts for the hydrogen bonding behavior and their ability to become associated with the molecules of water as shown in Fig. **1**.

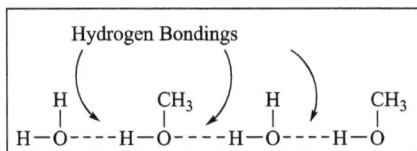

Fig. (1). Hydrogen bond between water and methanol.

Increasing the molecular weight by incorporating larger alkyl group(s) gives bulk to the alcohol and reduces its ability to form hydrogen bonds with water molecules and thus reduces its water solubility. The water solubility is reduced dramatically when the side chain contains more than five carbon atoms. Octanol, for example, is considered as the lipid phase (and thus a lipid solvent), which is immiscible in water, and is often employed to determine lipid-water partition coefficients of drugs. Increasing the number of hydroxyl groups on the other hand increases water solubility. Ethers, aldehydes and ketones may also become associated with water, but act only as hydrogen bond acceptors unlike hydroxyl group that works both as the hydrogen bond donor and acceptor. Solubility of the drug molecules is also affected by the strength and number of hydrogen bond(s) with the solvent. For example, nitrogen is less electronegative than oxygen and therefore forms weaker hydrogen bond; therefore it is slightly less soluble in water. However, the primary amines which can form three hydrogen bonds have comparable solubility as alcohols but reduced solubility with the inclusion of more alkyl groups as in secondary and tertiary amines. It is their salts which are considered water soluble through the ionization process, while the neutral amine drugs are soluble in nonpolar lipophilic solvents.

Forces Involved in Solvation

The key to solubility of the drugs is the ability of their functional groups to bond with water or lipid solvent molecules. The common phrase *"like dissolves like"* elucidates that chemical compounds (drugs in this case) are more likely to dissolve in solvents that have similar chemical properties to themselves. Irrespective of the type of solvent, the most important types of intermolecular forces involved in such solute-solvent bonding are the van der Waals force, hydrogen bonds, ionic bonds and ion-dipole bonds.

Van der Waals Force

Van der Waals force is defined as an attraction between molecules like diatomic free elements, and individual atoms. They differ from covalent and ionic bonding in that they are unstable. They are caused by momentary polarization of particles, very weak and short lived interactions, approximately 0.5 to 1.0 kcal/mole for each atom involved. *Keesom force* (between charges, dipoles, quadrupoles, and in general between permanent multipoles), *Debyes force* (arising from induction or polarization), which is the interaction between a permanent multipole on one molecule with an induced multipole on another, or *London dispersion* (attraction experienced by noble gas atoms) are called *van der Waals forces*. These types of forces are mainly observed in hydrocarbon or hydrocarbon portion of a molecule including aromatic systems that cause solubilization in lipid solvents and are of little significance in water.

Hydrogen Bond

The most important force involved in water solubility is the *hydrogen bond*, a special type of dipole-dipole force. It exists between an electronegative atom and a hydrogen atom bonded to another electronegative atom (Nitrogen, Oxygen, or Fluorine). This is expressed as an X-H…Y system where X and Y are F, N or O. The covalent X-H distance is typically 1.1 Å whereas H…Y (actual hydrogen bond or association) distance is approximately 1.6-2 Å. The hydrogen bond is much stronger than van der Waals force, but weaker than covalent, ionic and metallic bonds. A hydrogen atom attached to a relatively electronegative atom (N, O) is a hydrogen bond donor (which are also acceptors) and another electronegative atom such as F, O, or N is a hydrogen bond acceptor (but are not donors), regardless of whether it is bonded to a hydrogen atom or not. Thus alcohols are good hydrogen bond donors while ketones or diethyl ethers are good hydrogen bond acceptors. Hydrogen bonds can vary in strength from very weak (about 1 kcal/mol) to strong (about 10 kcal/mol) and typical values are summarized in Table **1**.

Table 1. Different Types of Hydrogen Bonds and their Bond Strengths.

Type of Hydrogen Bond	Bond Strength (Kcal/mole)
O-H N	6.9
O-H O	5.0
N-H N	3.1
N-H O	1.9
HO-H OH$_3^+$	4.3

Water is the most important pharmaceutical solvent where hydrogen bonding plays an important role in solubilizing drug molecules. Every water molecule is H-bonded with up to four other molecules; the exact number though depends on the temperature and time. The high boiling point of liquid water is due to strong intermolecular hydrogen bonding. Drugs containing alcohol, amine, carboxylic acid or their derivatives are usually solubilized by water to different degrees depending on their ability to form hydrogen bond with water.

Ionic Bond

An **ionic bond** (or **electrovalent bond**), formed by the electrostatic attraction between two oppositely charged ions, is often observed between metal and non-metal ions (or polyatomic ions such as ammonium). Ionic bonds are formed by complete transfer of electrons. For example, metal donates one or more electrons, forming a positively charged ion called "cation" with a stable electronic configuration. The donated electron by metal is accepted by the nonmetal, forming a negatively charged ion called "anion" with a stable electron configuration. The electrostatic attraction between the oppositely charged ions brings them together to form a bond, as observed in case of NaCl, where Na donates an electron to form cation while Cl accepts an electron to form anion. Compounds formed by ionic bonds generally have a high melting point and tends to be soluble in water. It is an important strategy of preparing water soluble salts of acidic or basic drugs. Salts are associated by ionic bonds between a cation and an anion that dissociate in water and are well solvated through another type of bond called ion-dipole bond discussed below. Ionic bonds are stronger than hydrogen bond, usually >5 kcal/mol, and is least affected by temperature and bond distance.

Ion-Dipole Bond

One of the most important type of bond involved in dissolving salts of acidic or basic drugs in water is the *ion-dipole bond* occurring between an ion (cation or anion) and a dipole found in water ($H-O^{\delta-}-H^{\delta+}$). The cations associate with the atom or region with rich electron density (*e.g.* O in water) while the anions associate with the atom or region with poor electron density (*e.g.* the H atom in water).

The hydrogen bonding, ionization, or the ion-dipole bonds are not observed in nonpolar solvents where induced dipole-induced dipole, permanent dipole-induced dipole, or van der Waals type of interactions drive the solubilization of nonpolar compounds. Because nonpolar solvents lack the aforementioned interactions (hydrogen bonding, ionization *etc.*) they are unable to dissolve ionic or polar compounds.

Irrespective of the type of solvent, polar or nonpolar, the more the associative interaction of the solute to the solvent, the more is the solubility of the solute in the solvent. In general, the solubility of drug in water increases with the increase in number of polar functional groups (like, hydroxyl, carboxylic acid *etc*.) and decreases with increase in the number of carbon atoms. . Functional groups such as alcohol, amine, amide, carboxylic acid, sulfonic acid and phosphorus oxy-acid groups either ionize or possess relatively strong intermolecular forces of attraction with water through hydrogen bonding which usually result in increased water solubility. On the other hand, the aromatic compounds are less soluble in water and more soluble in nonpolar solvents. The opposite is true regarding lipid solubility. Consequently, replacing polar functional groups by nonpolar functional groups significantly increase lipid solubility and decreases water solubility. The classic nonpolar groups include alkyl groups and halogens. The longer the alkyl group, the higher its contribution to lipid solubility. Branching, cyclization, and unsaturation applied to an alkyl chain will diminish the contribution to lipid solubility as compared to normal alkyl groups with the same number of carbons. Typically, the halogen contribution towards lipid solubility is proportional to its atomic weight; I > Br > Cl > F. Note that F can contribute to both lipid and water solubility. For example F in $-CH_2F$ can form hydrogen bonding with water, while F in $-CF_3$, which looks similar to an alkyl group $-CH_3$, lacks the ability to form hydrogen bonding with water.

Polarity, Lipophilicity and Related Parameters

As discussed in preceding section, the polarity is an important parameter that can predict the solubility. Following three parameters can aid in predicting the solute solubility.

1. Remember "*like dissolves like*". This concept can be illustrated by the following example. Polar solutes such as urea are highly soluble in polar solvent like water, less soluble in fairly less polar solvent compared to water like methanol and practically insoluble in nonpolar solvents like benzene. In contrast, nonpolar solutes such as naphthalene are insoluble in water, fairly soluble in methanol, and highly soluble in nonpolar benzene (Fig. **2**).
2. **Hydrophilicity** (water-loving) is the tendency of a molecule to be solvated by water and **hydrophobicity** (water-hating) is the association of non-polar groups or molecules in an aqueous environment which arises from the tendency of water to exclude non polar molecules.
3. **Lipophilicity** (lipid-loving), on the other hand, represents the affinity of a molecule or a moiety for a lipophilic environment. It is referred to as the fat-lovingness, or the ability of a chemical compound to dissolve in fats, oils, lipids, and non-polar solvents such as hexane or toluene. These non-polar solvents are themselves lipophilic in nature. Thus lipophilic substances tend to

dissolve in other lipophilic substances, while hydrophilic substances tend to dissolve in water and other hydrophilic substances. Lipophilicity, hydrophobicity and non-polarity, all essentially describe the same molecular attribute; the terms are often used interchangeably [5, 6].

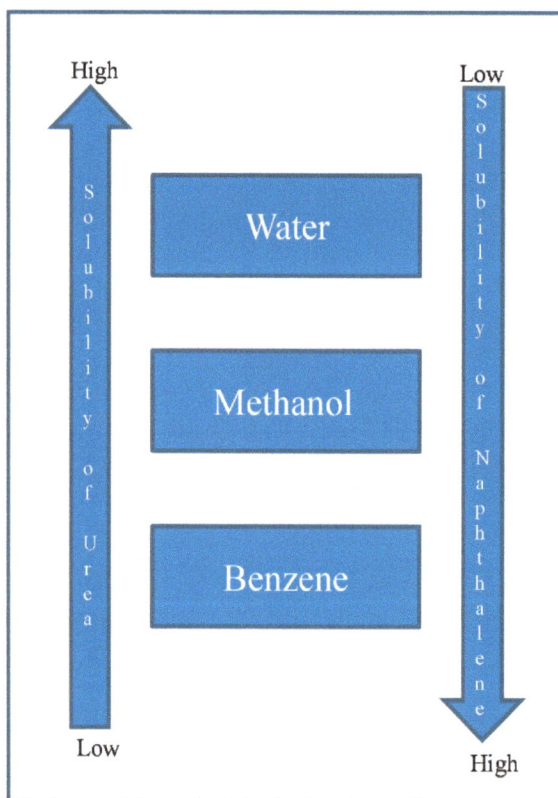

Fig. (2). Illustration of solubility of polar and nonpolar solutes into polar and nonpolar solvents.

Predicting Solubility of Drugs Based on Functional Groups

Lemke's Empirical Method [7]

The notion *"like dissolves like"* is a simplified expression of the interactions of functional groups of the drugs with the solvent. Several functional groups that are liked by water due to their hydrogen bonding potential (called polar groups) are shown in the Table **2**. These groups are usually present in drugs to a varying degree and therefore impart water solubility to varying degree. One can calculate the *water solubilizing potential (WSP)* of a drug using these functional groups and thus predict, fairly accurately, their solubility in water or lipid solvents. The water WSP for any such functional group is higher if it is present as the monofunctional group than if it is present as one of polyfunctional groups in a drug. This discrimination is due to the fact that most of these functional groups are capable of forming intra- and intermolecular hydrogen bonding, which decreases their ability or potential to associate with water molecules and therefore solubility. Another important factor that adversely affects the drug solubility is the formation of zwitterions, for drugs possessing both acidic and basic functional groups. Because in such cases the ionization of both the acidic and basic functional groups to negative and positive charges respectively causes intramolecular ion-ion interaction that diminishes their ability to bond with water and consequently dramatically reduces the solubility. Drugs with a single charge – positive or negative – exhibits very high water solubility due to its ability to form strong ion-dipole bonding with water [7].

Table 2. Water Solubilizing Potential (WSP) of Some Organic Functional Groups [7].

Functional Group	Monofunctional Drug	Polyfunctional Drug
Alcohol (R-OH)	5-6 Carbons	3-4 Carbons
Phenol (Ar-OH)	6-7 Carbons	3-4 Carbons
Amine (R_3N)	6-7 Carbons	3 Carbons
Carboxylic acid (R-COOH)	5-6 Carbons	3 Carbons
Ester (R-COOR)	6 Carbons	3 Carbons
Amide (R-CONR)	6 Carbons	3 Carbons
Ether (C-O-C)	4-5 Carbons	2 Carbons
Aldehyde (R-CHO)	4-5 Carbons	2 Carbons
Ketone (R-CO-R)	5-6 Carbons	2 Carbons
Urea (-NCON-)		2 Carbons
Carbonate (OCOOR)		2 Carbons
Carbamate (OCONR)		2 Carbons

Charge (+, or -) (present in salts): 1 charge can solubilize 20-30° C

Let us consider the case of the antidepressant drug *imipramine* (Fig. **3**). It possesses two amine functions with total WSP = 3 + 3 = 6 carbons. That means its two N atoms are capable of solubilizing only about 6 carbons. Since imipramine has 19 carbons in its structure it is insoluble in water, however, soluble in organic solvents like chloroform, ether or methanol. Imipramine hydrochloride salt on the other hand has a positive charge on the side chain nitrogen which is converted to the ammonium ($-NH^+(CH_3)_2$) cation. Thus one can easily predict that imipramine hydrochloride with only 19 carbons in total is water soluble as its WSP = 3 (for one N atom without a positive charge) + 30 (for one positive charge) = 33 carbons.

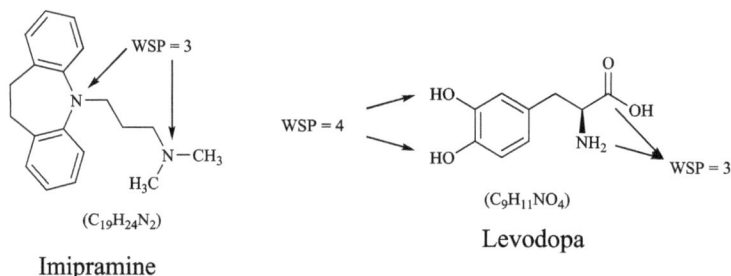

Fig. (3). WSP of Imipramine and levodopa.

Let us now consider the antiparkinsonian drug **levodopa** (Fig. **3**). Based on the more conservative calculation of WSP for polyfunctional drug, its total WSP = 3 + 3 + 4 + 4 = 14 carbons. It contains only 9 carbons while its functional groups are capable of solubilizing at least 14 carbons and it is supposed to be highly water soluble, but in practice it is only slightly soluble. This is due to the fact that it has a carboxylic acid (-COOH) and an amine (-NH$_2$) functions. Both of these functional groups are ionized in water producing a negative ion (-COO$^-$) and a positive ion (-NH$_3^+$), the zwitterions, that interact with each other reducing the water solubility dramatically driving it to only be slightly soluble in water.

It should be noted that this method of predicting water solubility of drugs based on the WSP is only an estimate which works well in many cases but not very accurate. It is suggested that the more conservative calculation using the polyfunctional drug's values is relatively more reliable than using the monofunctional values as most drugs are polyfunctional compounds.

Cate's Analytical Method [7]

This is another method of predicting water solubility employing the *hydrophilic-lipophilic values (π values)* of the functional groups or fragments of a drug structure, which is more useful. This in turn is based on the principle of drug partitioning between lipid and water layers *i.e.* the log *P* value of drugs (discussed in the next section), which is a sum of the π values of each fragments of the drug (Table **3**).

The following examples of **phenytoin** will illustrate this method of predicting water solubility (Fig. **4**). In its structure there are two amide fragments (-CONH-), two phenyl rings and one aliphatic carbon atom, which stands at the total π value that is the calculated log *P* value (Clog *P*) of (2.0 + 2.0 + 0.5 – 0.7 – 0.7) = +3.1. The Clog *P* value of > +0.5 signifies water insolubility thus phenytoin is predicted to be water insoluble, which is also the case in practice. It is slightly soluble in alcohol, chloroform and ether.

Table 3. π Values of Organic Functions (IMHB = Intramolecular Hydrogen Bonding) [5].

Fragment	π Value
Phenyl	+2.0
IMHB	+0.65
C (Aliphatic)	+0.5
Cl	+0.5
O_2NO	+0.2
S	0.0
O_2N (aromatic)	-0.28
O=C-O (carboxyl)	-0.7
O=C-N (not amine)	-0.7
O_2N (aliphatic)	-0.85
O (hydroxyl, phenol, ether)	-1.0
N (amine)	-1.0

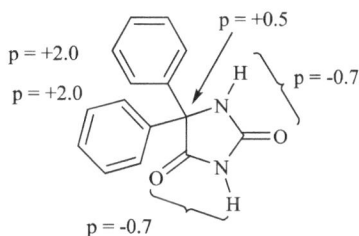

Phenytoin

Fig. (4). The Clog *P* value of phenytoin.

The Clog *P* value is based on the presence of the organic atoms like C, Cl, N, O, or S. However, other factors like intramolecular hydrogen bonding (IMHB), which has a π value of + 0.65, also plays an important role in decreasing water solubility. It is the combination of all these factors that gives the actual Clog *P* value and thus water solubility. The net Clog *P* value of -0.5 of less signifies water solubility; higher the negative value, more the water solubility.

PARTITION COEFFICIENT

In the field of organic and medicinal chemistry, a *partition* (*P*) is the ratio of concentrations of a compound in the two phases of a mixture of two immiscible solvents at equilibrium. Hence these coefficients are measures of differential solubility of the compound between these two solvents. *The partition coefficient (Log P) is a measure of differential solubility of a compound in a hydrophobic solvent (n-octanol) and a hydrophilic solvent (water).* The logarithm of these two values enables compounds to be ranked in terms of hydrophilicity (or lipophilicity). Lipophilic drugs with high partition coefficients are preferentially distributed to hydrophobic compartments such as lipid bilayers of cells while hydrophilic drugs (low partition coefficients) preferentially are found in hydrophilic compartments such as blood serum (Fig. **5**). The propensity of partitioning of a compound from octanol to water is directly proportional to the number of hydrogen bonds it can form with water [8].

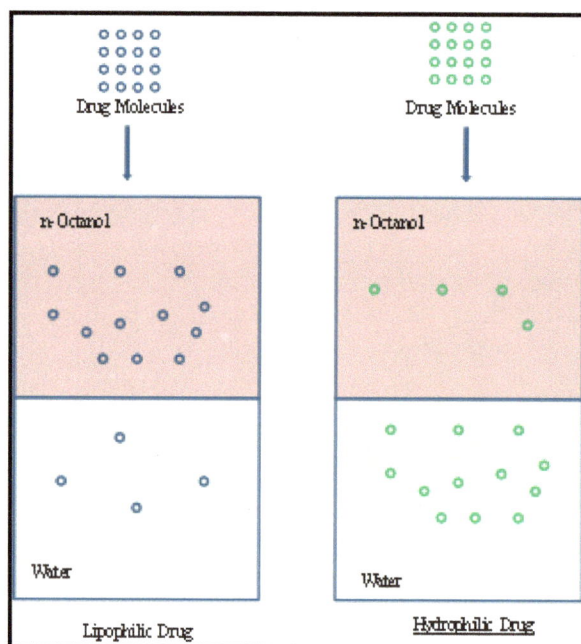

Fig. (5). Illustration of the partition coefficient of drug molecules.

Lipid-Water Partition Coefficient

The octanol-water partition coefficient is the ratio of the concentration of a chemical in octanol and in water at equilibrium and at a specified temperature (Fig. **5**). This is often termed as *lipid-water partition coefficient (LWPC).* Compounds with low log *P* values, that is high aqueous solubility, are good for dissolution but exhibit low permeability through lipid bilayer.

In 1899, Charles E. Overton demonstrated that the greater the lipid solubility of a compound, the greater is its rate of penetration through a plasma membrane. This correlation, sometimes referred to as *Overton's rule,* provided one of the earliest indications that lipids are a major component of the plasma membrane [9]. A second breakthrough with regard to drug activity came in the early 1960s when Hansch and co-workers published the concept that a drug's activity was really a function of two processes [10, 11]. The first being its *transportation* from point of entry to receptor site (*pharmacokinetics*) and the second being its *interaction* with the receptor (*pharmacodynamics*).

Hansch proposed that the first step is necessary for the drug to make its way from the outside of the cell to a particular place on the inside and saw it as being highly dependent upon drug structure. For the drug to travel to the receptor site, it must interact with both an aqueous and an organic environment, *i.e.* it would preferably be both hydrophilic and lipophilic – given that the cell is essentially a lipid-type structure bathed both inside and outside in a dilute solution of salts.

Moreover, the cell is protected by a membrane which is lipid, or carbohydrate in nature. One of the most protective membranes is the "*blood-brain barrier*" that protect the brain from entering polar molecules what nature considers being bad for that organ. Hansch believed that the ability of a drug to get through a membrane might be modeled by its partition coefficient between a lipid solvent and water. He used 1-octanol as a model for the lipid part of the membrane. The 1-octanol/water partitioning system seems to mimic the lipid membranes/water systems found in the body. It turns out that 1-octanol is not as non-polar as initially predicted. Water-saturated octanol contains 2.3 M water, whereas 1-octanol-saturated water contains little of the organic phase. The water in the 1-octanol phase apparently approximates the polar properties of the lipid bilayer, whereas the lack of octanol in the water phase mimics the physiological aqueous compartments, which are relatively free of nonpolar components. However, it should be remembered that the 1-octanol/water system is only an approximation of the actual environment found in the interface between the cellular membranes and the extracellular/intracellular fluids.

The classical and most reliable method of log *P* determination is the shake-flask method, which consists of dissolving a known amount of the solute in question in a measured amount of water-saturated octanol and octanol-saturated water, then measuring the concentration of the solute in each solvent. The aqueous phase will be buffered with a phosphate buffer at pH 7.4 to reflect physiological pH. The amount of chemical in one or both of the phases is determined by an appropriate analytical technique and the partition coefficient calculated from the equation below.

$$P = \frac{\text{Solubility in Octanol}}{(\text{Solubility in Water}) \times (1 - \alpha)}$$

Where α = degree of ionization of the compound in water, based on calculations from ionization constants. He wanted to run these studies with low α, *i.e.* drug is mostly unionized and experimentally determined log P for a large number of compounds by placing the compound in a separating funnel with varying volumes of 1-octanol and water, shaking for a particular period of time and determining the concentration of compound in each layer. Note that log P varies slightly with compound concentration and temperature, but for molarities of 0.01 or less, which is negligible. The most common method of measuring the distribution of the solute is by UV/VIS spectroscopy. A faster method of log P determination makes use of high-performance liquid chromatography where the log P of a solute can be determined by correlating its retention time with similar compounds with known log P values.

Large log P values indicate high lipid solubility while small log P values indicate higher water solubility. Thus large log P value implies a compound is non-polar while small log P value indicates a compound is polar. It is usually desired that drugs possess a hydrophilic and lipophilic balance. Drugs with very low log P (usually less than zero) value may not be considered suitable candidate for the purpose of drug development because they lack the ability to cross lipophilic biological membrane. On the other hand, drugs with very high log P value (usually more than 3.5) have very poor water solubility and will exhibit poor solubility in aqueous biological fluids. These drugs may also concentrate in lipid rich region of the body causing drug accumulation inside the body. Fig. **6** shows the preferred log P values for drug considered suitable for development. It should be noted that few drug candidates in spite of log P values outside the desirable range has been proved to be successful drugs. This suggests that the body is much more complicated system and decision based purely on log P values can be erroneous. However, log P value provides a direction in predicting drug's behavior inside the complex biological system.

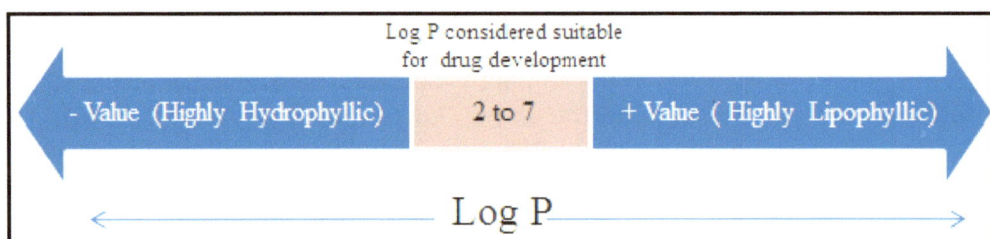

Fig. (6). Illustration of log P range with biological significance.

It should be noted that partition coefficient has no unit, it is always positive, and cannot be zero. Finally, log P does not represent the absolute solubility but only the ratio of the two solubility parameters.

Application of Partition Coefficient to Drugs

It has been shown that the absolute and relative solubility of drugs in aqueous and lipid phases of the body are the physical properties of primary importance in providing and maintaining effective concentrations of drugs at their sites of action. These physical and chemical properties are important variables in determining a drug's biological response. An excellent correlation has been obtained between chromatographic retention parameters and biological response [12]. In homologous series of undissociated or slightly dissociated compounds in which the change in structure involves only an increase in the length of the carbon chain, gradations in the intensity of action have been observed for a number of unrelated pharmacologic groups of compounds. The lower members show a low order of biological activity. With the increasing length of the carbon chain, the activity increases, passing through a maximum. The increase in activity roughly parallels the decrease in water-solubility and the increase in lipid solubility. As indicated by the partition coefficient, this increase in potency may be associated with the availability of the compound for the cell where the action occurs. Further increase in the length of the carbon chain results in a rapid decrease in the activity. The observed decrease in activity with further increase in length of the chain may be due to the diminishing solubility of the compounds in the extra-cellular fluid which serve as a medium to transport the drug to the cell surface.

The relationship of drug's water solubility and bioavailability is complex and various factors must be considered for selecting the right drug candidate that can provide desired bioavailability. All pharmaceutical compounds can be grouped into the following classes based on their solubility and permeability. It has also been illustrated in Fig. **7** [13, 14].

- Class I: <u>High solubility, high permeability compounds</u>. Because these compounds are highly permeable through the jejunum and readily go into solution, these are generally very well-absorbed after oral administration.
- Class II: <u>Low solubility, high permeability compounds</u>. These compounds exhibit dissolution rate-limited absorption. By improving the dissolution rate of these compounds the rate of absorption can also be increased. Chemically these may have too high Log P values that make them hydrophobic.
- Class III: <u>High solubility, low permeability compounds</u>. These compounds exhibit permeability rate-limited absorption. In general, compounds with low log P values (too high polarity) demonstrate such behavior.
- Class IV: <u>Low solubility, low permeability compounds</u>. These compounds have very poor oral bioavailability. Normally, compounds in this class are large molecules with high lipophilicity or form zwitterions.

Class I (Amphiphillic)	Class II (Lipophillic)
Highly soluble	Poorly soluble
Highly permeable	Highly permeable
Eg: Metoprolol	*Eg: Diclofenac*
Class III (Hydrophillic)	Class IV (Risk-Phillic)
Highly soluble	Poorly soluble
Poorly permeable	Poorly permeable
Eg: Atenolol	*Eg: Cyclosporine*

Fig. (7). Classes of pharmaceutical compounds based on their solubility and permeability.

Due to the complexity in the membrane structure, drug absorption is a complex process of solubilization and permeability, which is influenced by physicochemical factors such as molecular size (*negatively correlated*), lipophilicity (*positively correlated*), polarity (*negatively correlated*), and the intramolecular hydrogen bond formation [15]. The drug-water hydrogen bond strength is also important, which must be broken before the drug can pass through the biological membrane. Vast majority of orally administered drugs are absorbed by passive diffusion through a highly lipophilic membrane, which is governed by *Fick's laws of diffusion*. The drug's log P values have been found to be highly correlated with such diffusion ability. The optimum log P value for drug absorbability by passive diffusion is in the range of 2 to 7 as mentioned before [16]. The active transport, however, is highly compound specific and will not follow this generalization [17].

The FDA's Center for Drug Evaluation and Research (CDER) has provided the waiver of *in vivo* bioequivalence study requirements for high-solubility/hig--permeability drug products (Class I) based on *in vitro* dissolution data. This waiver is only applicable for immediate-release solid oral dosage forms if the study is conducted following the CDER guidelines [18].

CASE STUDIES

Case 1. Ryan, a second year pharmacy intern has recently joined your hospital pharmacy. He has a very inquisitive mind and always backs up his decision with a rationale. He is currently assisting a pharmacy technician in setting up an automated compounder to make total parenteral nutrient (TPN) admixtures (pH of TPN ranges from 5-7). All the micronutrients added to the TPN admixture are added as salts for example calcium (as the gluconate salt), phosphate (as the sodium salt) Potassium (as an acetate salt) *etc*. Ryan observed that there were two vials on the shelf for calcium administration (calcium chloride ($CaCl_2$) and calcium phosphate [$Ca(H_2PO_4)_2$, $CaHPO4$, or $Ca_3(PO_4)_2$ depending on pH of the medium]. The pharmacy technician hanged calcium gluconate vial, Ryan asked the pharmacy technician to why he hanged a calcium gluconate vial and not calcium chloride or calcium phosphate. The pharmacy technician was not sure of the answer so he directed Ryan to you about this concern.

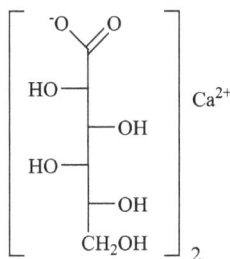

Calcium gluconate
(pKa 3.7)

1. What is the acid-base character of $CaCl_2$, calcium phosphate and calcium gluconate? Which calcium salt is the most dissociable and why? Does calcium gluconate dissociate in water? Why or why not?
2. Calcium phosphate can exist in three different ions depending on the pH of solution (monobasic biphosphate $H_2PO_4^-$, dibasic phosphate $H_2PO_4^{2-}$, and tribasic orthophosphate PO_4^{3-}). Which ions exist at what pH and which ions may exist in the TPN solution?

Tribasic calcium orthophosphate exists above pH 10 therefore only monobasic calcium biphosphate and dibasic and calcium phosphate can exist in the TPN admixtures. Furthermore, pH less than 6 favors the formation of monobasic calcium biphosphate and pH between 6 and 10 favors formation of dibasic calcium phosphate.

1. Why phosphate is generally added as sodium salts but not as calcium salt? Do you think Ryan had a legitimate concern and why? (Hints: Calcium phosphate is water insoluble and get precipitated if formed)

Yes, Ryan asked an excellent question. In preparing TPN admixtures the most dangerous physical incompatibility that occurs is through calcium phosphate precipitation. It is very important to use the right calcium salt because calcium salt and phosphate salt can react with each other (double displacement reaction) in the admixture and can lead to the formation of insoluble calcium phosphate which can precipitate in the admixture. It is recommended to use calcium gluconate salt over calcium chloride because calcium chloride is much more reactive than an equivalent amount calcium gluconate. In general, the degree of dissociation of calcium chloride is much higher than calcium gluconate. Therefore, more calcium will be available to form precipitate with calcium chloride than with calcium gluconate when added in same amount [19].

1. What should be recommended to follow to avoid precipitation of calcium phosphate?

The aqueous solubility of monobasic and dibasic calcium phosphates at 20°C is 18g/L and 0.04g/L, respectively. Since monobasic calcium biphosphate is 450 times more soluble than dibasic calcium phosphate the pH of TPN admixtures should be maintained below 6 to avoid precipitation [19].

STUDENT SELF-STUDY GUIDE

- What are the different factors that affect solubility?
- If given a drug structure you should be able to apply the Lemke's empirical method of determining its water or lipid solubility
- Understand the Cates analytical method of determining drug solubility. You have to solve it for a given drug structure.
- LWPC and its significance in drug transport and pharmacokinetics

STUDENT SELF ASSESMENT

Part I – Multiple Choice Questions:

Select the **best** lettered answer in each case.

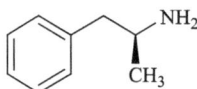

1. Calculate the π value and WSP value for and tell whether the drug is water soluble or not.
    A. $\pi = +3.0$; WSP = 6-7; water soluble
    B. $\pi = +3.0$; WSP = 3; not water soluble
    C. $\pi = +2.5$; WSP = 6-7; water soluble
    D. $\pi = +2.5$; WSP = 6-7; not water soluble
    E. $\pi = -2.5$; WSP = 3; not water soluble
2. Which statement is incorrect?
    A. Solubility ranges widely, from infinitely soluble such as ethanol in water (miscible), to poorly soluble, such as steroidal drugs in water.
    B. Van der Waals force is defined as attractions between molecules and

    differs from covalent and ionic bonding in Van der Waals forces are not stable and are the weakest interactions.

  C. Lemke's Empirical Method is a method that predicts water solubility by employing the log P (lipid-water partition coefficient) values of functional groups.

  D. Hydrophilicity is the tendency of a molecule to be solvated by water.

  E. Keesom force, Debyes force, and London dispersion forces are all Van der Waals forces.

3. The WSP of X, Y, Z is in order (Highest to lowest):

    X   Y   Z

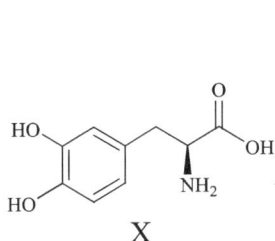

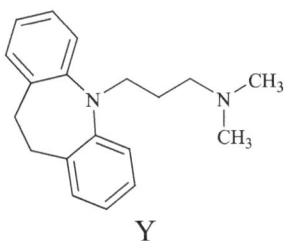

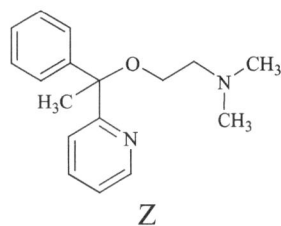

  A. X > Y > Z
  B. Y > Z > X
  C. Z > Y > X
  D. X > Z > Y
  E. Y > X > Z

4. The total π value of drugs X, Y, Z is in order (Highest to lowest):

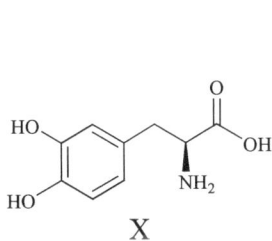

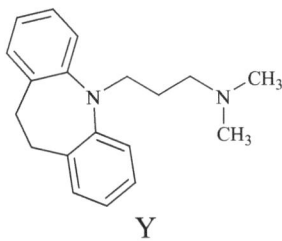

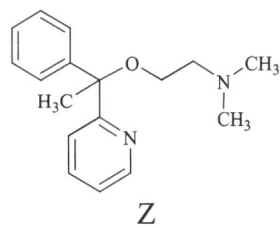

    X   Y   Z

  A. X > Y > Z
  B. Y > Z > X
  C. Z > Y > X
  D. X > Z > Y
  E. Y > X > Z

Part II – K-Type Questions:

Chose the answer

  A. If **I only** is correct
  B. If **III only** is correct
  C. If **I and II** are correct

D. If **II and III** are correct
E. If **I, II, and III** are correct

5. Which of the following statements are true?
   I. C log P values $> +0.5$ demonstrate water insolubility
   II. According to Cate's Analytical Method, negative π values signify more lipid soluble compound
   III. Zwitterions increase solubility in water by 2-fold

Part III – Matching Questions:

For each item, select the one lettered option that is most closely associated with the numbered items.

6. Match the drug structures I-V with the appropriate statements A-E.

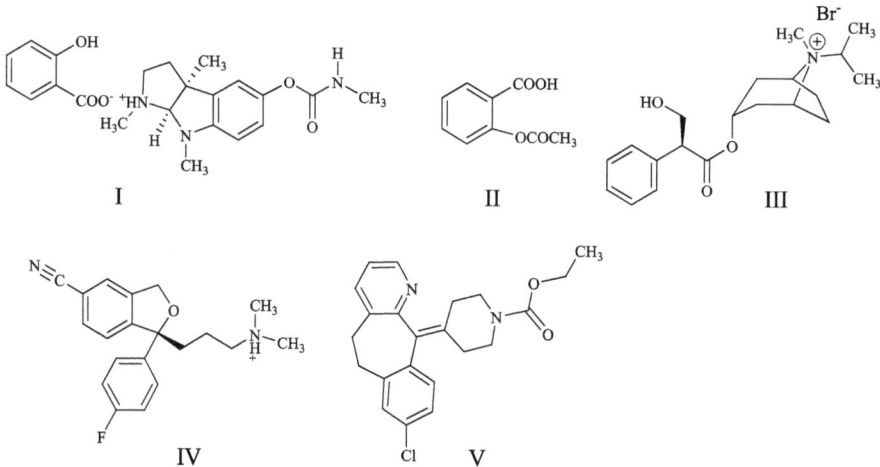

I   II   III

IV   V

A. This conjugated acid is water soluble and if mixed with $NaHCO_3$ solution will be precipitated out.
B. This drug is highly unionized in the gastric fluid and thus absorbed considerably from stomach.
C. This drug (not a salt) is highly ionized in the gastric fluid and thus is absorbed negligibly, if any, from stomach.
D. This one is a salt of organic acid and organic base and thus is only weakly soluble in water
E. This one is a neutral salt (neither acid nor base) and is water soluble.

CONSENT FOR PUBLICATION

Not applicable.

CONFLICT OF INTEREST

The authors declare no conflict of interest, financial or otherwise.

ACKNOWLEDGEMENT

Declared none.

REFERENCES

[1]   Sokoloski TD. Solutions and phase equilibria.Remington's Pharmaceutical Sciences. Easton, PA: Mack Publishing 1985; pp. 207-29.

[2]   Williamson KJ, Ed. Macroscale and Microscale Organic Experiments. 2nd ed., Lexington, MA: D. C, Heath 1994.

[3]   Jinno J, Kamada N, Miyake M, *et al.* Effect of particle size reduction on dissolution and oral absorption of a poorly water-soluble drug, cilostazol, in beagle dogs. J Control Release 2006; 111(1-2): 56-64.
[http://dx.doi.org/10.1016/j.jconrel.2005.11.013] [PMID: 16410029]

[4]   Hancock BC, Parks M. What is the true solubility advantage for amorphous pharmaceuticals? Pharm Res 2000; 17(4): 397-404.
[http://dx.doi.org/10.1023/A:1007516718048] [PMID: 10870982]

[5]   Clugston M, Fleming R, Eds. Advanced Chemistry. 1st ed., Oxford: Oxford Publishing 2000.

[6]   Compendium of Chemical Terminology 1997.http://goldbook.iupac.org

[7]   Lemke TL. Review of Organic Functional Groups: Introduction to Medicinal Organic Chemistry. 4th ed., Malvern, Pennsylvania: Lippincott Williams and Wilkins 2004.

[8]   Stenberg P, Luthman K, Artursson P. Prediction of membrane permeability to peptides from calculated dynamic molecular surface properties. Pharm Res 1999; 16(2): 205-12.
[http://dx.doi.org/10.1023/A:1018816122458] [PMID: 10100304]

[9]   Kleinzeller A. Ernest overton's contribution to the cell membrane concept: A centennial appreciation. Physiology (Bethesda) 1997; 12(1): 49-53.
[http://dx.doi.org/10.1152/physiologyonline.1997.12.1.49]

[10]   Hansch C, Maloney PP, Fujita T, Muir RM. Correlation of biological activity of phenoxyacetic acids with hammett substituent constants and partition coefficients. Nature 194: 178-80.
[http://dx.doi.org/10.1038/194178b0]

[11]   Hansch C. Quantitative approach to biochemical structure-activity relationships. Acc Chem Res 1969; 2(8): 232-9.
[http://dx.doi.org/10.1021/ar50020a002]

[12]   Breyer ED, Strasters JK, Khaledi MG. Quantitative retention-biological activity relationship study by micellar liquid chromatography. Anal Chem 1991; 63(8): 828-33.
[http://dx.doi.org/10.1021/ac00008a019] [PMID: 1877751]

[13]   Amidon GL, Lennernäs H, Shah VP, Crison JR. A theoretical basis for a biopharmaceutic drug classification: the correlation of *in vitro* drug product dissolution and *in vivo* bioavailability. Pharm Res 1995; 12(3): 413-20.

[http://dx.doi.org/10.1023/A:1016212804288] [PMID: 7617530]

[14] Löbenberg R, Amidon GL. Modern bioavailability, bioequivalence and biopharmaceutics classification system. New scientific approaches to international regulatory standards. Eur J Pharm Biopharm 2000; 50(1): 3-12.
[http://dx.doi.org/10.1016/S0939-6411(00)00091-6] [PMID: 10840189]

[15] Palm K, Luthman K, Ungell A-L, Strandlund G, Artursson P. Correlation of drug absorption with molecular surface properties. J Pharm Sci 1996; 85(1): 32-9.
[http://dx.doi.org/10.1021/js950285r] [PMID: 8926580]

[16] Seydel JK, Schaper KJ. Quantitative structure-pharmacokinetic relationships and drug design. Pharmacol Ther 1981; 15(2): 131-82.
[http://dx.doi.org/10.1016/0163-7258(81)90040-1] [PMID: 6124012]

[17] Stenberg P, Luthman K, Artursson P. Virtual screening of intestinal drug permeability. J Control Release 2000; 65(1-2): 231-43.
[http://dx.doi.org/10.1016/S0168-3659(99)00239-4] [PMID: 10699283]

[18] Martinez MN, Amidon GL. A mechanistic approach to understanding the factors affecting drug absorption: a review of fundamentals. J Clin Pharmacol 2002; 42(6): 620-43.
[http://dx.doi.org/10.1177/00970002042006005] [PMID: 12043951]

[19] Singh H, Dumas GJ, Silvestri AP, *et al.* Physical compatibility of neonatal total parenteral nutrition admixtures containing organic calcium and inorganic phosphate salts in a simulated infusion at 37 ° C. Pediatr Crit Care Med 2009; 10(2): 213-6.
[http://dx.doi.org/10.1097/PCC.0b013e31819a3bf4] [PMID: 19188866]

CHAPTER 5

Isosteric and Spatial Considerations of Drugs

M. O. Faruk Khan[1,*] and **Timothy J. Hubin**[2]

[1] *Department of Pharmaceutical Sciences and Research, Marshall University School of Pharmacy, Huntington, WV, USA*

[2] *Department of Chemistry, Southwestern Oklahoma State University, Weatherford, Oklahoma, USA*

Abstract: This chapter is a brief review of the isosterism and steriochemical principles and their applications in drug action. After study of this chapter, students will be able to:

• Define isosterism and bioisosterism of drugs and apply these concepts in drug discovery and drug action
• Explain how drugs' spatial factors control their activity
• Define important stereochemical parameters in drugs
• Assign drugs stereochemistry and apply those in drug action and receptor binding
• Explain the basis of the structure activity relationship and mechanism of drug action

Keywords: Bioisosterism, Drug receptor interaction, Isoterism, Receptor, Stereochemistry, Structure activity relationships.

ISOSTERISM AND BIOISOSTERISM

The concept of isosterism was developed by I. Langmuir in 1919 to correlate similarities in physical properties of nonisomeric molecules, and defined isosteres as those compounds or groups of atoms or radicals that have the same number and arrangements of electrons [1]. The rationale was that if molecules have similar outer electron configurations, they will have similar physicochemical properties, because they interact with other atoms primarily through their outermost electrons.

Grimm's Hydride Displacement Law (1925) [2] was an extension of the concept of isosterism. It states that atoms up to four places in the periodic table left of the

* **Corresponding author M. O. Faruk Khan**: Department of Pharmaceutical Sciences and Research, Marshall University School of Pharmacy, Huntington, WV, USA; Tel: 304-696-3094; Fax: 304-696-7309; E-mail: khanmo@marshall.edu

inert gases can unite with one to four hydrogen atoms to convert them into "pseudoatoms". These pseudoatoms behave like the elements in groups immediately to their right. For example, isosteric groups following Grimm's Law would include atoms, or pseudoatoms: N and CH; O, NH, and CH_2; F, OH, NH_2, and CH_3. As time progressed the definition of isosterism was broadened to include functional groups in which pseudoatoms take the place of atoms in simpler functional groups, such as O and NH in amide and imidine.

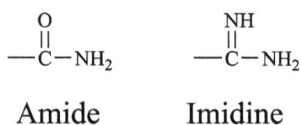

$$\underset{\text{Amide}}{\overset{\displaystyle O \atop \displaystyle \|}{-C-NH_2}} \qquad \underset{\text{Imidine}}{\overset{\displaystyle NH \atop \displaystyle \|}{-C-NH_2}}$$

Friedman [3] introduced the concept of *bioisosteres* to include all isosteres producing similar biological activity including antagonistic activities. He liberally defined bioisosteres as compounds resulting from the exchange of an atom or a group of atoms with another, broadly similar atom or group of atoms. The objective of a bioisosteric replacement is to create a new compound with similar biological properties to the parent compound. The bioisosteric replacement may be physicochemically or topologically based.

Alfred Burger (1970) classified and subdivided bioisosteres into "classical" and "non-classical" categories [4]. The classical isosteres are the groups of atoms which impart similar physical or chemical properties to a molecule due to similarities in size, electronegativity, or stereochemistry (this latter term will be defined later in this chapter). Examples are N_2 and CO, or CO_2 and N_2O or N_3^- and NCO^-. Similarities are observed most often in atoms that are in the same vertical columns of the periodic table because of the identical nature of the number and arrangement of valence electrons. This identical nature of valence electrons makes oxygen and sulfur atoms isosteres of each other. Similarly, all the halogen atoms (F, Cl, Br, I) are isosteres of each other. The common classical isosteres that occur in drug molecules are shown in Table **1** [5].

Another common set of classical isosteres are the aromatic and heteroaromatic ring equivalents: benzene, pyridine, thiophene, furan, and pyrrole. These structures can be considered classical isosteres due to their similar cyclic structures and there identical number of six π-electrons. Keep in mind that only certain heteroatom lone-pairs are included in the aromatic-required 6 π-electron count. If not included, lone-pairs are localized on the heteroatom [5].

Table 1. Commonly encountered classical isosteres in drug molecules [5].

Monovalent	Divalent	Trivalent	Tetravalent	Ring Equivalents
-H, -F, -Cl, -Br, -I, -C≡N -OH, -SH -OCH$_3$, -NH$_2$, -PH$_2$ -CF$_3$, -CH$_3$	-O-, -S- -Se-, -Te- -CH$_2$-, -NH-	-N=, -P=, -As=, -Sb= -CH= -CH=, -N=	=C=, =Si=, =N$^+$= =P$^+$=, =As$^+$=, =Sb$^+$=	-CH=CH-, -S-, -O-, -NH-, -CH$_2$-

Benzene   Pyridine   Thiophene   Furan   Pyrrole

Friedman [3] introduced the concept of *bioisosteres* to include all isosteres producing similar biological activity including antagonistic activities. He liberally defined bioisosteres as compounds resulting from the exchange of an atom or a group of atoms with another, broadly similar atom or group of atoms. The objective of a bioisosteric replacement is to create a new compound with similar biological properties to the parent compound. The bioisosteric replacement may be physicochemically or topologically based.

Unlike the classical isosteres, the *nonclassical isosteres* do not require the same number of atoms and do not necessarily fit the steric and electronic requirements of their classical counterparts. From a modern viewpoint, isosterism can include such parameters as basicity, acidity, electronegativity, polarizability, bond angles, size, molecular orbitals, electron density, and partition coefficient, all of which contribute to the overall physicochemical and biological properties of a molecule. Indeed, with increased knowledge of the structure of molecules, less emphasis has been placed on the number of electrons. Even when electron number is the same, variations in hybridization may lead to differences in bond angles, bond lengths and polarities. Typical nonclassical isostere references are instead likely to identify cyclic *vs.* noncyclic isosteres or functional group isosteres, *e.g.*, amides, imidines, sulfonamides, carboxylic acids, ethers, thioethers, *etc.* Some common examples of nonclassical bioisosteres are shown in Table **2**. Examples of some drugs that contain these isosteres are shown in the following section [5].

Isosterism in Drugs

The concept of isosterism is often used by pharmaceutical companies in an attempt to produce drugs with activities similar to the existing drugs. Isosteric analogues can often be produced inexpensively and result in additional active drugs. Unfortunately, there are no rules which will predict the activity of drugs resulting from isosteric replacement. Analogues of drugs made by isosteric replacement may in fact have similar, opposite, or even seemingly unrelated biological activity. Each series of isosteric drug candidates must be screened individually, for there is no general rule to predict if activity will increase or decrease. However, some common outcomes of isosteric replacement are known. When isosteric replacement modifies only the bridge connecting other groups necessary for activity, and is itself not essential, a graduation of similar activity results. However, when the isosteric replacement involves a part of the molecule essential for an interaction with a receptor, loss of activity or antagonism may result. Some of the important drugs obtained by isosteric replacements of a variety of the common isosteres of Tables **1** and **2** are discussed below.

Table 2. Some commonly occurring non-classical bioisosteres [5]

-CO-	-COOH	-H	-OH	R-S-R	$-CF_3$
$-CO_2-$	$-SO_2NH_2$	-F	$-CH_2O$	(R-O-R')	-CN
$-SO_2-$	-COOR				$-N(CN)_2$
$-SO_2NR-$	$-SO_3H$				$-C(CN)_3$
-CON-	-tetrazole				
-CH(CN)-	$-SO_2NHR$				
-ROCO-	$-NHCONH_2$				
-NHCO-					
-CONH-					

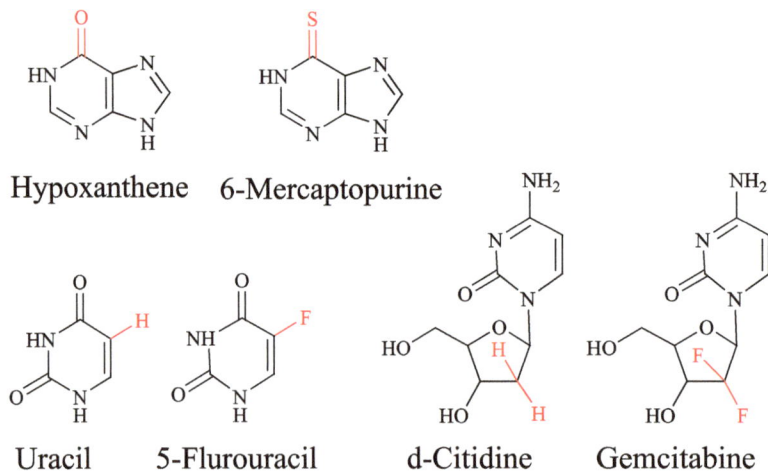

Hypoxanthene   6-Mercaptopurine   Uracil   5-Flurouracil   d-Citidine   Gemcitabine

6-Mercaptopurine is an isostere of hypoxanthine, a purine base that is an intermediate in the biosynthesis of nucleosides and nucleotides – the DNA and RNA bases. A classical isosteric replacement is when =O is replaced with the isostere =S, thus furnishing an analog that works as an antimetabolite for these bases and their nucleotides. This isostere acts as a cytotoxic agent used clinically for a variety of neoplasms.

The van der Waals radius for hydrogen (120 pm) and fluorine (147 pm) are similar. Therefore, the F atom has similar steric requirements as an H atom on a molecule. Enzymes can often bind isosteres with F replacement of H, but the reactivity is different, resulting in the inhibition of enzyme activity. The fact that the F atom cannot be removed as readily as the H atom, allows the fluorinated isosteres of the metabolites to inhibit the enzymes instead of acting as substrates. Moreover, the F atom makes the molecule more acidic, due to its electron withdrawing effect, thus binding the enzyme tighter than the natural substrate. Examples of drugs in clinical use that use the isosteric replacements of the H atom by the F atom and work in the same mechanism or enzyme systems include 5-fluorouracil, gemcitabine, and citidine, among many others.

The isosteric replacement of the -OH fragment by NH_2 is also frequently observed in drugs. Methotrexate, which is a bioisostere of dihydrofolate, is an example of a dihydrofolate reductase inhibitor (an antagonist) obtained by this classical isosteric replacement method.

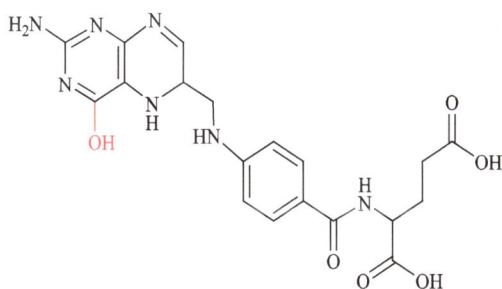

Dihydrofolic acid (DHFA)   Methotrexate

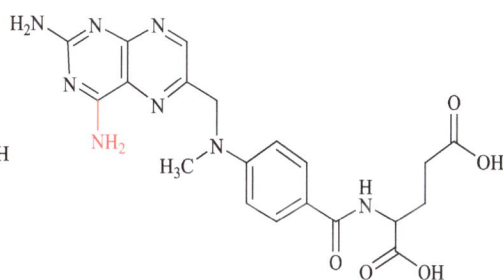

Carbachol is an isostere of acetylcholine where -$COCH_3$ of the neurotransmitter acetylcholine is replaced with -$CONH_2$ to provide an agonist.

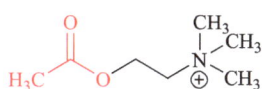

Acetylcholine   Carbachol

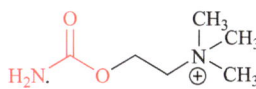

The isosteric replacement of the phenyl ring with the pyridine ring is an excellent method of getting potent antihistaminics. Chlorpheniramine is an example that has such a modification from diphenhydramine, in addition to other obvious modifications.

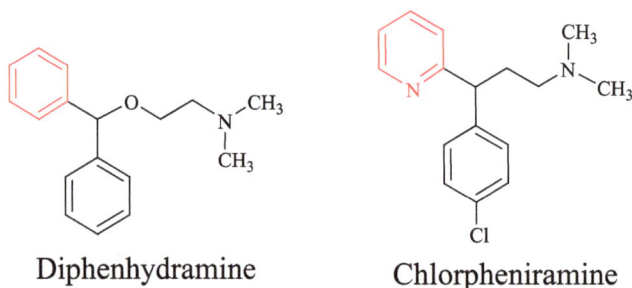

Diphenhydramine   Chlorpheniramine

Metiamide was found to be a potent H_2 blocker but possessed some disadvantages due to the presence of the thiourea (-NHCSNH-) function in the side chain. The clinically useful drug cimetidine was obtained by replacing this side chain thiourea function with a neutral, but polar, nitrile substituted guanidine moiety. Further modification of the heteroaromatic imidazole ring with a thiazole isostere provided nizatidine. All of these are excellent examples of bioisosterism in drug discovery.

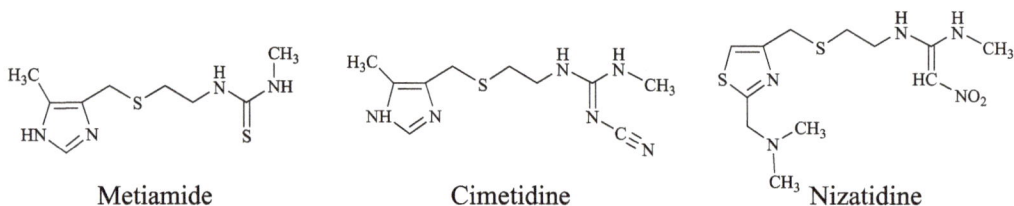

Metiamide   Cimetidine   Nizatidine

Other interesting examples of ring equivalents are the clinically useful drugs chlorpromazine and clomipramine. While chlorpromazine, which contains a -S-atom in the ring system, is an antipsychotic drug, clomipramine, which contains the isostere -CH_2CH_2- in place of the –S-, is an antidepressant.

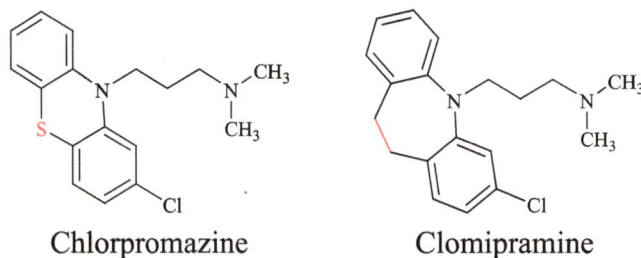

Chlorpromazine   Clomipramine

The isosteric replacement of the ring =CH- of norfloxacin with =N- resulted in enoxacin, both of which are potent second generation flouroquinolone antimicrobial agents.

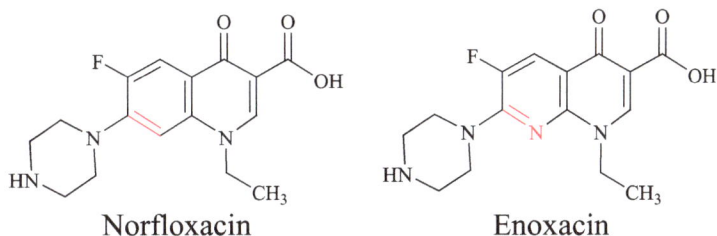

Norfloxacin   Enoxacin

Carboxylic acid (-COOH) and tetrazole are both weakly acidic functional groups and are considered bioisosteres. The polar –COOH present in EXP7711, when replaced with the lipophilic tetrazole, furnished the clinically useful angiotensin II inhibitor losartan. Most of the newly discovered antihypertensive angiotensin II inhibitors possess tetrazole as a weakly acidic function that is needed for their inhibitory activity, but with more lipophilic character and thus better oral bioavailability.

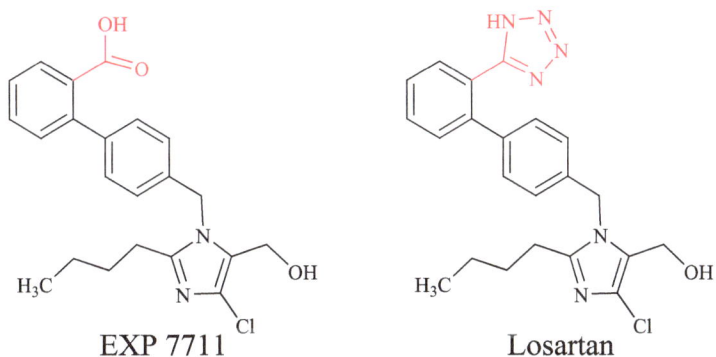

EXP 7711   Losartan

Sulfanilamide and related drugs (also known as congeners) are the result of the isosteric replacement of the –COOH function of p-aminobenzoic acid, a cofactor for dihydropteroate synthase, with –SO_2NH_2. The resulting "sulfa drugs" are competitive inhibitors of the enzyme dihydropteroate synthase, and thus are antibacterials.

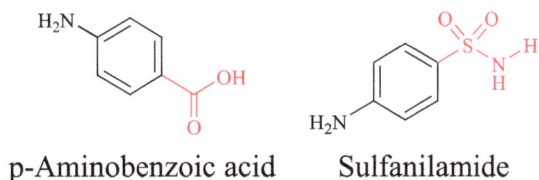

p-Aminobenzoic acid   Sulfanilamide

SPATIAL CONSIDERATION OF DRUGS

The classic statement below by Louis Pasteur, and the following timeless title of an article by J. H. van't Hoff, the father of stereochemistry, contributed enormously to the development of the modern concepts of stereochemistry and drug action:

"Most organic products, the essential products of life, are asymmetric and possess such asymmetry that they are not super-imposable on their images. This establishes perhaps the only well-marked line of demarcation that can at present be drawn between the chemistry of dead matter and the chemistry of living matter" – Louis Pastor, 1858 [6].

"Proposal for the extension of the structural formulae now in use in chemistry into space, together with a related note on the relationship between the optical active power and the chemical constitution of organic compounds – J. H. van't Hoff, 1974; English translation, 1901 [6].

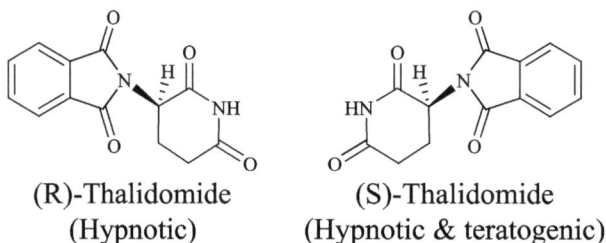

(R)-Thalidomide
(Hypnotic)

(S)-Thalidomide
(Hypnotic & teratogenic)

Tremendous advances in asymmetric synthesis and the corresponding improvements in the analytical techniques of chiral drugs over the past two decades have generated an explosion of research on drug chirality. The understanding of the significance of the pharmacodynamic and pharmacokinetic differences between the enantiomers of chiral drugs has increased the concern over the use of racemates or mixtures of diastereomers in therapeutics. The Food and Drug Administration (FDA) now requires the evaluation of both enantiomers of a stereoactive prospective drug. The *thalidomide tragedy* is particularly important to cite in this regard, and is frequently used by the popular press to support the arguments for developing chirally pure drugs [7, 8]. Racemic thalidomide was developed in the 1950s and was used to treat morning sickness. Fortunately, its use was limited and did not occur in the USA, as FDA approval was not given. Tragically, the drug had serious side effects, as it was found to be teratogenic and caused fetal abnormalities. It was later discovered that the (*S*)-enantiomer possessed the teratogenic and hypnotic effects while the (*R*)-enantiomer possessed only the hypnotic effects. Subsequent studies revealed that the enantiomers racemise under physiologic conditions. Instructive of the necessity for full exploration of a drug's properties, the FDA has recently approved racemic thalidomide for the treatment of leprosy under the strictest of guidelines.

The Significance of the Study of Stereochemistry

The effect of a drug's chirality on its pharmacodynamic and pharmacokinetic properties arises from the fact that most drugs must bind or otherwise interact with receptor sites and the reactive sites of enzymes, which are chiral themselves and have chiral topographical (surface structural) features. Most of these biological receptors or enzymes, *e.g.*, proteins, glycolipids and polynucleotides, are composed of the chiral building blocks of L-amino acids and D-carbohydrates. Many proteins are coiled into an α-helix where each peptide is H-bonded to the 3rd peptide away from itself. The distance between two consecutive peptide bonds is 3.6 Å, while the distance between two consecutive turns of the α-helix is 5.4-5.5 Å. It is then, not likely only coincidence that a considerable number of drugs have functional groups believed to be involved in receptor complexation that are separated by distances which are multiples of 5.4 Å, (*e.g.*, diphenhydramine, carbachol, procaine and adiphenine) and/or 3.6 Å (*e.g.*, quaternary cationic N⁺ drugs like carbachol and decamethonium). Thus, the 3D spatial arrangements of both the drug molecules and the receptor surface are important.

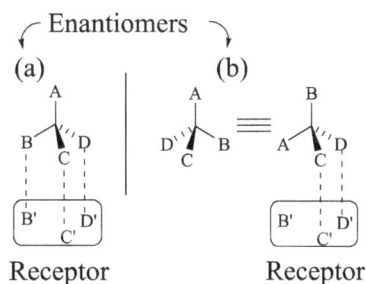

Fig. (1). The enantiomer (**a**) exhibits complimentary fit with receptor active site making all three interactions, while the enantiomer (**b**) shows two-site interactions only.

The complimentary fit of an enantiomeric pair of drugs with the receptor active site is illustrated in Fig 1. As shown in this figure, the enantiomer (a) is the active enantiomer since it interacts with all three bonding areas, whereas the enantiomer (b) is considered the inactive one as it makes only two-site interactions. The enantiomer of a chiral compound that is less potent for a particular action, including all the effects and side effects, is called a distomer. The more potent one of the pair is the eutomer. The ratio of the potency of the eutomer relative to that of the distomer is the eudismic ratio. Enantiomers may differ in a variety of ways regarding how they influence drug activity:

- The enantiomers may possess different pharmacological indications. For example, (+)-propoxyphene, with the (2S, 3R)-configuration, is an analgesic marketed as Darvon™. The (–) isomer, with the (2R, 3S)-configuration, is used as a pure antitussive called Novrad™ and lacks the classical opioid characteristics of its enantiomer.
- The enantiomers may possess different potency. For example, (S)-(-)-propranolol (eutomer) is about 130 times more potent than (R)-(+)-propranolol (distomer).
- Sometimes, one enantiomer is therapeutic but the other enantiomer is toxic or harmful. For example, dextromethorphan is a cough suppressant while levomethorphan is a narcotic analgesic. Further, (S,S)-ethambutol is tuberculostatic while (R,R)-ethambutol may cause blindness.
- Enantiomers may also possess opposite biological activity. For example, (+)-dobutamine is an α-adrenergic antagonist, but (-)-dobutamine is an agonist to the same receptor.

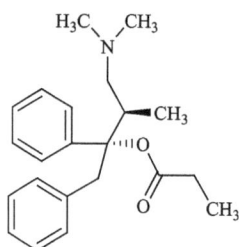

(+)-Propoxyphene   (-)-Propoxyphene

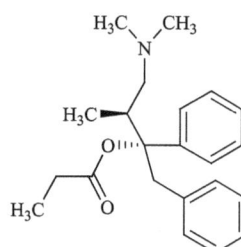

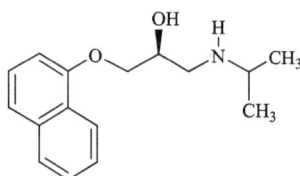

(*S*)-(-)-Propranolol   (*R*)-(+)-Propranolol

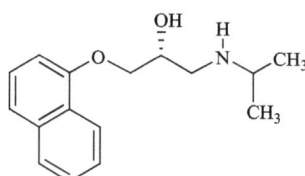

The difference in pharmacokinetic profiles of enantiomers are the result of several processes, including the interaction of the drugs with chiral biological macromolecules, such as active transport processes, and plasma protein binding, as well as processing by chiral enzymes during drug metabolism. Although difference in lipid solubility, pK_a, and size – factors important in drug transport through lipid membranes – are not expected between the enantiomers, diastereomers may differ in these parameters leading to differences in absorption through the membranes. For example, the oral bioavailability of D-methotrexate is only 2.5% that of the L-isomer, which may account for its differential absorption, protein binding and metabolism. Thus, the chirality or stereochemistry of drugs is an important factor to be considered for therapeutic agents. The following sections consider different stereochemical features of drugs. Definitions and illustrations are included to aid in understanding the different conformational and configurational aspects of drugs.

Stereochemical Definitions and Illustrations

Isomers are molecules with identical molecular formulas but different structural formulas or different stereochemical formulas and hence different physical and/or chemical properties. Molecules having identical molecular formulas but different structural formulas are termed *structural isomers*. For example, ethanol (CH_3CH_2OH) and methyl ether (CH_3OCH_3) are structural isomers as both have the same molecular formula (C_2H_6O), but different connectivity of atoms. Structural isomers would be expected to have very different physical/chemical characteristics (*i.e.* color, melting point, solubility, *etc.*).

Stereoisomers come about from chirality or from restricted rotation such as in a ring or olefinic (double) bond. There are two obsolete terms that have been used in the past with respect to stereoisomers. The first, *optical isomers,* is used to describe stereoisomers with different optical properties. These optical properties arise from chirality and will be discussed under that topic. The second, *geometric isomers,* is used to describe *cis-trans* isomerism.

Chirality is a geometric property whereby a molecule (or any rigid object for that matter) is not superimposable on its mirror image. Chirality arises either from chiral centers or from restricted rotation, referred to as axial chirality. Stereoisomers have identical molecular and structural formulas but they have different spatial arrangements of the same groups. Physical or chemical characteristics of steroisomers are usually different (except see enantiomers).

A chiral center is an atom bonded to a set of other moieties in a spatial arrangement which is not superimposable on its mirror image. The resulting molecule must have exactly zero planes of symmetry. Atoms lacking this property are said to be achiral. The word "chiral" comes from the Greek word cheir meaning "hand" because left and right hands are chiral (Fig. **2**). For most drug molecules, the chiral center is based on a carbon atom, classically referred to as an asymmetric carbon atom.

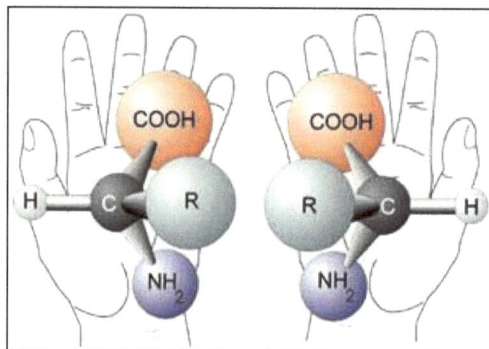

Fig. (2). The chirality or handedness of amino acids.

A carbon atom bonded to four different substituents will generally be chiral due to the tetrahedral arrangement of the sp^3 hybrid orbitals. Nitrogen and sulfur, among other atoms, can also form chiral centers. Not all asymmetric carbon (or other) atoms are chiral (See meso compounds below as an example). Nitrogen forms chiral centers that can be isolated only under limited circumstances. The lone-pair of electrons possessed by most nitrogen atoms form the fourth part of the tetrahedral structure, taking the place of one of carbon's four bonded moieties. With tertiary amines, the inversion rate is often too fast to separate the isomers. Ammonia has an inversion rate of ~2 X 10^{11} Hz with tertiary amines slower but still too fast to isolate. However, if the nitrogen is in a 3-membered ring and is connected to an atom with at least one unshared pair of electrons, then the isomer is stable and can be isolated as is seen in 1-chloro-2,2 dimethylaziridine. Quaternary amines (with four bound moieties, just like carbon) can be chiral and the isomers can be isolated.

Chloro-2,2
dimethylaziridine

Axial chirality is a term used to refer to stereoisomerism resulting from the non-planar arrangement of four groups in pairs about a chiral point in space. It is seen in two types of common organic chemicals (see Fig. **3**). Axial chirality results from restricted rotation around a single bond, as is seen with the *ortho* substituted biphenyls, which are referred to as atropisomers. Axial chirality is also seen in allenes, where the central atom lies on a chiral point of the molecule. Even though it is only disubstituted, this central carbon's two substituents are each disubstituted, producing an elongated, distorted tetrahedron that is still chiral.

(Fig. **3**). Comparatively few organic molecules, and virtually no drug molecules, derive their chirality from these structural types, so we will not concentrate on them.

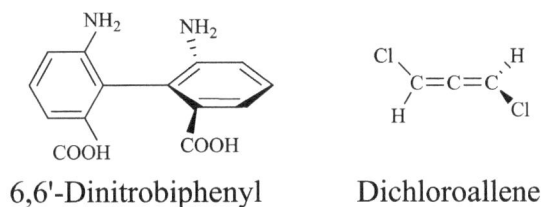

6,6'-Dinitrobiphenyl   Dichloroallene

Fig. (3). Axial chirality in biphenyls and allenes.

Light, or electromagnetic radiation, is composed of electrostatic and magnetic fields perpendicular to each other and to the direction of propagation (Fig. **4**). Ordinary light is unpolarized. That is, the electric and magnetic fields may have any orientation. When light is passed through long needle-shaped crystals, only the light with an electric field perpendicular to the crystals can pass. If all the electric fields lie in one plane, the resulting light is known as *plane polarized light*. Molecules with chiral centers have the ability to rotate plane polarized light in either clockwise or counterclockwise directions. The equipment used to polarize light with the use of such crystals is called a polarimeter.

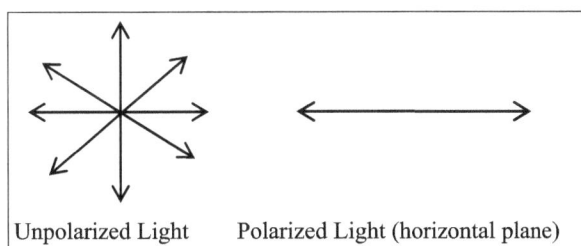

Unpolarized Light   Polarized Light (horizontal plane)

Fig. (4). Schematic diagram of an unpolarized and plane polarized light.

This process, call polarimetry, was one of the earliest methods used to distinguish between isomers of drugs. Rotation is dependent upon wavelength, temperature, concentration, and length of the path the light travels through the substance and is measured with a polarimeter. Rotation can be counterclockwise (CCW), to the left if facing the light source (levorotatory, levo, l-, –), or clockwise (CW), to the right if facing the light source (dextro, dextrorotatory, d-, +). However, (+) and (–) are preferred. When the root dextro- or levo- is found in a drug name it refers to optical rotation of light, not the D, L system described later.

Stereoisomers which are non-superimposable mirror images of each other are termed *enantiomers*. By definition, enantiomers are pairs of compounds related as an object to its mirror image. Compare the Fischer projections of D-glucose and L-glucose in Fig. **5**. In Fischer projections, vertically drawn bonds recede away from the viewer, below the page. Horizontal bonds rise above the page toward the viewer. Thus, D-glucose and L-glucose are mirror images of each other, which are presented in the shorthand notation of Fisher projections.

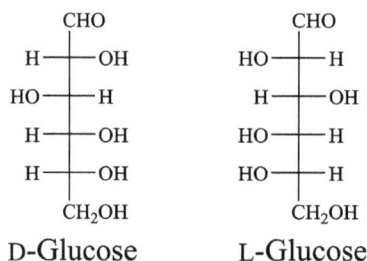

$$
\begin{array}{cc}
\text{CHO} & \text{CHO} \\
\text{H}-\!\!-\text{OH} & \text{HO}-\!\!-\text{H} \\
\text{HO}-\!\!-\text{H} & \text{H}-\!\!-\text{OH} \\
\text{H}-\!\!-\text{OH} & \text{HO}-\!\!-\text{H} \\
\text{H}-\!\!-\text{OH} & \text{HO}-\!\!-\text{H} \\
\text{CH}_2\text{OH} & \text{CH}_2\text{OH} \\
\text{D-Glucose} & \text{L-Glucose}
\end{array}
$$

Fig. (5). Enantiomeric pair of glucose.

Enantiomers rotate plane polarized light the same amount in opposite directions. When there are equal amounts of a pair of enantiomers in solution, it is called a racemic mixture. The enantiomers have identical molecular and structural formulas. Their physicochemical characteristics are more or less identical; however, enantiomers will interact differently with other chiral compounds. This is extremely important for understanding drugs, since many drugs bind to macromolecules which contain chiral components, *i.e.* proteins.

The term diastereoisomers (or diastereomers) refers to all other stereoisomeric compounds, regardless of their ability to rotate plane polarized light, and include both optical and geometric isomers. Compounds with more than one chiral center can form diastereoisomers, which are optically active but are not mirror images of each other. Diastereomers may differ in physicochemical properties and may be relatively readily separable by standard chemical techniques.

When a compound possesses a single chiral center, it can produce a single pair of enantiomers. But the presence of n such centers gives rise to 2^n stereoisomers and half that numbers of enantiomeric pairs. The non-enantiomers are termed diastereomers. An epimer is a diastereoisomer differing at only one chiral center. Compare the Fischer projection of D-glucose to D-galactose and D-mannose. All three are diastereoisomers. D-Glucose and D-galactose are also epimers (differing only at C4), which is also true for D-mannose and D-glucose (differing only at C2). However, D-galactose and D-mannose are not epimers (they differ at C2 and C4) but are diastereomers.

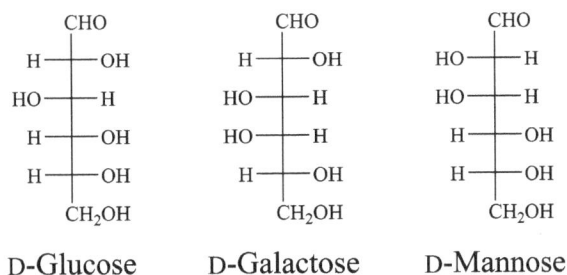

D-Glucose   D-Galactose   D-Mannose

The antimicrobial agent, chloramphenicol (Fig. **6**), has two chiral centers, thus 2^2 = 4 stereoisomers are possible. Only A is the active isomer with (1R, 2R)-configuration. The pairs A, B and C, D are enantiomers to one another, but A is the diastereomer of both C and D, and B is the diastereomer of both C and D. Since chloramphenicol has two chiral centers, A/C and B/D differ in only one chiral center, so they are also epimers to one another. The active isomer A is the (+)-isomer when determined in ethanol but the (-)-isomer when determined in ethyl acetate.

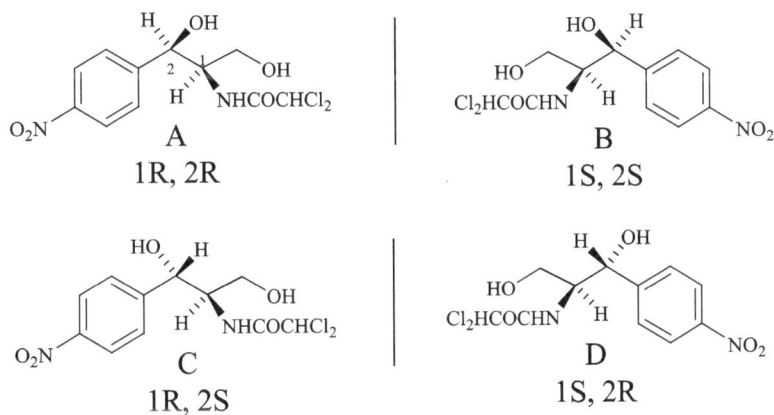

Fig. (6). Stereoisomers of chloramphenicol.

Meso Compounds. "Meso" is a term for the achiral member(s) of a set of diastereoisomers which also includes one or more chiral members. Meso compounds are molecules that have asymmetric carbons, yet are not chiral. To identify a meso compound, look for a mirror plane that produces equivalent halves of the molecule. Fig. **7** shows (+)-, (-)-, and meso-tartaric acid. A plane of symmetry is perpendicular to the C2-C3 bond in the meso- compound.

not mirror images

	COOH		COOH		COOH	
	H—C—OH		HO—C—H		H—C—OH	mirror images
	HO—C—H		H—C—OH		H—C—OH	
	COOH		COOH		COOH	mirror images

(+) Tartaric acid   (-) Tartaric acid   *meso*-Tartaric acid

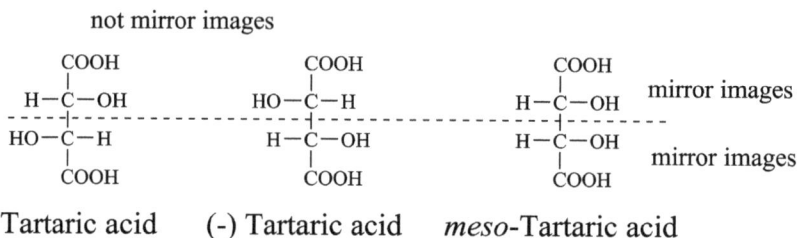

Fig. (7). Diastereoisomers and meso form of tartaric acid.

The D and L System. The D and L system is one of the two older systems used in describing the absolute orientation of the atoms attached to a chiral center. The chiral center is compared to the center found in glyceraldehyde (Fig. **8**). Glyceraldehyde was chosen since the early work in optical isomerism and chirality was done on sugars and this is the simplest chiral sugar.

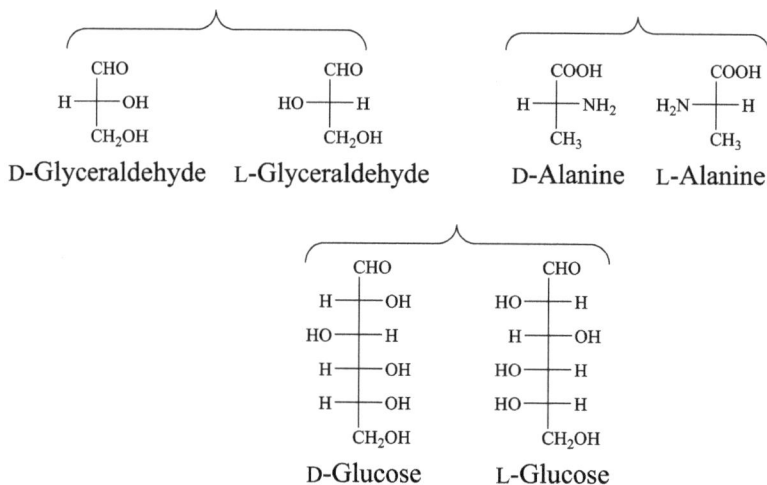

CHO	CHO	COOH	COOH
H—OH	HO—H	H—NH$_2$	H$_2$N—H
CH$_2$OH	CH$_2$OH	CH$_3$	CH$_3$
D-Glyceraldehyde	L-Glyceraldehyde	D-Alanine	L-Alanine

CHO	CHO
H—OH	HO—H
HO—H	H—OH
H—OH	HO—H
H—OH	HO—H
CH$_2$OH	CH$_2$OH
D-Glucose	L-Glucose

Fig. (8). D and L Configuration using glyceraldehyde as the reference.

The symbols used in this system are D- (for *dextro* or right) and L-(for *levo* or left). The reference for D- and L- are obtained by drawing glyceraldehyde in the proper Fischer projection with the aldehyde functional group at the top. Glyceraldehyde, or any other sugar, is in the D-form if the -OH functional group attached to the furthest chiral carbon (second from the last with numbering going from top to bottom) is on the right. Conversely, if it is on the left it is the L- form. Caution must be taken to prevent confusion with the symbols for optical activity, d- and l-, for dextrorotatory and levorotatory. To minimize confusion, the preferred symbols for dextro- and levo-rotatory are (+) and (−). There is no relationship between D- & L- and d- & l-.

The Erythro and Threo System. This second old system of descriptors comes from carbohydrate nomenclature for two simple sugars, *erythrose and threose*. It is used to describe the relationship of pairs of chiral atoms in diastereoisomers. In the *erythro* form, the same groups (for example the –OH on C2 and C3 in Fig. **9**) are on the same side of the Fischer projection. In the *threo* form, these groups are on opposite sides of the Fischer projection.

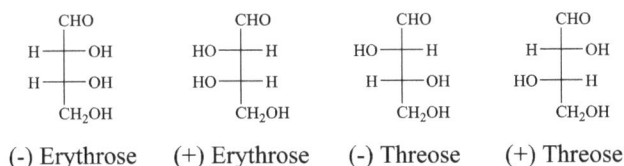

(-) Erythrose   (+) Erythrose   (-) Threose   (+) Threose

Fig. (9). Erythrose and threose to show erythro and threo system.

S and R Absolute Configuration (Preferred). There is no relationship between S and R and D- and L-, or between S and R and d- and l-. The rules for assigning absolute S and R configuration were published by Cahn, Ingold and Prelog, commonly known as the CIP rule or the sequence rule and is discussed in the following section [9].

Cahn-Ingold-Prelog (CIP) Rules [9]

To determine whether a chiral center is Rectus (*R, right*) or Sinister (*S, left*), we assign the priority of the four different moieties attached to a chiral carbon atom by the Cahn-Ingold-Prelog (CIP) rules, also known as the sequence rules. The determination of the absolute configurations *R* or *S* of the chiral carbon is performed by following the sequence rules as illustrated below.

Sequence Rule I: First, label the 4 groups "a, b, c, d" in order of decreasing priority according to the atomic number (atomic mass must be used when working with isotopes) of the atoms attached directly to the chiral atom. The highest group is assigned "a" and goes to the highest atomic number, the lowest atomic number is assigned "d" (Fig. **10**). These atoms are considered the first shell. In determining the configuration of most drug molecules, only the following atoms are encountered: H=1, C=6, N=7, O=8, F=9, P=15, S=16, Cl=17, Br=35, I=53. Unshared (lone-pair) electrons are given 0.

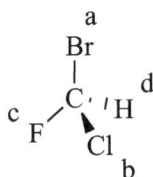

Fig. (10). Illustration of Sequence Rule I.

Sequence Rule II: When the atoms directly attached to the chiral atom are the same, the priority is established by the next atoms in the group. Work outward along the chains until a point of difference is reached (see Fig. **11**).

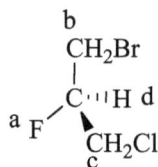

Fig. (11). Illustration of Sequence Rule II.

Sequence Rule III: When groups are equivalent at a branch point, proceed along those branches which start with the heaviest atoms. If these branches are the same, proceed along those branches which start with the next heaviest atom until a point of difference is reached. Priority of groups is established based on the atoms at the first point of difference (Fig. **12**).

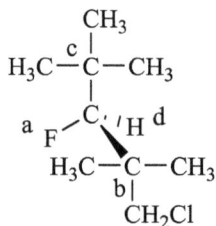

Fig. (12). Illustration of Sequence Rule III.

Sequence Rule IV: When two atoms are joined by a multiple bond, both atoms are considered to be multiplied in determining the priority of the group. These atoms are called phantom atoms and must be considered as part of a shell, but have no higher shell attached (Fig. **13**).

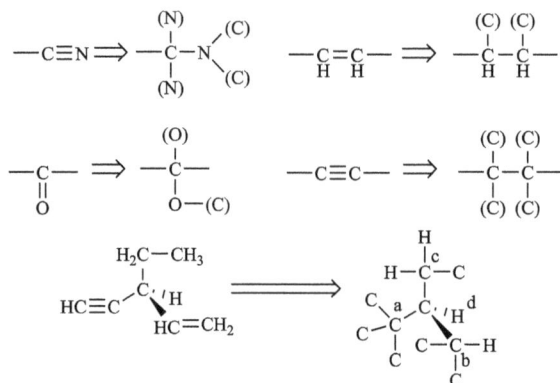

Fig. (13). Illustration of Sequence Rule IV showing phantom atoms.

Once the groups are assigned their priorities, the molecule is viewed in a way so that the lightest group points directly back. The convention accepted in chemistry is that bonds oriented above the plane of the paper, toward the reader are signified by solid wedge bonds while hashed bonds, dashed bonds, or hashed wedge imply the substituent is oriented below the plane of the paper, away from the reader. Finally, if tracing a path from groups "a" to "b" to "c" is clockwise, it is designated as (R)- and if it is counterclockwise it is designated as (S)-configuration. Consider the following compounds:

The second step is often hard to do without the aid of models. On two-dimensional paper the best that can be done is to draw the structure with one bond going up from, and one bond going down from the plane of the paper (forward or back). This is illustrated below.

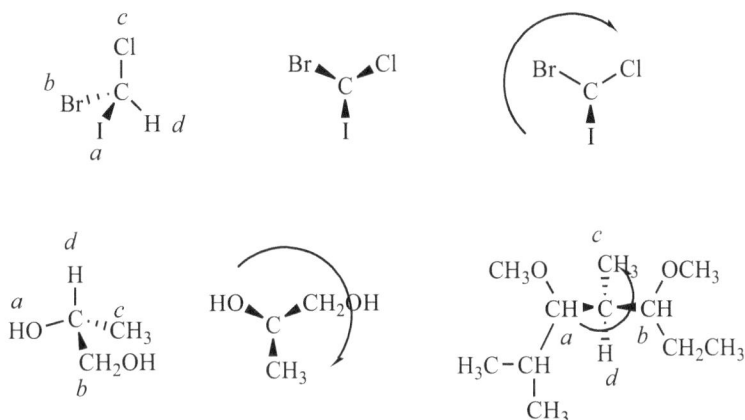

As drug structures become more complicated, the structure may not show the orientation of the lowest priority group. However, the configuration can still be deduced if the orientation of at least one group is known. Eliel [10] has devised a scheme for assigning configurations in such cases. The steps are as follows:

Obtain the orientation forward (front) or back (rear) for the group whose orientation is known, the "reference group". Ignore the reference group and determine if tracing a path of the remaining groups (in decreasing priority) is clockwise (CW) or counterclockwise CCW). Assign the proper configuration to the rotation by using Table **3**.

To illustrate this method, the assignment of chiral centers X and Y are shown below.

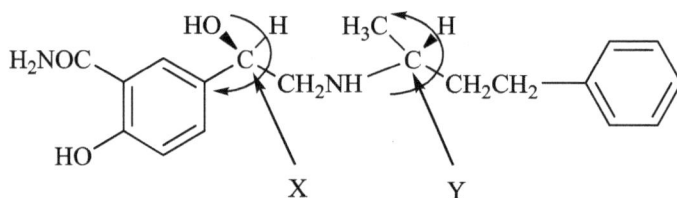

In case X, the orientation of the -OH group is known (forward) and its priority is "a". The rotation from "b" to "c" to "d" is clockwise. The chart associates forward for group "a" with a "+" meaning this configuration is R. In case Y, the orientation of the –H is known (forward) and its priority is "d". The rotation from "a" to "b" to "c" is counterclockwise. The chart associates forward for group "d" with "–". Therefore, the configuration is R.

Table 3. Eliel Modification in the R/S System Assignment.

Reference Group Orientation	Reference Group Priority[1]			
	a	b	c	d
Forward	+	–	+	–
Back	–	+	–	+

[1] + means CW is *R* and CCW is *S*, – means CW is *S* and CCW is *R*

Also consider the following example:

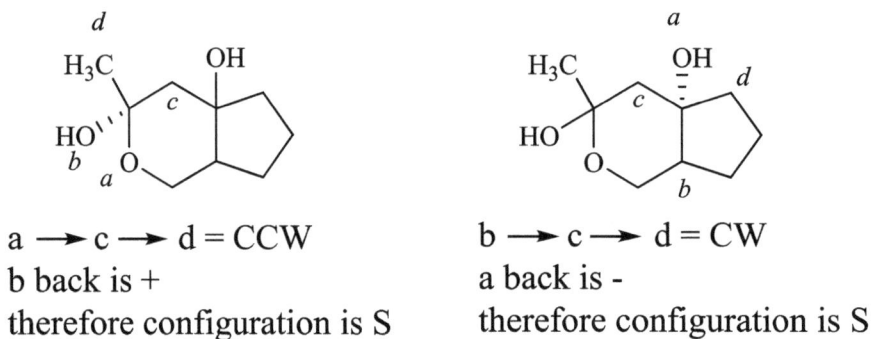

a ⟶ c ⟶ d = CCW
b back is +
therefore configuration is S

b ⟶ c ⟶ d = CW
a back is -
therefore configuration is S

If a chiral carbon is drawn with only three substituents, assume the fourth is a hydrogen atom and the hydrogen atom's orientation is *opposite* of the dash/wedge bond which is shown. For example, in pravastatin the topmost hydroxyl is oriented up. Therefore, its hydrogen is oriented down. Since hydrogen is priority *d,* determining the configuration is simplified.

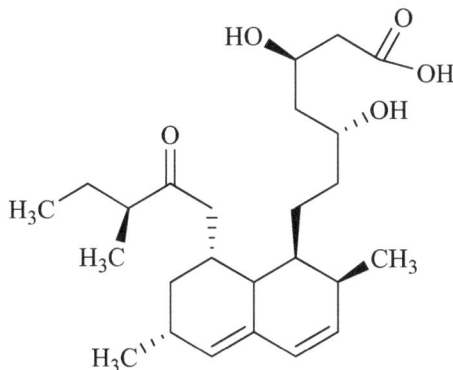

Pravastatin

Esomeprazole has a chiral sulfur. Three of the groups are shown below. The fourth group is a lone-pair of electrons. These *non-bonding electrons have an atomic number of zero*, thus have a lower priority than hydrogen. Since the methylene group is oriented back, it is assumed the n-electrons are oriented up. Verify that the configuration is *S*.

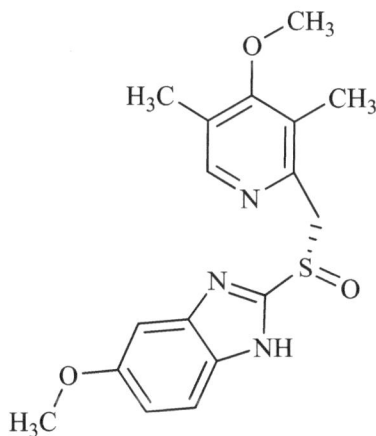

Esomeprazole

In summary, the following flowchart can be used to determine the configuration at atom 13 of the structure shown.

Cis-Trans Isomerism

This type of isomerism is found with olefinic bonds or alicyclic compounds; an older term for this phenomenon is *geometric isomerism*. These isomers have different physicochemical properties which result in different biological activity.

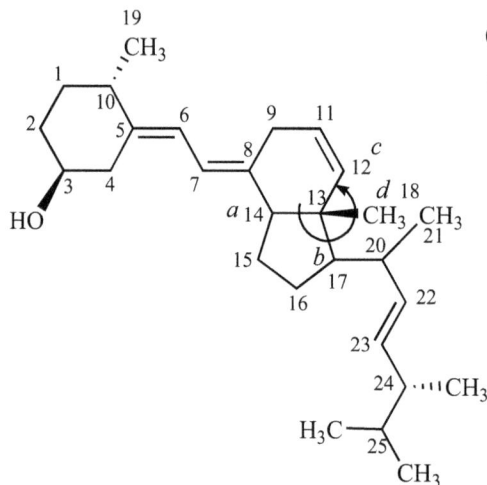

Configuration at atom 13
CCW with —R

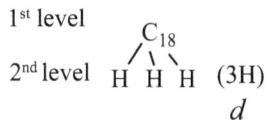

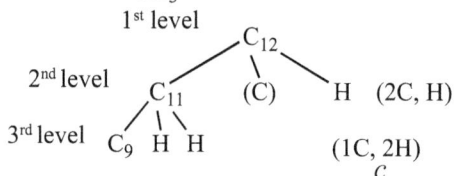

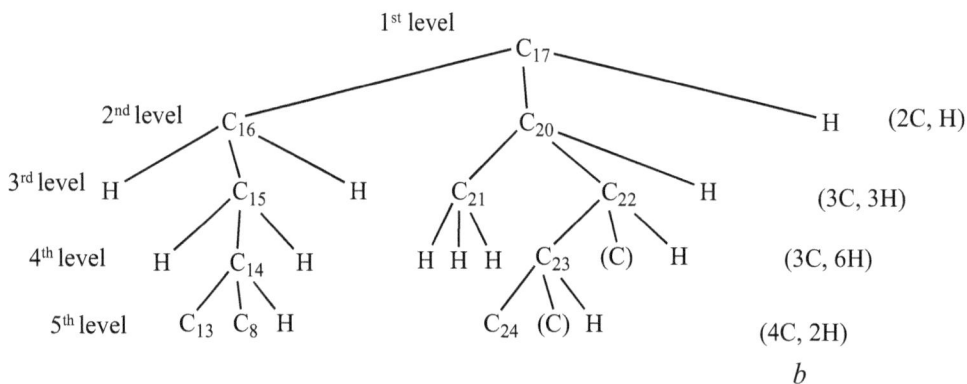

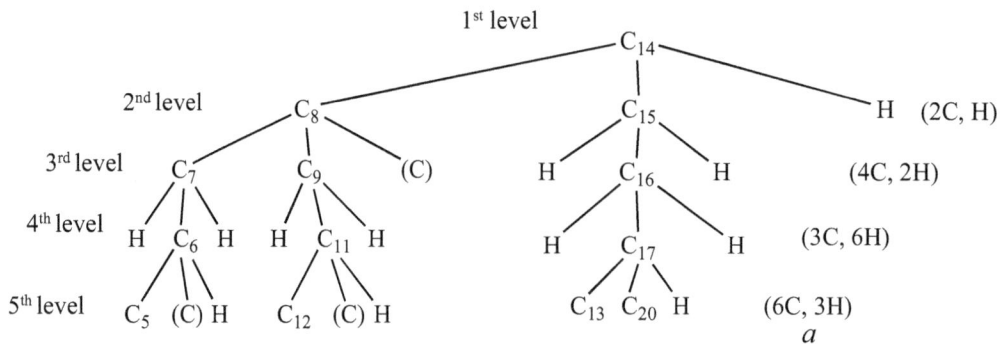

1st level $\quad$ C$_{18}$

2nd level $\quad$ H H H $\quad$ (3H)

d

1st level $\quad$ C$_{12}$

2nd level $\quad$ C$_{11}$ $\quad$ (C) $\quad$ H $\quad$ (2C, H)

3rd level $\quad$ C$_9$ H H $\qquad$ (1C, 2H)

c

1st level $\quad$ C$_{17}$

2nd level $\quad$ C$_{16}$ $\qquad$ C$_{20}$ $\qquad$ H $\quad$ (2C, H)

3rd level $\quad$ H $\quad$ C$_{15}$ $\quad$ H $\qquad$ C$_{21}$ $\quad$ C$_{22}$ $\quad$ H $\quad$ (3C, 3H)

4th level $\quad$ H $\quad$ C$_{14}$ $\quad$ H $\qquad$ H H H $\quad$ C$_{23}$ $\quad$ (C) $\quad$ H $\quad$ (3C, 6H)

5th level $\quad$ C$_{13}$ C$_8$ H $\qquad$ C$_{24}$ (C) H $\qquad$ (4C, 2H)

b

1st level $\quad$ C$_{14}$

2nd level $\quad$ C$_8$ $\qquad$ C$_{15}$ $\qquad$ H $\quad$ (2C, H)

3rd level $\quad$ C$_7$ $\quad$ C$_9$ $\quad$ (C) $\qquad$ H $\quad$ C$_{16}$ $\quad$ H $\quad$ (4C, 2H)

4th level $\quad$ H C$_6$ H $\quad$ H C$_{11}$ H $\qquad$ H $\quad$ C$_{17}$ $\quad$ H $\quad$ (3C, 6H)

5th level $\quad$ C$_5$ (C) H $\quad$ C$_{12}$ (C) H $\qquad$ C$_{13}$ C$_{20}$ H $\quad$ (6C, 3H)

a

Olefinic Bonds. When carbon atoms are joined by double bonds, there is no longer free rotation around the bond. Substituents tend to lie in a plane on either side of the bond. When there is one non-hydrogen substituent on each carbon, then these substituents will lie either on the same side of a line parallel to the double bond (*cis*) or on opposite sides (*trans*) of this line.

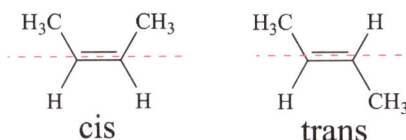

cis   trans

As the number and complexity of the substituents increases, there is ambiguity as to which of the multiple substituents are being referred to as being on the same or opposite sides of the double bond. Therefore, the *Z* and *E* system was devised using the CIP rules to designate the orientation around a double bond. *Z* (German, *zusammen* = together) or *E* (German, *entgegen* = opposite) is used to designate the orientation around an olefinic bond, which replaces the inadequate system of *cis* and *trans*. No absolute correlation between *cis*/*trans* and *E*/*Z* is evident.

To determine *Z* or *E*, using the Cahn-Ingold-Prelog sequence rules, a/b priorities are assigned to the groups attached to each end of the double bond, working on each alkene carbon independently. Then the priority on each end of the double bond is compared. If the two groups of higher priorities are on the same side of the plane defined by the double bond, the configuration is designated "*Z*". If the two groups of higher priorities are on opposite sides of the plane of the double bond, the configuration is designated "*E*". Conveniently, if lines are physically traced on the written molecular structure from priority "a" to priority "b" of alkene C1, then to priority "a" and lastly to priority "b" of alkene C2, a sideways "Z" is drawn for the "Z" isomer. Similarly, an "E" (missing the middle of its three parallel lines) is drawn for the "E" isomer.

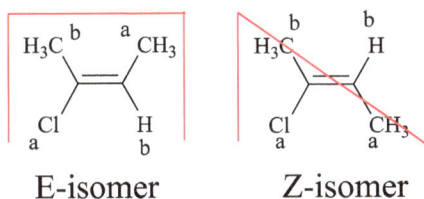

E-isomer   Z-isomer

Alicyclic Compounds. Aliphatic ring structures limit free rotation around single carbon bonds and once again *cis* and *trans* isomerism occurs. For example, *cis* and *trans* 1,2-dichlorohexanes.

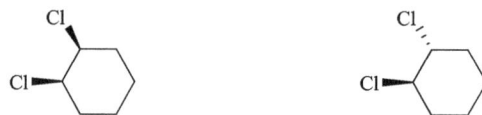

cis(1,2)-Dichlorohexane   *trans*(1,2)-Dichlorohexane

Ring Fusion. Some drugs have multiple fused rings. "Fused" rings have 2 adjacent carbons in common (see Fig. **14**). Often how the rings are fused determines their pharmacological properties. The best examples can be found among the steroid drugs. Depending on the ring fusion, steroid drugs can function as hormones, cardiac glycosides, or bile acids.

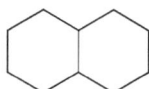

Fused rings

Fig. (14). Red bond showing the two adjacent carbons in common.

The orientation of substitution on a cyclohexane ring is the basis for determining if two rings are fused *cis* or *trans*. Substituents on the cyclohexane chair conformer are either axial (ax) or equatorial (eq). These names derive from the Earth's "axis" and "equator". Bonds parallel to the axis of a "globe" mentally superimposed on the cyclohexane in Fig. **15** are designated "axial". Bonds around the perimeter are perpendicular to the axis of the superimposed globe, but parallel to the circle that represents the equator and are thus called "equatorial".

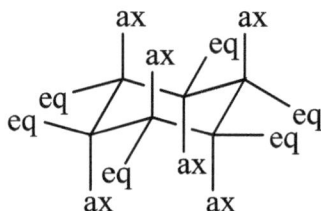

Fig. (15). Cyclohexane showing its axial and equatorial bonds.

Fig. **16** shows that substituents on adjacent carbon atoms may be *cis* or *trans* depending on their relative orientation. If their orientation is axial–equatorial they are *cis*, but if the orientation is either axial–axial or equatorial–equatorial, the substituents are *trans* to each other. Since carbon forms four bonds, each atom in the ring has four substituents. Two are adjacent carbons in the ring and the other two comprise one axial-equatorial pair. If hydrogen is one of these two, it is often not drawn. Nevertheless, its orientation can be predicted as illustrated in Fig. **16**.

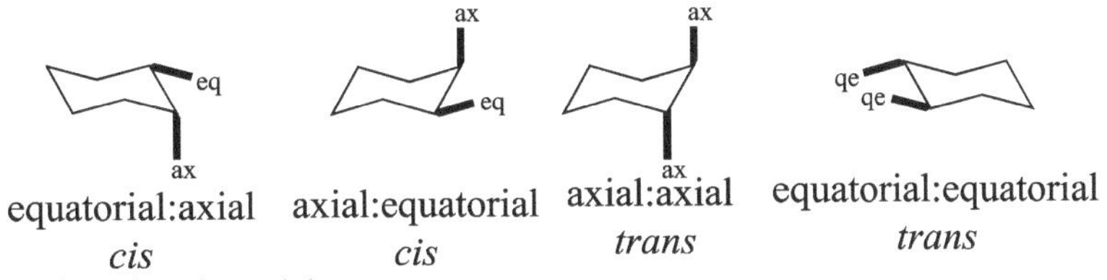

ax   ax

eq   eq

ax   ax

qe
qe

equatorial:axial   axial:equatorial   axial:axial   equatorial:equatorial
cis   *cis*   *trans*   *trans*

Fig. (16). Bond orientations in cyclohexane.

There are several additional methods besides direct comparison to Fig. **17** to determine if a ring fusion is *cis* or *trans*. One method uses the orientation of the substituents at the fusion carbons. Choosing one of the two rings as a reference ring is selected. The second step is to add the missing hydrogens at the fusion carbons. This addition of the missing hydrogens demands knowledge of the orientation of all the bonds as illustrated in Fig. **17**.

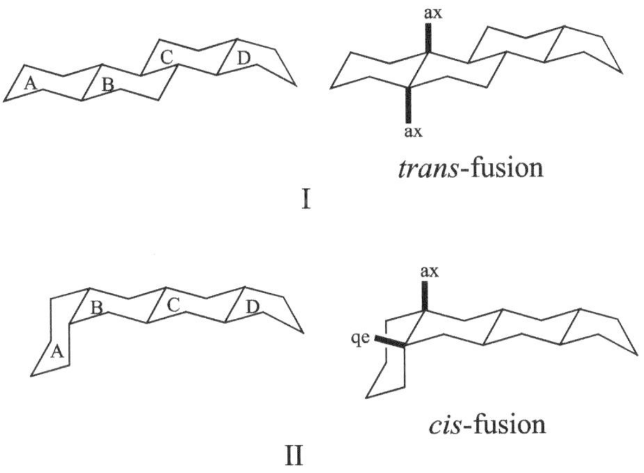

C D

A B

ax

ax

trans-fusion

I

B C D

A

ax

qe

cis-fusion

II

Fig. (17). The *trans* and *cis* fusions in steroid nucleus.

In situation I, both rings A and B are drawn in the familiar chair conformation so either one can be chosen as the reference ring. Adding the missing hydrogens to the reference ring (ring A in this example) one can get the two hydrogens in *axial:axial* conformations thus they must be *trans* to each other, therefore, the ring fusion is said to be *trans*.

In situation II, ring B is drawn in the familiar chair conformation so is used as the reference ring. Adding the missing hydrogens to the reference ring (ring A/B in this example), one can get the two hydrogens in *axial:equatorial* conformations. Thus, they must be *cis* to each other. Therefore, the ring fusion is said to be *cis*.

Steroid hormones like testosterone, hydrocortisone and progesterone have unsaturation in ring A which prevents a true A/B *cis* fusion. But, the ring fusion is similar to a *trans* fusion. Reduction of the unsaturation produces an inactive metabolite with a *cis* A/B fusion. 5α-Dihydrotestosterone, a metabolite of testosterone, has a *trans* A/B as well as C/D fusions and is biologically active as a steroid hormone. Bile acids have a *cis* A/B fusion similar to the above mentioned metabolite and thus lack hormonal activity. Cardiac glycosides have a *cis* A/B fusion as well as a *cis* C/D fusion and thus have biological activities different from the other steroids (Fig. **18**).

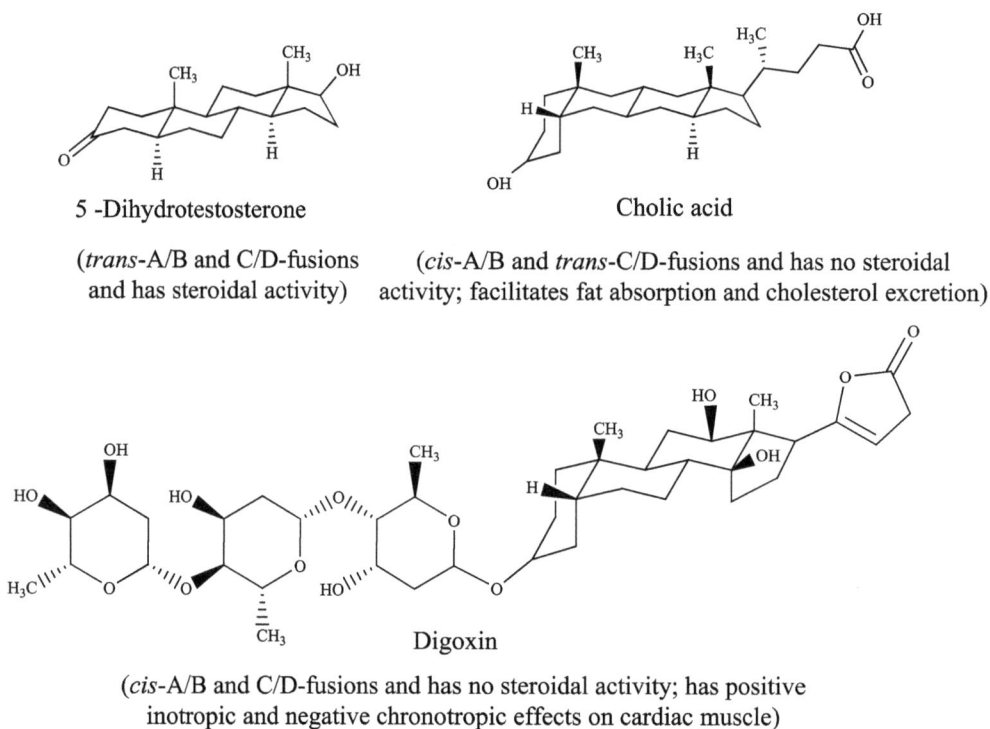

5 -Dihydrotestosterone
(*trans*-A/B and C/D-fusions
and has steroidal activity)

Cholic acid
(*cis*-A/B and *trans*-C/D-fusions and has no steroidal
activity; facilitates fat absorption and cholesterol excretion)

Digoxin
(*cis*-A/B and C/D-fusions and has no steroidal activity; has positive
inotropic and negative chronotropic effects on cardiac muscle)

Fig. (18). Few important steroid structures showing their ring fusions as well as biological activity patterns.

STUDENT SELF-STUDY GUIDE

1. What is isosterism and bioisosterism? What are ring equivalents? Identify the isosteres in given functional groups as well as in drug structures.
2. Which stereoisomer of thalidomide is hypnotic and which one is teratogenic?
3. Significance of stereochemistry in drug action with examples of propranolol, propoxyphene, methorphan, ethambutol, and dobutamine.
4. What is *cis-trans* isomerism? How do you determine the *cis-trans* fusion type in steroids? Which fusion type is important for biological action of steroids? What is a meso compound?
5. Configuration and conformations.

6. Determining *R/S* and *E/Z* from drug structures.
7. What is the sequence rule?

STUDENTS SELF ASSESMENT

Part I: Directions: Each of the numbered items or incomplete statements in this section is followed by answers or by completions of the statements. Select the **one** lettered answer or completion that is **best** in each case.

1. Which one is <u>not</u> a classical isostere of -S-?
   A. –NH–
   B. –O–
   C. –F
   D. –CH$_2$CH$_2$–
   E. –CH=CH–
2. See the following structure and find the <u>INCORRECT</u> statement from the keys A – E afterwards:

   A. The fusion type between ring A and ring B is a *trans* fusion
   B. Steroid hormones have unsaturation in ring A and thus they possess *trans* type of A/B ring fusion
   C. Saturation of steroid hormones produces metabolite with a *cis* fusion which is active
   D. Bile acids have *cis* A/B fusion
   E. Cardiac glycosides have *cis* A/B and C/D fusions
      Part II: Directions: Each item below contains three suggested answers of which **one or more** is correct. Chose the answer
         A. if **only I** is correct
         B. if **only III** is correct
         C. if **I and II** are correct
         D. if **II and III** are correct
         E. if **I, II, and III** are correct
3. Which statements regarding the configuration / optical activity of structures A-C are <u>true</u>?

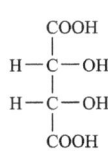 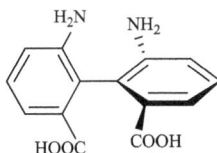 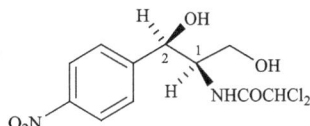

   A   **B**   **C**

I. "A" is optically active
II. "B" has R_A configuration
III. "C" has 1R configuration

Part III: Directions: The group of items in this section consists of lettered options followed by a set of numbered items. For each item, select the **one** lettered option that is most closely associated with it. Each lettered option may be selected once, more than once, or not at all.

1. For each pair of molecules (I-V), select the letter (A-E) that best describes the relationship.
   A. Geometric isomers
   B. Enantiomers
   C. Axial isomers
   D. Isosteres
   E. Conformational isomers

   I.

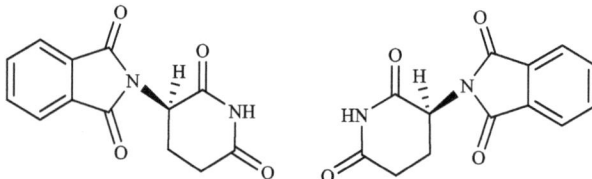

   II.

   III.

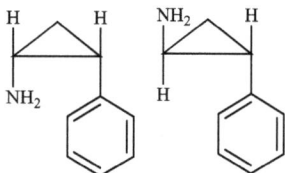

   IV.

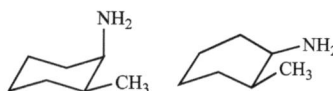

   V.

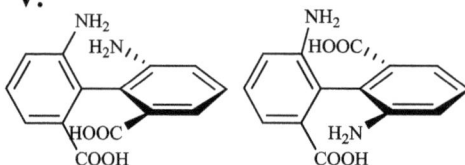

CONSENT FOR PUBLICATION

Not applicable.

CONFLICT OF INTEREST

The authors declare no conflict of interest, financial or otherwise.

ACKNOWLEDGEMENT

Declared none.

REFERENCES

[1]   Langmuir I. Isomorphism, isosterism and covalence. J Am Chem Soc 1919; 41: 1543-59.
        [http://dx.doi.org/10.1021/ja02231a009]

[2]   Grimm HG. Structure and Size of the Non-metallic Hydrides Z. Electrochem 1925; 31: 474-80.

[3]   Friedman HL. Influence of Isosteric Replacements upon Biological Activity, Washington, EUA,
        National Academy of Science, 1951; No.206, p. 295.

[4]   Burger A. Medicinal Chemistry, 3rd Ed., NY, EUA, Wiley, 1970,p. 127.

[5]   Lima LM, Barreiro EJ. Bioisosterism: a useful strategy for molecular modification and drug design.
        Curr Med Chem 2005; 12(1): 23-49.
        [http://dx.doi.org/10.2174/0929867053363540] [PMID: 15638729]

[6]   Pasteur L. Memoirs by Pasteur, van't Hoff, Le Bel and Wishcensus. New York: American Book
        Company 1901; p. l-33.

[7]   Kim JH, Scialli AR. Thalidomide: the tragedy of birth defects and the effective treatment of disease.
        Toxicol Sci 2011; 122(1): 1-6.
        [http://dx.doi.org/10.1093/toxsci/kfr088] [PMID: 21507989]

[8]   Fintel B, Samaras AT, Carias E. The thalidomide tragedy: lessons for drug safety and regulation
        http://scienceinsociety.northwestern.edu/content/articles/2009/research-digest/thalidomide/title-tba

[9]   Cahn RS, Ingold CK, Prelog V. Specification of molecular chirality. Angew Chem 1966; 78: 413-47.
        [http://dx.doi.org/10.1002/ange.19660780803]

[10]   Eliel EL. The *R/S* system: a new method for assignment and some recent modifications. J Chem Educ
        1985; 62: 223-4.
        [http://dx.doi.org/10.1021/ed062p223]

Further Reading

Rakoff H, Rose, NC. Rakoff H and Rose NC eds., Organic Chemistry, New York, NY: The Macmillan
Company; 1966.

Preferred IUPAC Names. Chapter 9: Specification of configuration and conformation. Available at:
old.iupac.org/reports/provisional/abstract04/BBprs310305/ Chapter9.pdf. Accessed on September 12, 2011.

CHAPTER 6

Fundamentals of Drug Action

M. O. Faruk Khan[1,*] and Taufiq Rahman[2]

[1] *Department of Pharmaceutical Sciences and Research, Marshall University School of Pharmacy, Huntington, WV, USA*

[2] *Department of Pharmacology, University of Cambridge, Cambridge, UK*

Abstract: This chapter is a brief review of the mechanistic aspects of drug action and discusses the concepts of receptors and drug receptor interactions critical for pharmacological responses of drugs. After study of this chapter, students will be able to:

• Discuss about receptors from historical perspectives
• Explain the mechanistic principles of drug action in light of receptors
• Summarize different theories of drug action such as:

- Occupancy Theory
- Rate Theory
- Induced-Fit Theory
- Macromolecular Perturbation Theory
- Occupation-Activation Theory of "Two-State" Model

• Understand drug receptor interactions

Keywords: Affinity, Agonist, Antagonist, Drug receptor interaction, Efficacy, Intrinsic activity, Partial agonist, Receptor, Theories of drug action.

HISTORICAL PERSPECTIVE

Since ancient times, various substances or preparations have been used to treat ailments based on empirical knowledge, as well as traditional practices without the knowledge of how these substances worked against certain diseases. It was not until the early eighteenth century that our present knowledge about generic mechanisms and dogmas of drug action was discovered by the seminal studies of some notable physiologists, pharmacologists, biochemists and biophysicists.

* **Corresponding author M.O. Faruk Khan**: Department of Pharmaceutical Sciences and Research, Marshall University School of Pharmacy, Huntington, WV, USA; Tel: 304-696-3094; Fax: 304-696-7309; E-mail: khanmo@marshall.edu

Theories of drug action evolved, initially from studies enquiring how endogenous ligands like acetylcholine work, followed by the introduction and expansion or modifications of the receptor concept. Subsequent enrichment of the receptor concept came from the parallel work of the biophysicists who deciphered how enzymes work. Therefore, it is pertinent to discuss the historical perspective of the mechanism of drug action under two broad aspects – pharmacological and biophysical.

Claude Bernard's (1813-1878) classical experiments with curare in the mid-1800s showed that this neuromuscular blocking agent could only prevent contraction of skeletal muscle that was induced by nerve stimulation, but was ineffective against a directly-stimulated muscle [1]. This provided, for the first time, evidence that a drug can act locally and more importantly, suggested the presence of a junction between nerve and a tissue. However, the precise mechanism underlying such information transfer remained elusive.

Paul Ehrlich (1854–1915), a German scientist, was the first to conduct a thorough and systematic study of the relationship between the chemical structure of organic molecules and their biological actions. His studies involved selective staining of cells by dyes, as well as the remarkably powerful and specific actions of bacterial toxins, which convinced him that biologically active molecules had to be bound in order to be effective [2]. In his own words, '*corpora non agunt, nisi fixata*', which literally tells us that 'entities do not act, unless bound'. Whilst Ehrlich was the first to develop a receptor concept, **John Newport Langley** (1852–1925), a Cambridge physiologist, was the first to propose a *receptor theory* for the action of drugs and transmitter substances in the body based on his work with nicotine and curare [3]. Nicotine contracts skeletal muscle, which is blocked by the actions of curare by blocking the response of the muscle to nerve stimulation. From these studies in 1905, Langley inferred that the muscle must possess a "*receptive substance*" for which Ehrlich coined the term 'receptor'. Today, both Ehrlich and Langley are equally recognized for the introduction of the receptor concept. Alternative theories had been proposed by other scientists, *e.g.*, **Walther Straub's** (1874-1944) '*physical theory of drug action*' or '*poison-potential theory*', which were not well-accepted. In favor of this theory, he showed that a drug or poison acted only as long as it diffused across the cell membrane due to a 'concentration potential' between the outside and the inside of the cell [4].

The very first attempt to provide a quantitative basis for drug action on the receptor was made in 1909 by **Archibald Vivian Hill** (1886–1977), a research student of Langley [5]. At Langley's suggestion, Hill conducted a mathematical analysis of the size and time course of the action of nicotine and curare on the isolated *rectus abdominis* muscle of the frog. These curves reflected a gradual combination of the drug with Langley's receptive substance within the muscle. His observations supported the receptor theory, but not the physical theories of drug action. For the first time, he applied the *law of mass action* to the relationship between ligand concentration and receptor occupancy at equilibrium and to the rate at which this equilibrium is approached. He derived what later became to be known as Langmuir's adsorption isotherm (although Langmuir's work came several years later).

Hill's evidence for a reversible, unimolecular chemical reaction between a drug and its receptor was followed up in the 1920s by another of Langley's student, **Alfred Joseph Clark** (1885–1941). Clark, in 1933, experimented with acetylcholine on isolated ventricular strips from the frog heart, in which it diminished the force of contractions, and on the *rectus abdominis* muscle of the frog, in which it increased contractions [6]. From both experiments, like Hill, Clark observed a sigmoidal nature of the concentration-response curves. Clark calculated the minimal amount of acetylcholine binding to a heart muscle cell that would be sufficient to produce a demonstrable action. He arrived at about 20,000 acetylcholine molecules per cell – an amount that was not enough to form a continuous layer over the heart cells or to cover a larger area inside them. The drug molecules would therefore be fixed only to a 'very small fraction' of the cell's surface. Drawing these observations together, Clark suggested that the simplest way to explain the concentration-response curve was to assume that a reversible, monomolecular reaction occurred between the drug and some receptor in the cell or on the cell's surface. He thus pioneered the '**occupancy theory**' of drug action (see later).

In the 1920s, **John Henry Gaddum** (1900-1965) produced new experimental evidence supporting the receptor model of drug antagonism. He examined the actions of adrenaline on isolated uterus of the rabbit, arriving at a sigmoid curve which plotted the intensity of contraction against the logarithm of the drug concentration. He introduced the concept of *competitive antagonism* [7].

During the 1940s, the concept of competitive antagonism, especially *competitive inhibition*, gained new support due to research on antibiotic properties of sulfonamides. In particular, it was demonstrated that sulfanilamide inhibited the growth of bacteria by competing with the metabolite, para-aminobenzoic acid (PABA) for an active site on a bacterial enzyme. Subsequent research revealed the paradoxical phenomenon that some of the PABA derivatives could have both metabolite-like (that is, growth stimulating) and inhibitory effects on the same microorganism, depending on the experimental conditions. In the early 1950s, the Dutch pharmacologist **Everhardus Jacobus Ariëns** (1918-2002) investigated this problem, showing that a similar '*dual action*' also applied to some muscle paralyzing agents and sympathomimetics, which was not explained by Clark's *receptor theory*. He extended this theory [8] by introducing the concept of '*intrinsic activity*' as a new factor that determined the effects of a drug. **Robert P. Stephenson** (1925–2004), a colleague of Gaddum at the Edinburgh department of pharmacology, proposed a similar extension of the receptor theory in 1956 [9]. Stephenson coined the term 'efficacy' which is comparable to 'intrinsic activity' of Ariëns. Their work introduced the concept of *full and partial agonism* while Stephenson also introduced the concept of '*spare receptors*'.

While physiologists and pharmacologists had been chasing the possible mechanisms by which molecules trigger biological responses, biochemists and biophysicists had been involved in finding out the precise mechanism of enzyme catalysis. Thus, various theories regarding the physicochemical nature of enzyme-substrate interactions at the molecular level emerged. For example, **Hermann Emil Fisher** (1852-1919), a notable German chemist of his time, introduced the famous '*Lock and Key theory*' of enzyme-substrate interaction in 1894 [10]. This theory alludes to the structural specificity required for a drug to interact with its receptor, considering both the interacting partners to be rigid. This theory had been the cornerstone for understanding enzyme-substrate (and drug-receptor) interaction for many years and remains attractive in general; however, there are examples it fails to explain. Later in 1958, studies on the enzyme hexokinase led an American biochemist, **Daniel E. Koshland, Jr**. (1920-2007), to replace the "*lock and key*" theory with the "*induced-fit theory*" (see later) [11].

DRUG RECEPTORS

A small number of clinically useful drugs work in a structurally non-specific way, *i.e.* they do not require a target receptor to evoke their biological response. For example, antacids work simply by chemically neutralizing the hydrochloric acid secreted within stomach. Other examples of such drugs include osmotic diuretics (*e.g.*, mannitol), antidotes for heavy metal poisoning, and most laxatives (*e.g.*, lactulose, polyethylene glycol). However, most drugs work in a *structurally specific* way, *i.e.* their therapeutic as well as toxic effects stem from their interactions with specific cellular target molecules, collectively known as **receptors**. This concept, as discussed above, evolved from the pioneering work of Langley and Ehrlich in early nineteenth century. From a pharmacological point of view, a **receptor** is a cellular macromolecule that is present mostly on the cell membrane, but also within the cytoplasm or cell nucleus that binds to a specific molecule (ligand) such as a neurotransmitter, hormone, metabolite, or a drug molecule and thereby initiating cellular response. Drug-induced changes in the biochemical and biophysical properties of the receptor result in physiological changes that constitute the biological actions of the drugs. Most drug receptors are **proteins** in nature and fall into one of the following categories –

- Enzymes (*e.g.*, dihydrofolate reductase, the receptor for the antineoplastic drug methotrexate)
- Ionotropic receptors or ion channels (ligand gated channels and voltage gated channels)
- Metabotropic receptors (G-protein coupled receptors that bind to endogenously produced hormones, neurotransmitter, *etc.*)
- Kinase linked and related receptors (*e.g.*, receptors for various growth factors and thus for some anticancer drugs)
- Nuclear receptors (*e.g.*, receptors for thyroid hormone, some fat-soluble vitamins and steroids)
- Cytoskeletal or structural proteins (*e.g.*, tubulin, the receptor for colchicine, an anti-inflammatory agent)
- Transporters or carrier proteins (*e.g.*, Na^+-K^+ ATPase, the receptor for cardiac glycosides)

Many of the above receptor subfamilies, notably the G-protein coupled receptor (GPCR) superfamily and growth factor receptors, can be coupled to respective executioner or **effector** components that orchestrate diverse cellular effects which may occur over a wider time scale. For example, the effects produced by the ion channels are very fast (milliseconds), those produced by steroid and thyroid hormones are very slow (several minutes to several hours), and those induced by the GPCRs occupy an intermediate time scale (seconds to minutes) [12].

In addition to the above-mentioned protein-receptor families, there are also **non-protein receptors** or drug target sites including nucleic acids (DNA, RNA), membranes, and fluid compartments. One also needs to remember that many drug molecules can bind to some plasma proteins and a few other carrier proteins (*e.g.*, low density lipoproteins, transferrin *etc.*). But these proteins, though capable of binding drugs, are not considered as receptors in the context of pharmacology and medicinal chemistry since they are physiologically inert and the binding of drugs does not influence their functional properties. In the more general context of cell biology, the term receptor is used to describe various cell surface molecules (such as T-cell receptors, integrins, Toll receptors, *etc.*) involved in immunological responses to foreign proteins and the interaction of cells with each other and with the extracellular matrix. These receptors play a critical role in cell growth and migration and are also emerging as drug targets. These receptors differ from conventional pharmacological receptors in that they respond to proteins attached to cell surfaces or extracellular structures, rather than to soluble mediators [13].

MECHANISM OF DRUG ACTION ON RECEPTOR LEVEL

One important notion is that drugs only modify the biological activities transduced through specific receptor proteins but they do not generate any new activity *per se*. As mentioned above, most of the drugs physically interact with their receptors and thus effectively are exogenously administered (or endogenously made available) **ligands** of these proteins. Ligands are endowed with a property called **affinity***i.e.* the ability to interact with a receptor protein.

Upon binding at specific binding site(s) on their receptors, drugs can either mimic the action of endogenously produced ligands as **agonists** or can oppose the biological effects of the endogenous ligands as **antagonists**.

From a mechanistic point of view, when a drug acts as an agonist of a receptor, it **activates** the receptor, *i.e.* the binding of a drug induces changes in the structure of that receptor in such a way that a biologic response is elicited. The ability of a ligand to initiate receptor activation is termed as **efficacy**. Agonists therefore *have both affinity and efficacy* for cognate receptors. Agonists can be of following subtypes:

Full agonists: Such drugs mimic the physiologic agonist. Example: isoproterenol (β-adrenergic agonist).

Partial agonists: Such drugs activate receptors but are unable to elicit the maximal response of the receptor system. The extent of response is intermediate between an agonist and antagonist. Example: Dobutamine (a partial agonist at β-adrenergic receptor).

Inverse agonists: Inverse agonists cause constitutively active targets to become inactive. Example: Ro15-4513 is an inverse agonist of $GABA_A$ receptor. Antihistamines are now considered as inverse agonists of H_1 receptor.

Conversely, drugs that act as antagonists of receptors generally are capable of physically interacting with them but incapable of triggering receptor activation and thus any biological response. Antagonists thus *have affinity but lack efficacy* for their target receptors. A drug can antagonize a receptor in several ways:

- **Competitive or reversible inhibition**: A drug occupies the active site and prevents binding of the physiological (*i.e.* endogenously produced) ligands. Examples: ACEI, rennin inhibitors, angiotensin receptor inhibitors.
- **Non-competitive or irreversible inhibition**: Drugs covalently bind at the active site of the enzyme and irreversibly inhibit it. Examples: inhibitors of acetylcholine esterase such as physostigmine, neostigmine *etc.*, cyclooxygenase

(COX) inhibition by aspirin, phenoxybenzamine antagonism of α-adrenergic receptor.

- **Allosteric inhibition**: Drugs bind at sites other than the active site, causing a conformational change in the enzyme that prevents it from binding to its physiological substrate. Examples: nonnucleotide reverse transcriptase inhibitors, Na^+/K^+-ATPase inhibition by digoxin, antihistamines binding to histamine H_1 receptor. Some enzymes as well as ion channels and metabolic receptors exhibit allosteric behavior, which in the absence of physiological agonists, typically prefer the inactive conformation; channels will be closed, and metabolic receptors will not be stimulated.

In addition to the above mentioned types of antagonisms, there are certain types of drugs that act as antagonists of particular receptors (more appropriately the signals transduced by these receptors), but do not require direct interaction with these receptors. These include:

- **Physiological antagonists**: In this case, two drugs acting on different cognate receptors have opposing pharmacological actions. For example, histamine acts on receptors of the parietal cells of the gastric mucosa to stimulate acid secretion, while omeprazole blocks this effect by inhibiting the proton pump; the two drugs can be said to act as physiological antagonists.

- **Pharmacokinetic antagonists**: This type of antagonists reduce bioavailability, and thus the concentration of an agonist at its site of action. This can happen in a number of ways, through induction of drug metabolizing enzymes in the liver (*e.g.*, Phenobarbital reducing the anticoagulant effect of warfarin this way), impaired rate of absorption of the drug from gastrointestinal tract or enhancement of renal excretion of the drug.

THEORIES OF DRUG ACTION

As mentioned earlier, the actions of most drugs are mediated through binding or interactions with their receptors. Different theories incorporating various models and equations have been proposed to understand the mechanisms of drug action in a quantitative way. Although these drug-receptor theories can be envisaged as largely adapted from enzyme kinetics, there can be a significant difference in the mechanism of drug action *via* a non-enzymatic receptor protein (*i.e.* G-protein coupled receptors) from the classical enzyme-substrate interactions. For example, unlike enzyme substrates, most drugs are not chemically altered by their receptors, but rather they interact in a reversible manner to produce a change in the state of the receptor. It should be noted that none of these theories can provide an explanation for the mechanism of action of all drugs.

Occupancy Theory

The classic occupancy theory of drug-receptor interaction was proposed by A.J. Clark and Gaddum (1926) [14]. It was actually an adaptation of the Langmuir isotherm theory dealing with adsorption of gases onto metal surfaces. The occupancy theory capitalizes on the laws of mass action and thus rests on several assumptions. First, the drug (D) and receptor (R) bind reversibly to form a drug-receptor complex (DR) in 1:1 stoichiometry. The DR complex generates a response (E) (eq. (1)).

$$D + R \leftrightarrow DR \rightarrow response\ (E) \tag{1}$$

Second, the magnitude or intensity of the response is directly proportional to the amount or concentration of the DR complex ([DR]) formed (eq. (2)). Third, the maximum possible response (E_{max}) is elicited when all the receptors (R_T) become occupied by the drug, forming the DR complex (eq. (3)). Since the biological response E emanates only from the DR complex (where D is an agonist, see later), E is a function of the fractional occupancy (eq. (4) and Fig. **1**) [15].

$$E \propto [DR] \tag{2}$$

$$when\ [DR] = [R_T], E = E_{max} \tag{3}$$

$$Fractional\ Occupancy = \frac{[DR]}{[R_T]} = \frac{[DR]}{[R+DR]} = \frac{E}{E_{max}} \tag{4}$$

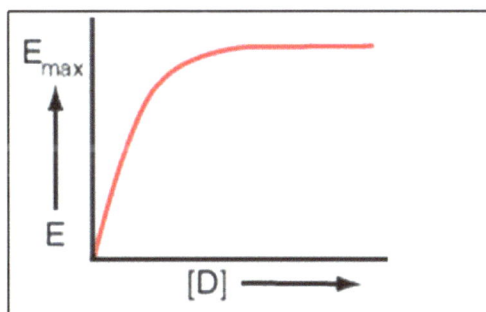

Fig. (1). Drug concentration *vs* response curve according to occupancy theory.

Using the law of mass action for the simple bimolecular reaction shown in eq.1, one can derive the equilibrium dissociation constant (K_D) and therefore quantify the fractional occupancy by eq. (5) [15], which is comparable to the *Hill-Langmuir adsorption isotherm.*

$$Fractional\ Occupancy = \frac{[DR]}{R_T} = \frac{[D]}{[D]+K_D} \tag{5}$$

The term K_D is a concentration and it quantifies the 'the affinity' for a particular drug (more appropriately, a ligand) for its receptor and as such is a very useful parameter. It is the concentration of the drug that produces a fractional occupancy of 50%. Since concentration of receptor is finite within a tissue, a large increase in drug concentration will not necessarily produce higher DR complexes (*i.e.* the receptor pool will be saturated) but will only lead to secondary, less affinity binding to various non-specific sites other than the receptor protein. This may create unwanted side effects.

Modification of Occupancy Theory

Clark's occupancy theory is fairly general; his equations describe many phenomena and have laid the foundation of modern treatises of drug-receptor interaction. It is generally compatible with in vitro radioligand binding experiments with purified receptor proteins and also with some drugs that are able to trigger a full response if given at adequate concentration in certain systems. However, it fails to reconcile with results from many ex vivo or in vivo studies of drugs. The problems seem to lie mainly within two basic assumptions of the occupancy theory. First, according to the theory, the maximal response to the drug is equal to the maximal tissue response, leading to the expectation that all agonists would produce the same maximal response. However, this is not true for some drugs, known as the partial agonists, for which a maximum response can never be achieved even at extremely high doses. Second, the theory assumes that the relationship between occupancy and response is linear and direct. Therefore, a 50% receptor occupancy will result in a half-maximal response and thus K^D equals to EC_{50} (*i.e.* the concentration of drug producing 50% of E_{max}). This is rarely seen in biological systems. **Nickerson** (1956) first showed that agonists such as histamine could produce a maximal tissue response at extremely low receptor occupancies (far less than maximal) [16]. Such data suggested the existence of nonlinear hyperbolic relationships between receptor occupancy and response. In most cases, tissues rather have '**spare receptors or receptor reserve**' and drugs need to occupy only a minor proportion ($\leq$10%) of the total receptor population to evoke a maximum response.

Ariens (1954) and **Stephenson** (1956) independently modified Clark's occupancy theory by introducing a proportionality factor, α in eq. (4), forming eq. (6) [8, 9].

$$Response = \frac{E}{E_{max}} = \frac{\alpha[DR]}{[R+DR]} \tag{6}$$

Ariens termed this factor as the intrinsic activity (α) [17] whereas Stephenson termed this as efficacy (e) [9]. They envisaged two basic requirements for any molecule to elicit a biological response – the ability to form the DR complex (affinity) and then the ability of the DR complex to produce the biological response (intrinsic activity or efficacy). Thus agonists and antagonists have similar affinities for a particular receptor but only the agonists have the efficacy to trigger a response. In Clark's original equation, $\alpha = 1$, and this is true for full agonists which have the highest intrinsic activities. For partial agonists, α is less than 1, but greater than 0. Antagonists elicit no response, so they have α value of 0. All of them have comparable affinities.

Clark's original theory, followed by Ariens and Stephenson's modification, enabled the recognition of two major factors in the definition of agonist activity: the drug-specific parameters of affinity and intrinsic efficacy. Some years later, **Furchgott** introduced further modification of the occupancy theory, recognising Stephenson's efficacy as a composite of both the agonist-specific component (i.e. affinity and intrinsic activity) and the tissue (or receptor-system related) component (i.e. parameters such as receptor density and the efficiency of coupling between receptors and the effector mechanisms of the tissue) [18]. The separation of these components allowed for the measurement of parameters that are independent of the tissue and characterize the molecular interaction between the agonist and the receptor. Furchgott introduced intrinsic efficacy (ε) and defined this as the unit stimulus per occupied receptor, using the relationship $e = \varepsilon \times [R_T]$.

Rate Theory

Although most of the observed pharmacological responses are compatible with the notion that a drug's action is being elicited while actually occupying a site (i.e. occupancy theory of Clark) on the target receptor, this cannot account for all such phenomena. As mentioned above, the occupation theory has been modified in several ways, especially with the introduction of the concept of intrinsic activity to account for the fact that occupancy of a receptor can have either an agonistic or an antagonistic effect. It still remains unclear how some drugs (e.g., nicotine) first stimulate and then block a response and also why many drugs initially exert an effect that subsequently declines, or "fades," even though the drug remains in contact with the tissue.

In 1961, **Paton**, inspired by his interest in the biphasic effects of nicotine on ganglia, presented a theory to account for such issues, which he called rate theory [19]. According to this theory, activation of a receptor only occurs in a quantal manner as a result of its initial encounter or collision with the drug molecule. Further effects are dependent on the drug's ability to dissociate from the receptor so that the receptor becomes free to be re-stimulated. Thus, it is the rate of stimulation or total number of successful collisions of the drug per unit time with its receptor that governs a drug's ability to elicit a response.

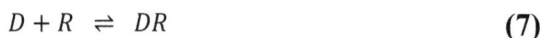

$$D + R \; \rightleftharpoons \; DR \qquad\qquad (7)$$

In terms of association (k_{+1}) and dissociation (k_{-1}), rate constants of the above bimolecular reaction between the drug and its receptor, an effective agonist, must have a high k_{-1} along with a high k_{+1}. However, an antagonist may associate quickly (*i.e.* a high k_{+1}) but dissociates slowly (*i.e.* have a low k_{-1}) and will thus limit the probability of fresh encounters. Partial agonists are likely to have dissociation constants intermediary between those of full agonists and the antagonists. The k_{-1} or rate of dissociation of the DR complex in Patton's rate theory thus determines what in occupation theory is called the intrinsic activity, whereas the affinity remains equivalent to the reciprocal of the dissociation constant k_{-1}/k_{+1}. Stimulation followed by blockade or fading of a response is possible due to the difficulty of a drug in dissociating from the DR complex following the initial successful binding that produces an effect.

The rate theory clearly offers some added insight into the mechanism of drug action but it has its own limitations. One of the major premises of the rate theory is that the stable drug-receptor complex is not required for pharmacological activities. However, in real experiments, the drugs often seem to form stable complexes with their receptor proteins. This phenomenon on a molecular level cannot be explained by Patton's theory. The question regarding why two similar compounds are antagonistic to each other is also left unanswered.

Induced-Fit Theory

Interaction of a drug molecule with its receptor is generally specific and analogous to that of an enzyme-substrate interaction. For the latter, Emil Fischer proposed the '*lock and key theory*' which considers the enzyme active site as a rigid and sturdy lock, with an exact fit to only one substrate (key). Such a simplistic assumption is key to the specificity of enzyme action, which had been accepted as the universal mechanism for enzyme ligand/substrate binding for more than half a century until challenged by an alternative mechanism of '*induced fit*' by **Koshland** in 1958 [11]. According to the induced fit theory, enzymes (or receptors) need not be rigid and may not exist in the appropriate conformation required to bind with the substrate or drug. The latter *induces a conformational change in the receptor* to cause a complementary fit and thus the initiation of a biological response and once the drug is released, it returns to its original shape. Only agonists induce a change in conformation and elicit responses. Under this scheme, drug molecules can also undergo conformational changes upon approaching a specific binding site.

Both the rigid and flexible views of molecular associations have since been widely accepted. These modes can be predicted by comparing the structures of the free, unbound, protein molecule (Apo form) with the structure when complexed with its ligand. If the structures are similar, the binding mode is likely to belong to the rigid, 'lock and key' type mechanism. However if the structural comparison reveals rather a large conformational change around the vicinity of the ligand-binding domain, it is more likely that the binding mode falls into the 'induced fit' category.

Macromolecular Perturbation Theory

This evolved from the induced–fit theory by Belleau [20]. It mainly suggests that interaction of a drug with a receptor results in two general types of perturbations: specific conformational perturbations that make possible the binding of certain molecules (agonists) that produce a biological response or nonspecific conformational perturbations that accommodate molecules (antagonists) that do not elicit a response. If a drug contributes to both possible perturbations, then a mixture of the two complexes exists and results in a partial agonist.

Occupation-Activation Theory of "Two-State" Model

This is essentially an extension of the macromolecular perturbation theory and the receptor is envisaged to be in dynamic equilibrium between two conformational states, 'resting' (R) and 'activated' R*, which exist in equilibrium (Fig. 2). Normally, when no ligand is present, the equilibrium lies far to the left, and few receptors are found in the R* state. For constitutively active receptors, an appreciable proportion of receptors adopt the R* conformation in the absence of any ligand. Agonists have higher affinity for R* than for R, so the equilibrium shifts towards R*. The greater the relative affinity for R* with respect to R, the greater the efficacy of the agonist. Partial agonists favor both states to different extents. An inverse agonist has a higher affinity for R than for R* and so shifts the equilibrium to the left. A 'neutral' antagonist has equal affinity for R and R* so does not affect the conformational equilibrium but reduces the binding of other ligands [21].

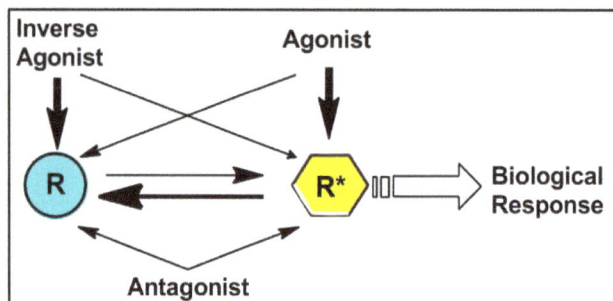

Fig. (2). Occupation-Activation Theory of Two-State Model.

This theory allows for the agonist binding site in the R* state to be different from the antagonist binding site in the R state and it also provides an understanding of large structural differences which may occur between agonists and antagonists. A major problem with this theory is that since it is known that receptors are not actually restricted to two distinct states but have much greater conformational flexibility, there is more than one inactive and active conformation. The different conformations that can be adopted may be preferentially stabilized by different ligands, and may produce different functional effects by activating different signal transduction pathways. Redefining efficacy for such a multistate model is difficult, however, and will require a more complicated state transition theory than that described in the two-state model [21].

DRUG RECEPTOR INTERACTIONS

Each drug-binding site on the target receptor protein has unique chemical characteristics that are determined by the specific properties of the amino acids that make up the site. The three-dimensional (3D) structure, shape, and reactivity of the site and the inherent structure, shape, and reactivity of the drug determine the orientation of the drug with respect to the receptor, and govern its specificity and affinity for the receptor protein. Rarely a drug-receptor interaction involves a single type of interaction. Rather multiple chemical interactions occur between the two molecules, some of which are fairly weak (such as van der Waals' forces) and some of which are extremely strong (such as covalent bonding). The ensemble of these interactions provides the specificity of the overall drug–receptor interaction. Affinity, which is usually experimentally reported by the K_D value, is a measure of the favorability of a drug–receptor interaction. Even a minor variation in the functionalities of the drug molecules can significantly alter the binding interactions and thus the affinity that eventually contributes to the overall potency, efficacy, and duration of drug action. The DR interaction of imatinib with the tyrosine kinase active site is shown as an example in Fig. (**3**).

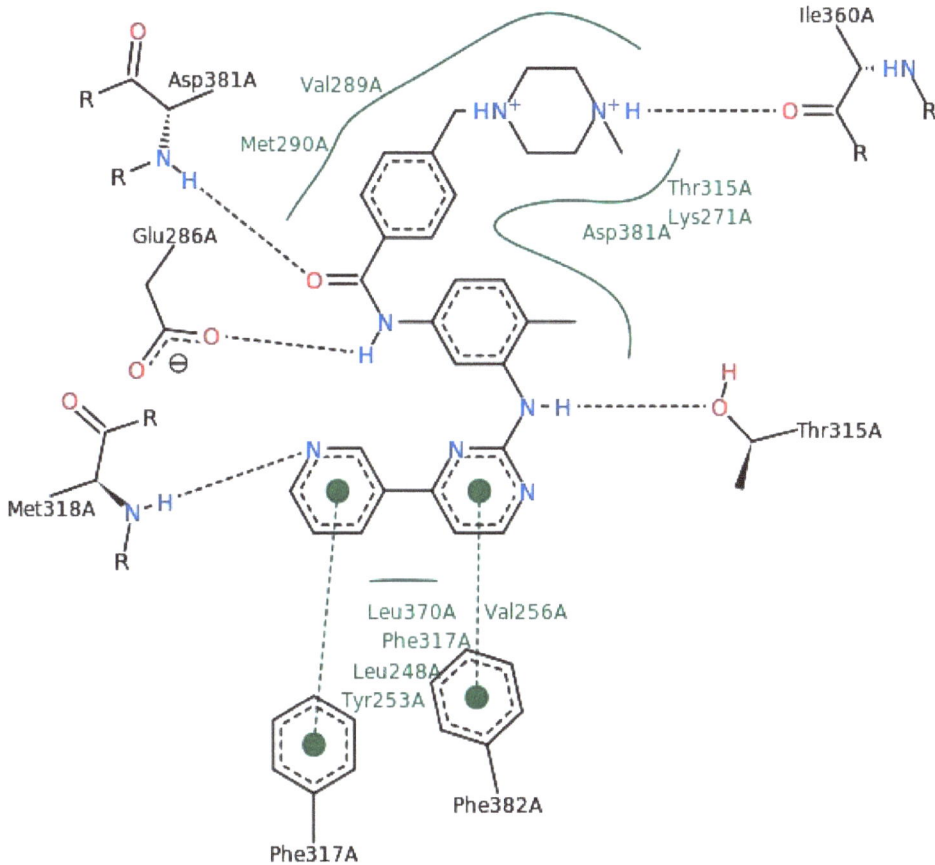

Fig. (3). Interaction of imatinib with the ATP-binding site of the BCR-Abl tyrosine kinase (pdb: 2HYY). Arrows indicate hydrogen bonding and only those residues in proximity enough to participate in van der Waals' interactions are shown. The sum total of these forces creates a strong (high affinity) interaction between this drug and its receptor.

The Interaction (Bond) Types

1. Covalent Bonds
2. Non-covalent Bonds
    a. Ionic Bonds
    b. Dipole Interactions
    c. Hydrogen Bonds: a specialized type of Dipole-Dipole bond
    d. van der Waals Interactions
    e. Hydrophobic Bonding

    f. Chelation and Complexation
    g. Charge Transfer Interactions

Covalent Bonds

Covalent bonds are formed by interactions of electrons from different atoms. The two participating atoms donate one electron each, forming an electron pair which is shared by both of them. This type of covalent bonding is the simplest and is known as a single or sigma (σ) bond. In addition to this, two atoms can use extra electrons (π electrons) to have an additional one or two covalent bonds (known as π bonds) leading to double or triple bonding. Covalent bonds are the strongest bonds (irreversible) which can be possibly made between a drug and its receptor (Fig. **4**) with bond strength (measured by the thermal energy required to break them) ranging from 40 to over 200 kcal/mol.

The selectivity of drug action is important for all drugs but it is extremely crucial for covalently bound drugs. Since covalently bound drugs cannot be easily removed, the side effects associated with such drugs are likely to be long lasting. To regain activity, the cell must synthesize a new receptor molecule to replace the inactivated protein. This is why only a few drugs, including the anticancer alkylating agents and acetylcholine esterase inhibitors, work by covalent bond formation (Fig. **4**). Among the types of covalent bonds employed by various drugs, alkylation, acylation and phosphorylation are common.

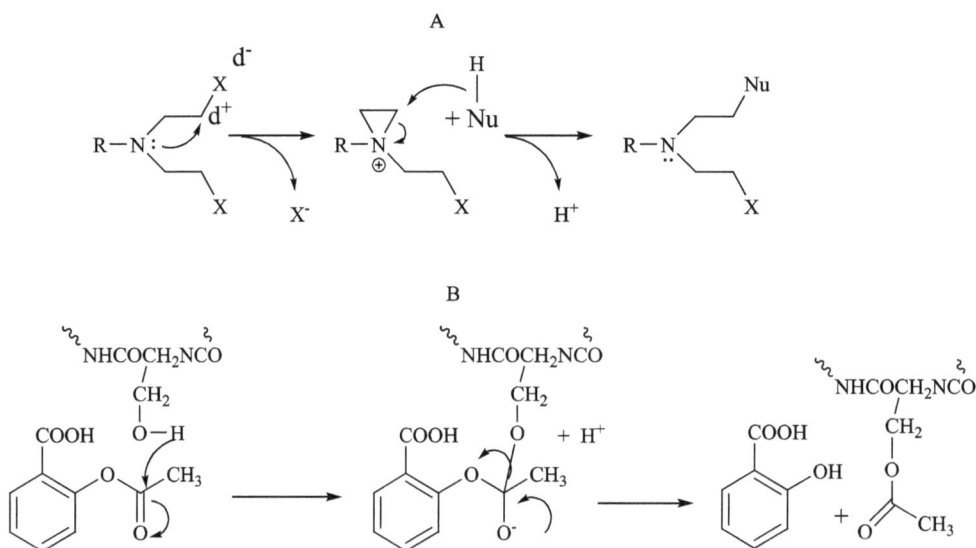

Fig. (4). Covalent interactions; A) Anticancer nitrogen mustard with nucleic acid nucleophile (alkylation), and B) Aspirin with COX active site serine (acylation).

Non-covalent Bonds

The non-covalent bindings are the most common type of interaction of drugs with receptors, enzymes, and other macromolecules. Such bonds are much weaker and reversible unlike covalent bonds. There are a large number of such interactions with bond energies ranging from 0.5-10 kcal/mol. In most cases, drugs bind through the use of multiple non-covalent bonds, which collectively give relatively strong interaction for desirable therapeutic effects. The interactions may be electrostatic and often involve ions (species possessing discrete charges), permanent dipoles (having a permanent separation of positive and negative charge), and induced dipoles (having a temporary separation of positive and negative charge induced by the environment). The drugs acting by non-covalent interaction need to come within reasonable proximity to its receptor to allow any such interaction to occur and thus to give therapeutic effects. The distance requirements for drugs are in order (greatest to shortest): ionic bond > ion-dipole bond > dipole-dipole interaction > hydrophobic bond [22].

Ionic Bonds

The attractive forces between the negatively charged and positively charged functions result in ionic bonds. Thus the negatively charged carboxylate groups will cause ionic interaction with the positively charged ammonium groups. The pK_a of the functional groups and their ionization at physiological pH dictates the ability to form these bonds.

Ionic bonds are the strongest among all the non-covalent interactions with bond strengths of 5-10 kcal/mol. The strength of such interactions is highly dependent on the nature of the interacting species and the distance, r, between them. Between two ions, the energy falls off as $1/r$. Since ionic bonds can form over the greatest distance (when compared to other interaction forces), they are often the initial bonding attraction between a drug and its receptor. Many drugs are acids or amines, easily ionized at physiological pH, and able to form ionic bonds by the attraction of opposite charges in the receptor site. For example, the ionic interaction between the protonated amino group on salbutamol or the quaternary ammonium on acetylcholine and the dissociated carboxylic acid group of its receptor site (Fig. **5**). Similarly, the dissociated carboxylate group on the drug can bind with ionized amino groups on the receptor.

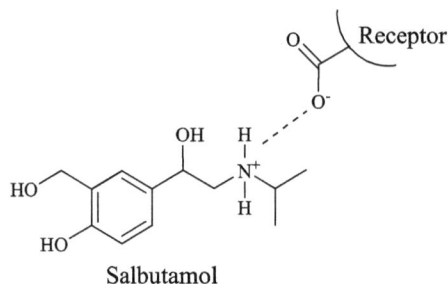

Salbutamol

Fig. (5). Ionic interaction of salbutamol with adrenergic receptor.

A **salt bridge** is an important phenomenon in a receptor site that allows ionization of weakly basic amino acid such as histidine (pKa 6.5), in a **friendly environment** (where the donor and acceptor atoms are arranged in a complimentary fashion), which would otherwise be only 10% ionized at physiologic pH (7.4). For example, in the acetylcholine esterase active site, the pyrrole like NH atom of the imidazole ring of histidine forms a hydrogen bond with the ionized carboxylate of nearby glutamic acid, resulting in a stronger dipole (Fig. **6**). This increases the nucleophilicity of the imine-like ring N atom and thus enhances its basicity and ionizability. Thus, the three amino acids (Glu, His, Ser) form a triad to cause a very fast hydrolysis of acetylcholine. However, in a **hostile environment** (where a nonpolar group is arranged in the vicinity of a polar group), when they encounter a hydrophobic region instead, the ionization of acidic and basic functional groups decreases.

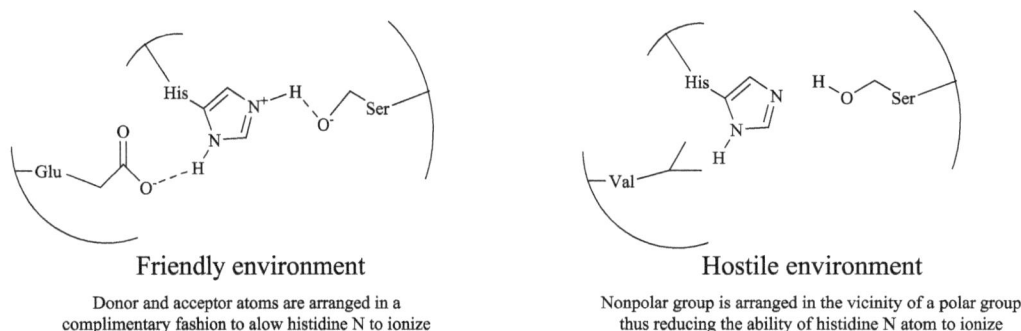

Friendly environment

Donor and acceptor atoms are arranged in a
complimentary fashion to alow histidine N to ionize

Hostile environment

Nonpolar group is arranged in the vicinity of a polar group
thus reducing the ability of histidine N atom to ionize

Fig. (6). The salt bridge formation in acetylcholine esterase active site, in friendly environment (the triad). In hostile environment it is not possible.

Dipole Interactions

The difference in electronegativity between atoms comprising functional groups causes partial charge separation resulting in dipoles. When the dipoles of drug and receptor come in close proximity (more stringent distance requirement), they produce dipole interactions. The more electronegative oxygen atom attached to carbon atom in carbonyl groups (ketones, esters, amides) results in such dipoles, and thus dipole interactions, with the biological receptors when they occur in drugs. Such interactions are very common since most drugs possess carbonyl or other functional groups exhibiting partial charge separation and dipoles. Such interactions have bond strengths between 1-7 kcal/mol.

There are two common types of dipole interactions, **ion-dipole** and **dipole-dipole**, with distance requirements of $1/r^2$ and $1/r^3$, respectively. Fig. **7** shows the ion-dipole and dipole-dipole interactions of chlorpheniramine and captopril with receptor site amino acid serine.

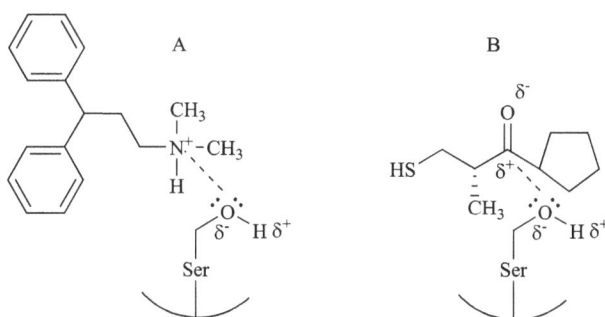

Fig. (7). Dipole interactions of active site serine residue with, A) chlorpheniramine (ion-dipole), and B) captopril (dipole-dipole).

Hydrogen Bonds

Hydrogen bonds are a subtype of dipole interactions that occur with the aid of a bridged hydrogen atom between two electronegative atoms – hydrogen bond donor and hydrogen bond acceptor (Fig. **8**). It has a bond strength of 3-7 kcal/mol. These bonds tend to be highly directional, forming straight bonds between donor, hydrogen, and acceptor atoms. Hydrogen bonds are possibly the most common type of drug receptor interactions. Such bonds are also very important for the maintenance of secondary and tertiary structures of proteins and nucleic acids and thus are very crucial for their conformation and functions. Breaking the hydrogen bonds of proteins causes their denaturation. Hydrogen bonds are crucial to imparting water solubility and sometimes ionic bond strength and are also important in enhancing the drug-protein binding. In Fig. **8**, the bond formed with the O-H-O orientation in 180° is stronger than the one formed with a curved orientation (N-H-O).

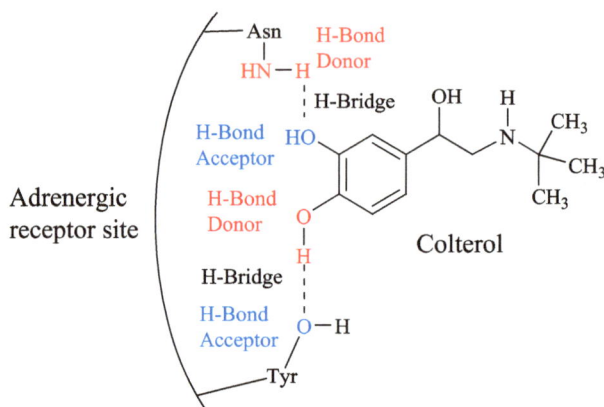

Fig. (8). Hydrogen bond between colterol and adrenergic receptor site showing the hydrogen bond donor and acceptors.

Sometimes drugs can form both the ionic and hydrogen bond within the same set of functional groups of the receptors. For example, the carboxylate function of the antidiuretic drug furosemide can form both a hydrogen bond as well as an ionic bond with the side chain amine of lysine in the receptor site (Fig. **9**). This type of bond is stronger than simple ionic bonds or simple hydrogen bonds and is known as reinforced ionic bonds through hydrogen bonds or reinforced hydrogen bonds through ionic bonds.

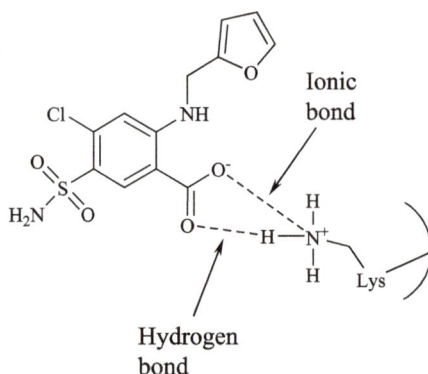

Fig. (9). Reinforced ionic bond through hydrogen bond.

The hydrogen bond donors (designated as X) are O, N, or S, and the hydrogen bond acceptors (designated as Y) are O, N, F or a double bond. The O and N containing functional groups, including hydroxyl, carbonyl, amino, and amido, are the most frequently observed hydrogen bonding groups in drugs and other biological systems. Some hydrogen bond donor and acceptor functional groups are shown in Table **1**.

Table 1. Some hydrogen bond donors and acceptors.

Donors	Acceptors
R–C(=O)–O–H ⟶	R₂C=O (R above and below, two acceptor arrows)
R–C(R)(R)–O–H ⟶	R₂O (two acceptor arrows)
R–O–H ⟶	R–O–H (acceptor arrow)
R–N(H)(H) (two donor arrows)	R–N=R (donor/acceptor arrows)
R–N(R)(H) (donor arrow)	–P=O (two acceptor arrows)

Van der Waals Interactions

Some nonpolar functional groups, including aromatic rings and alkyl and alicyclic chains, when in close proximity (critical distance requirement) can produce induced electrical interactions through the instantaneous fluctuation of their electron clouds. These interactions result in an attraction of the positively charged nuclei and the electrons of the nearby atom, known as Van der Waals forces. The dipole–dipole interactions discussed above are a type of Van der Waals interaction. The Debye forces also known as dipole-induced dipole interactions are also Van der Waals forces, which have a distance requirement of $1/r^5$. Other Van der Waals forces include induced dipole-induced dipole interactions (dispersion or London dispersion forces), which falls off as $1/r^6$. Although the van der Waals force is very weak, calculated to be about 0.7 kcal/mol arising from a pair of methylene groups, it can be significant when occurring between groups of sufficient size. For example, a phenyl ring, containing 6 carbon atoms, can provide a total of about 6 x .7 = 4.2 kcal/mol and an n-butyl side chain can provide about 4 x .7 = 2.8 kcal/mol bond strengths [23].

Hydrophobic Interactions

Hydrophobic interactions occur among nonpolar or hydrophobic molecules (and also among the hydrophobic regions or groups within a macromolecule such as a drug receptor) as they tend to associate with one another while being in a polar solvent (usually water). Unlike Van der Waals interactions, no actual bond is formed during the process. The polar water molecules preferably interact with each other and therefore exclude the nonpolar hydrophobic substances. This preferential interaction of water molecules drives the nonpolar molecules to coalesce and cause phase separation. The water molecules arrange in an orderly fashion around nonpolar drug and/or receptor molecules. The hydrophobic molecules then squeeze the water molecules out and then fewer water molecules are rearranged around the coalesced hydrophobic phase. This process raises the entropy of water, the energy that is utilized in hydrophobic and/or Van der Waals interactions between nonpolar drugs and a receptor (Fig. **10**) [22]. The non-polar groups of drug molecules (*e.g.*, alkyl or aryl groups) often interact with the non-polar amino acid residues of the receptor protein including leucine, isoleucine, lysine, valine, tryptophan and phenylalanine in this manner.

Chelation

Chelation is a process through which the electron donating groups (O, N, and S) form a ring structure by binding to metal ions. Chelates exist in either 4-, 5-, or 6-membered rings. The 4-member rings occur only if sulfur is involved. The two most important aspects regarding chelation are:

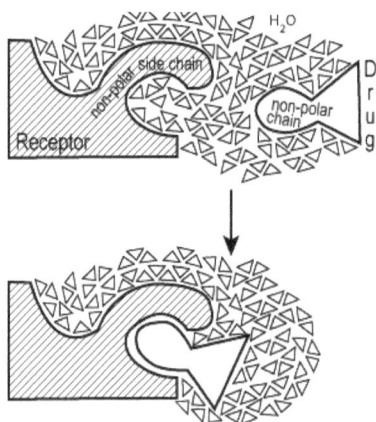

Fig. (10). Illustration of the hydrophobic interaction between drug and receptor.

A. Binding to enzymes (or their cofactors) that require metallic catalysis. For example, the enzyme lipid peroxidase requires the presence of iron atom to cause lipid oxidation. This is an important mechanism for doxorubicin's cardiotoxicity, which is minimized by coadministration of dexrazoxane that can chelate iron (Fig. **11**).

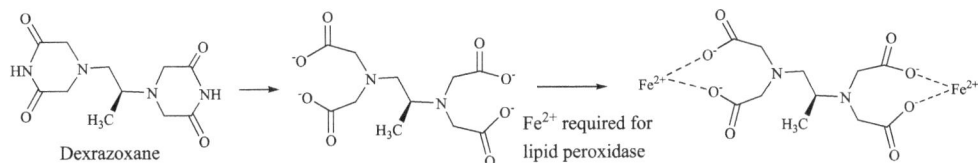

Fig. (11). Chelation of iron by dexrazoxane metabolite.

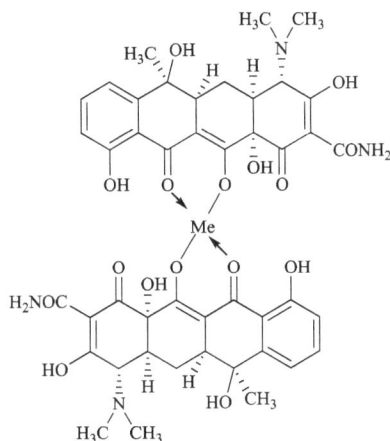

Fig. (12). Tetracycline metal chelate; Me = Fe^{3+}, Fe^{2+}, Cu^{2+}, Ni^{2+}, Co^{2+}, Zn^{2+}, Mn^{2+}, Mg^{2+}, Ca^{2+}, Be^{2+}, Al^{3+}.

B. Binding of drug molecules to metals in the GI tract. For example, tetracyclines will chelate with a variety of metals including calcium, magnesium, aluminum, and iron in the GI tract (Fig. **12**). Its B and C rings are more susceptible to form the metal complexes, or chelates. These chelates have very poor water solubility and therefore poor absorption and bioavailability [24].

Sometimes a metal ion interacts with a **single atom** without forming a ring by a process known as **metal complexation**. Some enzyme inhibitors containing -SH, -COO⁻ or -PO₄⁻ functional groups may cause this type of interaction. The angiotensin converting enzyme (ACE) inhibitors are a good example of such drugs that contain one of these functional groups and bind to the zinc ion for activity (Fig. **13**).

Charge Transfer Interactions

Charge transfer (CT) interactions are electrostatic attractions between relatively **electron-rich (donor)** molecules and relatively **electron-deficient (acceptor)** molecules. The donor molecules may include alkenes, alkynes, aromatics carrying electron donating groups and π-electron-rich heterocyclic rings such as thiophene, furan and pyrrole; the acceptor groups may include thiols, alcohols, amines, aromatic rings carrying electron withdrawing groups and π-Electron-deficient hetrocyclic rings such as purines and pyrimidines (Table **2**). Most drugs are carrying aromatic and heterocyclic rings that bind to the receptors consisting of heterocyclic rings, *e.g.*, proteins and nucleic acids. Thus, the CT interactions between electron-rich and electron-poor aromatic rings are most common in drug-receptor interactions. The CT interaction is similar to the dipole interaction but is stronger and not usually higher than 7 kcal/mole. Molecules within the proximity of 3-3.5 Å distance can produce such interaction. Note that all the six-membered heterocyclic rings systems are electron-poor and the five-membered heterocyclic rings are electron-rich rings.

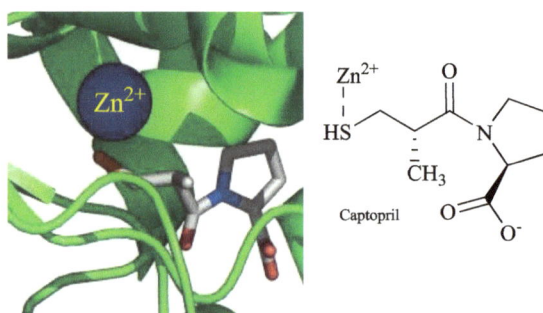

Fig. (13). Metal complexation of captopril in the active site of angiotensin converting enzyme (ACE).

Table 2. Some Common electron-rich and electron-poor heterocyclic rings occur in drugs and biological system.

Electron-poor Rings	Electron-rich Rings
Six-membered heterocyclic rings	Five-membered heterocyclic rings
Pyrimidine Pyridazine Pyrazine	Imidazole Oxazole Thiazole
Six-six fused heterocyclic rings	Five-six fused heterocyclic rings
Quinoline Isoquinoline	Indole Benzoxazole

STUDENTS SELF-STUDY GUIDE

- The mechanisms of drug action: reversible, irreversible and allosteric mechanisms
- The theories of drug action: occupancy, rate, induced fit, macromolecular perturbation, occupation-activation theory of two state model

- What is a full agonist, partial agonist and antagonist?
- All the different types of chemical interactions of drug and receptors with examples
- What is a charge transfer interaction? Identify the electron rich and electron poor rings in drug structures.
- What is chelation/metal complexation and its importance in drug action?

STUDENTS SELF ASSESMENT

Part I: Directions: Each of the numbered items or incomplete statements in this section is followed by answers or by completions of the statements. Select the **one** lettered answer or completion that is **best** in each case.

1. The intrinsic activity of a drug is defined as _____ .
   - A. its ability to bind to its biological target
   - B. its property that enables to produce responses
   - C. the maximal stimulatory response induced by a drug in relation to that of a given reference compound
   - D. the dose that is required to produce a specific effect
   - E. the maximum activity that an antagonist possesses
2. All of the following regulatory proteins serve as primary drug targets <u>except</u>:
   - A. Enzymes
   - B. Carrier molecules
   - C. Ion-channels
   - D. Receptors
   - E. Glucocorticoid
3. Which one of the following Bond types represents the right order in terms of binding strength (<u>lower to higher</u>)?
   - A. Ionic → Ion-dipole → Hydrogen → Hydrophobic → van der Waals
   - B. van der Waals → Hydrophobic → Hydrogen → Ion-dipole → Ionic
   - C. van der Waals → Ion-dipole → Ionic → Hydrogen → Hydrophobic
   - D. Hydrophobic → Ion-dipole → Ionic → Hydrogen → van der Waals
   - E. Hydrogen → Ion-dipole → Ionic → Hydrophobic → van der Waals
4. Which statement is <u>not true </u>about drug receptor interaction?
   - A. Interaction of drug with a biological receptor site produces biological response.
   - B. The order of forces (lowest to highest) involved in different types of bonding is: van der Waals → Hydrophobic → Hydrogen → Covalent → Ionic.
   - C. Covalent bonds between drugs and receptors are less common.
   - D. Most useful drugs bind through the use of multiple weak bonds.
   - E. A high degree of specificity in binding drug to its receptor site is desirable

for getting a potent and safer drug.

5. Based on induced-fit theory of drug receptor interaction, which statement is NOT TRUE?
   A. The receptors exist in appropriate lock and key conformation required to bind with drugs for response
   B. The drugs may induce a conformation change in the receptor
   C. The drug could undergo a conformation change to fit with the receptor
   D. Agonist induce a change in the receptor
   E. Antagonist bind without a change in the receptor

6. Which statement about Occupation-Activation Theory of "Two-State" Model of drug-receptor interaction is TRUE?
   A. Agonist shifts equilibrium to the inactive state
   B. Antagonists shift equilibrium to the active state
   C. Partial agonist favors both inactive and active states to different extents
   D. Both A and B
   E. A, B and C

7. According to modified occupancy theory drugs must have a(n) _____ _____ that is the ability of the drug-receptor complex to produce a biological effect.
   A. Intrinsic activity
   B. Receptor occupancy
   C. Binding site
   D. Receptor affinity
   E. Functional group

8. Which is a noncovalent type drug receptor interaction?
   A. Alkylation
   B. Acylation
   C. Phosphorylation
   D. Chelation and Complexation
   E. Oxidation and reduction

9. The following diagram is an example of:

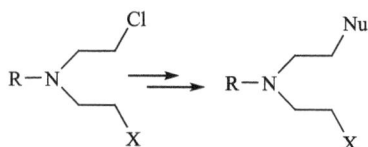

   A. irreversible inhibition
   B. reversible competitive inhibition
   C. noncovalent inhibition
   D. chelation

Questions 10 and 11 are based on following drug receptor interactions

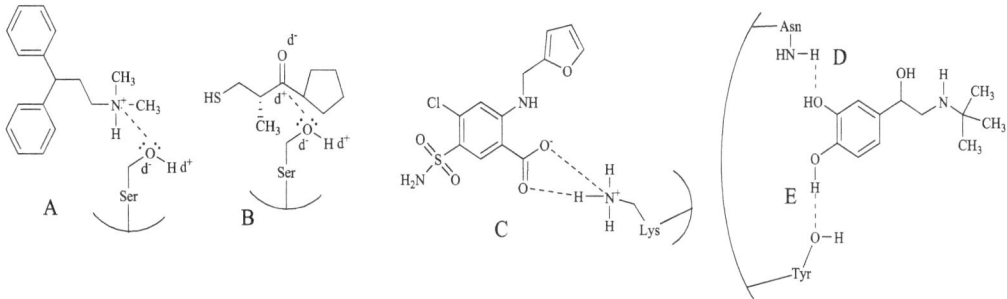

10. Which of the bond types is the strongest?
11. Which one represents the weakest bond?

1. Which statement is <u>not</u> true?
    A. Chelation of metals by tetracyclines increases its activity
    B. Chelation of iron by lipid peroxidase is important for its activity
    C. Dexrazoxane that chelates iron is important for its inhibitory activity of lipid peroxidase
    D. Captopril chelates Zn^{+2} at enzyme active site which is important for its activity
    Part II: **Directions:** Each item below contains three suggested answers of which **one or more** is correct. Chose the answer
    A. if **I only** is correct
    B. if **III only** is correct
    C. if **I and II** are correct
    D. if **II and III** are correct
    E. if **I, II, and III** are correct
2. Examples of allosteric inhibitors include:
    I. nonnucleotide reverse transcriptase inhibitors
    II. rennin inhibitors
    III. angiotensin converting enzyme inhibitors
3. Examples of nonprotein drug receptors are:
    I. an ion channel
    II. DNA
    III. a fluid compartment
4. A drug that dissociates very slowly but associate quickly to its receptor may be:
    I. an agonist
    II. a partial agonist
    III. an antagonist
5. Which of the following drug-receptor interactions is/are <u>not</u> considered actual bond(s)?
    I. A dipole-dipole interaction

    II. A van der Waals interaction

    III. A hydrophobic bond

6. The drug(s) that may cause ionic interaction at the receptor site include(s):

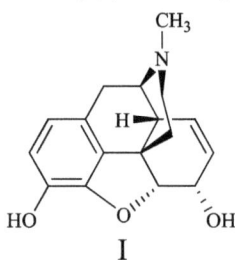

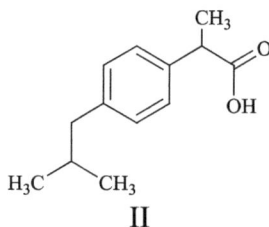

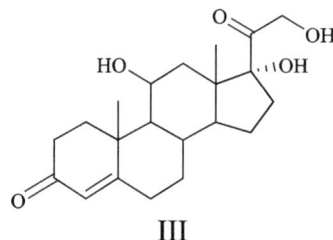

         I   II   III

7. Tryptophan when occurs in receptor may cause charge transfer interaction with the following ring systems when occur in a drug molecule:

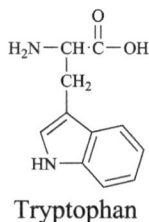

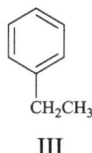

    Tryptophan   I   II   III

CONSENT FOR PUBLICATION

Not applicable.

CONFLICT OF INTEREST

The authors declare no conflict of interest, financial or otherwise.

ACKNOWLEDGEMENT

Declared none.

REFERENCES

[1]   Olmsted JMD. Claude Bernard Physiologist. London: Cassell & Co. 1939.

[2]   Ehrlich P, Morgenroth J. In: Himmelweit F Ed. The Collected Papers of Paul Ehrlich in Four Volumes Including a Complete Bibliography. Vol. II Immunology and Cancer Research (English translation). London: Pergamon; 1957.

[3]   Langley JN. Croonian Lecture, 1906 — on nerve endings and on special excitable substances in cells. Proc R Soc Lond 1906; B78: 170-94.
[http://dx.doi.org/10.1098/rspb.1906.0056]

[4]   Straub W. Zur chemischen Kinetik der Muskarinwirkung und des Antagonismus Muskarin-Atropin. Pflügers Arch Ges Physiol 1907; 119: 127-51.
[http://dx.doi.org/10.1007/BF01680084]

[5]   Hill AV. The mode of action of nicotine and curari, determined by the form of the contraction curve and the method of temperature coefficients. J Physiol 1909; 39(5): 361-73.
[http://dx.doi.org/10.1113/jphysiol.1909.sp001344] [PMID: 16992989]

[6]   Clark AJ. The mode of action of drugs on cells. London: Edward Arnold & Co. 1933.

[7]   Gaddum JH. The quantitative effects of antagonistic drugs. J Physiol 1937; 89: 7-9.

[8]   Ariens EJ, De Groot WM. Affinity and intrinsic-activity in the theory of competitive inhibition. III. Homologous decamethonium-derivatives and succinyl-choline-esters. Arch Int Pharmacodyn Ther 1954; 99(2): 193-205.
[PMID: 13229430]

[9]   Stephenson RP. A modification of receptor theory. Br J Pharmacol Chemother 1956; 11(4): 379-93.
[http://dx.doi.org/10.1111/j.1476-5381.1956.tb00006.x] [PMID: 13383117]

[10]   Fischer E. Einfluss der Configuration auf die Wirkung der Enzyme. Ber Dtsch Chem Ges 1894; 27: 2984-93.

[11]   Koshland DE. Application of a theory of enzyme specificity to protein synthesis. Proc Natl Acad Sci USA 1958; 44(2): 98-104.
[http://dx.doi.org/10.1073/pnas.44.2.98] [PMID: 16590179]

[12]   Rang HP, Dale MM, Ritter JM, Flower RJ, Eds. Rang & Dale's Pharmacology. 5th ed.,

[13]   Krauss G, Ed. Biochemistry of Signal Transduction and Regulation. 3rd ed., Wiley-VCH 2003.
[http://dx.doi.org/10.1002/3527601864]

[14]   Maehle AH, Prüll CR, Halliwell RF. The emergence of the drug receptor theory. Nat Rev Drug Discov 2002; 1(8): 637-41.
[http://dx.doi.org/10.1038/nrd875] [PMID: 12402503]

[15]   Kenakin T, Ed. Pharmacologic analysis of drug-receptor interactions. 3rd ed., New York: Lippincott-Raven 1997.

[16]   Nickerson M. Receptor occupancy and tissue response. Nature 1956; 178(4535): 697-8.
[http://dx.doi.org/10.1038/178697b0] [PMID: 13369505]

[17]   Ariëns EJ. Affinity and intrinsic activity in the theory of competitive inhibition. I. Problems and theory. Arch Int Pharmacodyn Ther 1954; 99(1): 32-49.
[PMID: 13229418]

[18]   Furchgott RF. The use of -haloalkylamines in the differentiation of receptors and in the determination of dissociation constants of receptor–agonist complexes. Adv Drug Res 1966; 3: 21-55.

[19]   Paton WDM. A theory of drug action based on the rate of drug-receptor combination. Proc R Soc Lond B Biol Sci 1961; 154: 21-69.
[http://dx.doi.org/10.1098/rspb.1961.0020]

[20]   Belleau B. A molecular theory of drug action based on induced conformational perturbations of receptors. J Med Chem 1964; 7: 776-84.
[http://dx.doi.org/10.1021/jm00336a022] [PMID: 14262809]

[21]   Gringauz A, Ed. Introduction to Medicinal Chemistry: How Drugs Act and Why. Wiley-VCH 1997.

[22]   Garrett R, Grisham CM, Eds. Eds., Biochemistry; 2009; Brooks/Cole: Cengage Learning, Boston, MA.

[23]   Nogrady T, Weaver DF, Eds. Medicinal Chemistry- A Molecular and Biochemical Approach. New York: Oxford University Press Inc. 2005.

[24]   Dürckheimer W. Tetracyclines: chemistry, biochemistry, and structure-activity relations. Angew Chem Int Ed Engl 1975; 14(11): 721-34.
[http://dx.doi.org/10.1002/anie.197507211] [PMID: 812385]

CHAPTER 7

Drug Metabolism

Rahmat Talukder[1], Ashok Philip[2] and M. O. Faruk Khan[3,*]

[1] *Department of Pharmaceutical Sciences, University of Texas at Tyler College of Pharmacy, Tyler, TX, USA*

[2] *Department of Pharmaceutical Sciences, Union University School of Pharmacy, Jackson, TN, USA*

[3] *Department of Pharmaceutical Sciences and Research, Marshall University School of Pharmacy, Huntington, WV, USA*

Abstract: This chapter is a detailed account of drug metabolism, prodrugs and related terminology that are critical knowledge base for pharmacist and pharmacy education. After study of this chapter, students will be able to:

• Comprehend the fundamental concepts of drug metabolism
• Describe the significance of drug metabolism
• Identify key enzymes involved and the sites of drug metabolism
• Explain phase I and phase II metabolic pathways, including:
♦ Phase I (Functionalization)
- Oxidation of aromatic moieties, olefins, benzylic & allylic C atoms and α-C of C=O and C=N, aliphatic and alicyclic C, C-heteroatom system, C-N (N-dealkylation, N-oxide formation, N-hydroxylation), C-O (O-dealkylation), C-S (S-dealkylation, S-oxidation, desulfuration), alcohols and aldehydes, and miscellaneous oxidative reactions
- Reduction of aldehydes and ketones, Nitro and azo compounds, and miscellaneous reductive metabolisms
- Hydrolytic reactions of esters and amides, epoxides and arene oxides by epoxide hydrase
♦ Phase II (Conjugation)
- Glucuronic acid conjugation, sulfate conjugation, glycine and other amino acid, glutathione or mercapturic acid, acetylation, methylation
- Define and differentiate between prodrug, soft drug and antedrugs
- Discuss metabolic routes of some individual drugs

Keywords: Antedrug, Conjugation, Drug interactions, Drug metabolism, Enterohepatic circulation, First pass effect, Phase I metabolism, Phase II metabolism, Prodrug, Soft drug.

* **Corresponding author M.O. Faruk Khan**: Department of Pharmaceutical Sciences and Research, Marshall University School of Pharmacy, Huntington, WV, USA; Tel: 304-696-3094; Fax: 304-696-7309; E-mail: khanmo@marshall.edu

INTRODUCTORY CONCEPTS

Roles Played by Drug Metabolism

Metabolism is one of four pharmacokinetic parameters, *i.e.*, absorption, distribution, metabolism and excretion (ADME), with metabolism and excretion together considered as elimination. Kidney, the major excretory organ, primarily excretes polar compounds and with respect to drugs, those that are extensively ionized at urinary pH. On the other hand, the kidney is unable to excrete drugs with high lipid water partition coefficient (LWPC), drugs that are unionized at urinary pH. In general, as a result of metabolism, drugs become more polar, ionizable and thus more water soluble with enhanced elimination. It also effects deactivation and thus detoxification. In some instances, the drugs are metabolically activated, referred to as Prodrugs, especially by phase I mechanism. Overall, metabolism may result in one of the following outcomes: 1) drug inactivation or detoxification, 2) metabolite(s) with similar activity, 3) metabolite with different activity, 4) drug intoxication, or 5) drug activation.

1. Drug Inactivation or Detoxification

When metabolism of a drug or a toxic substance results in the formation of an inactive metabolite, it is termed as drug inactivation or detoxification, respectively. For example, the glucuronide conjugation of the *p*-amino group (N4) or the N atom of the sulfonamide group (N1) of sulfa drugs leads to inactivation. On the other hand, the sulfate conjugation of the toxic phenol leads to detoxification (Fig. **1**). In general, CYP [1], UDP-glucuronosyltransferases [2] and glutathione *S*-transferases [3] play important roles in drug inactivation and detoxifications.

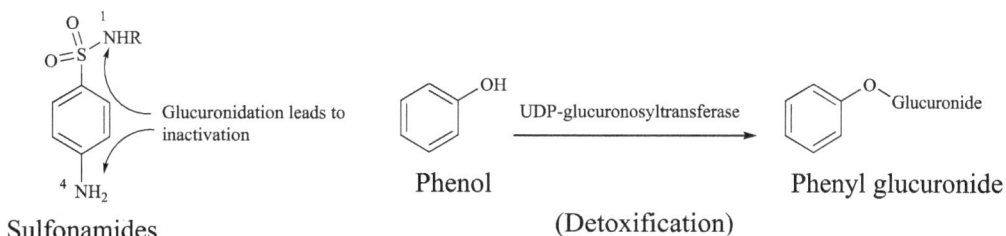

Fig. (1). Inactivation and detoxification reactions.

2. Similar Activity

Metabolism may result in a metabolite with similar activity, but with different potency and/or pharmacokinetic and safety profiles. For example, diazepam is an anxiolytic agent with longer duration of action compared to its active metabolites temazepam and oxazepam (Fig. **2**) [4]. Acetaminophen is an active metabolite of phenacetin with improved pharmacological and toxicological profiles, and so are fexofenadine and desloratadine [5].

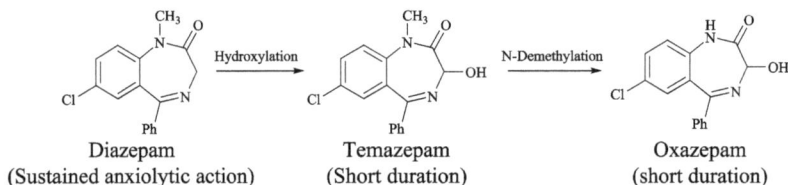

Diazepam
(Sustained anxiolytic action)

Temazepam
(Short duration)

Oxazepam
(short duration)

Fig. (2). Metabolism of diazepam leading to similarly active metabolites.

3. Different Activity

Drug metabolism may also result in a metabolite with entirely new pharmacologic activity. For example, the hydralazine derivatives of monoamine oxidase inhibitors were developed based on the structural modification of isoniazid. The observation that isoniazid made patients euphoric led to develop its isopropyl (lipophilic) analog to deliver it into the brain better. The isopropyl derivative of isoniazid was named iproniazid (Fig. **3**) that was approved in 1958 as an antidepressant agent [6].

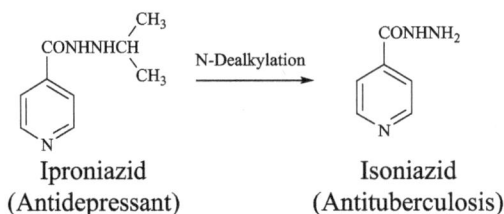

Iproniazid
(Antidepressant)

Isoniazid
(Antituberculosis)

Fig. (3). Iproniazid is antidepressant but its metabolite isoniazid is antitubercular agent.

4. Intoxication

Sometimes metabolism may lead to the formation of toxic products, generally as a result of phase I metabolism. For example, acetylhydrazine and isopropylhydrazine, the metabolites of isoniazid and iproniazid, respectively are hepatotoxic. The CYP450 enzymes are responsible for the production of these toxic metabolites, which are extremely reactive acetylating and alkylating agents (Fig. **4**) [7]. Although the phase 2 metabolism (conjugation) is considered to be involved in detoxification reaction, it can also lead to toxic metabolites. For example, some *N*-glucuronides of arylamines are involved in bladder and colon cancers.

Similarly, GSH conjugation, the most important detoxification mechanism in the humans against xenobiotics, may also cause renal toxicity due to the high level of γ-glutamyl transpeptidase activity in kidneys. For example, the GSH conjugate of 2-bromohydroquinone (in the liver) is converted to a nephrotoxic metabolite, 2-bromo-(di-cystein-*S*-yl) hydroquinone, into the kidneys by the action of γ-glutamyl transpeptidase [8].

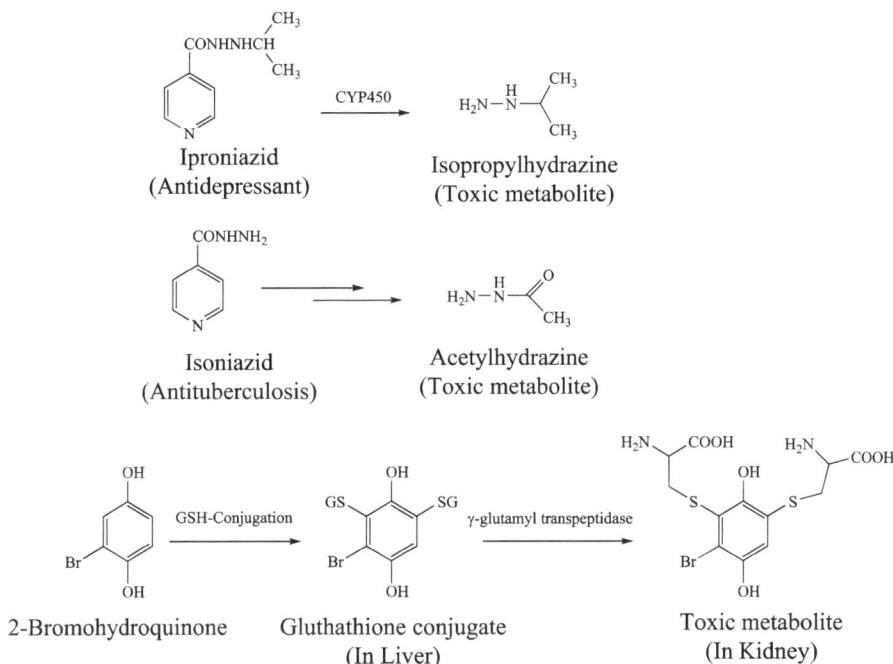

Fig. (4). Drug metabolism producing toxic metabolites.

5. Activation

In some instances, drugs need activation before they exhibit the intended pharmacological action, which is especially true for prodrugs. See the prodrug section later in this chapter for details.

Enzymes Involved in Drug Metabolism

Oxidases

Oxidases are flavoproteins that catalyze oxidation-reduction reactions. In this process molecular oxygen (O_2) is utilized as the electron acceptor forming water (H_2O) or hydrogen peroxide (H_2O_2) as by products. Some of the most important oxidases are *cytochrome c oxidase, cytochrome P450 (CYP), monoamine oxidase, NADPH oxidase*, and *xanthine oxidase*. When an oxidase utilizes two atoms of O_2

to oxidize two different substrates, for example saturated fatty acyl-CoA and NADPH, simultaneously then it is known as *mixed function oxidase*. Hepatic microsomal flavin contains monooxygenases (MFMO or FMO) Oxidize S and N functional groups by a different mechanism than CYP enzyme system. FMOs do not work on primary amines and will not oxidize substrates with more than a single charge nor will they oxidize polyvalent substrates. Monoamine oxidases (MAO) are located in the outer membrane of mitochondria and dehydrogenases are found in cytoplasm, all of which are flavin containing enzymes.

Two MAOs have been identified: MAO–A and MAO–B. Equal amounts are found in the liver, but the brain contains primarily MAO–B. MAO–A is found in the adrenergic nerve endings and shows preference for serotonin, catecholamines, and other monoamines with phenolic aromatic rings. MAO–B prefers non–phenolic amines. MAO metabolizes both 1° and 2° amines if N is attached to α-carbon, and both C & N must have at least one replaceable H atom. 2° amines are metabolized by MAO if the substituent is a methyl group. β–Phenylisopropylamines such as amphetamine and ephedrine are not metabolized by MAOs, and are potent inhibitors of MAOs.

Aldehyde dehydrogenase metabolizes 1° and 2° alcohols and aldehydes containing at least one "H" attached to α-C; 1° alcohols typically go to the aldehyde then acid; 2° alcohols are converted to ketone, which cannot be further converted into the acid. The aldehyde is converted back to an alcohol by alcohol (keto) reductases (reversible), however, it goes forward as the aldehyde is converted into carboxylic acid; 3° alcohols and phenolic alcohols cannot be oxidized by this enzyme; no "H" attached to adjacent carbon.

Cytochrome P450 System

Cytochrome P450 system is the most important enzyme family involved in drug metabolism. It accounts for about 75% of the total number of different metabolic reactions [9]. The endoplasmic reticulum functions to place the various components of this system in the proper 3D configuration. **Cytochrome P450 (CYP)** is a **P**igment that, with CO bound to the reduced form, absorbs maximally at **450**nm and hence is named so. Cytochromes are hemeproteins that function to pass electrons by reversibly changing the oxidation state of the Fe in heme between the 2+, 3+ and 4+ coordinated states (Fig. **5**). CYP is probably best classified as a "heme-thiolate" protein, by the Enzyme Commission. It *serves as an electron acceptor–donor*. CYP is not a singular hemoprotein but rather a family of related hemoproteins. Over 20,000 distinct CYP proteins have been identified in nature with ~57 functionally active in humans, *this may explain its broad substrate specificity* [10].

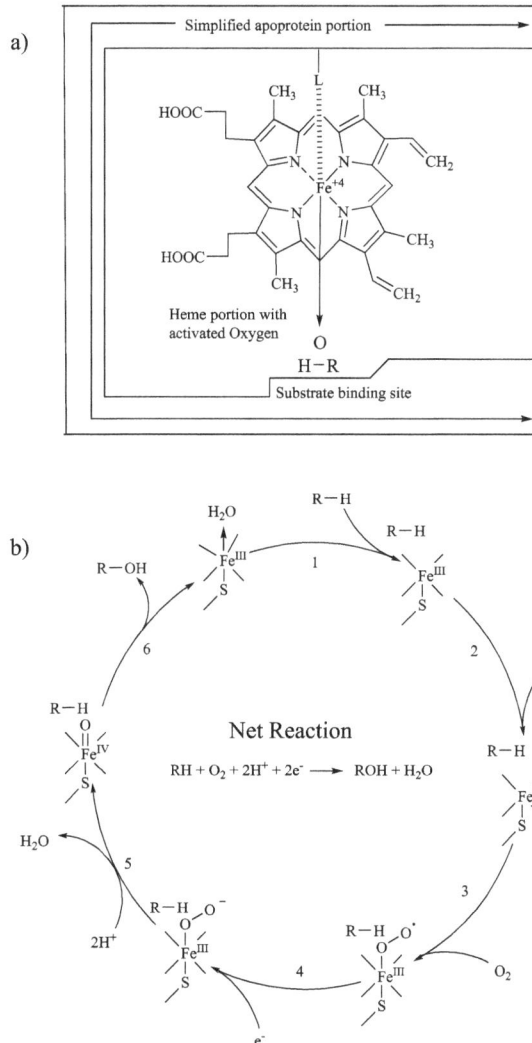

Fig. (5). 1. A) Simplified diagram of CYP showing the substrate binding site and the reactive oxygen species at the heme-iron. B) The mechanism of CYP mediated chain reactions in drug and other substrate metabolism (6-steps): 1. The drug (RH) binds to the substrate binding site of the enzyme displacing a water molecule and inducing a conformational change of the active site; 2. The electron transfer chain favors the transfer of an electron from NAD(P)H reducing the ferric heme iron (Fe^{3+}) to the ferrous state (Fe^{2+}).; Molecular oxygen binds covalently to the distal axial coordination position of the heme iron of the heme-thiolate proteins (CYP) activating the oxygen; 3. The electron-transport system of either cytochrome P450 reductase, ferredoxins, or cytochrome b5 favors the transport of a second electron reducing the dioxygen adduct to a negatively charged, short-lived, peroxo group; 4. The peroxo group is rapidly protonated twice releasing one water molecule, and forming a highly reactive species commonly referred to as **P450 Compound 1** (or Compound I), most likely an iron(IV)oxo (or ferryl) species, which catalyzes a wide variety of reactions; 5. A hypothetical hydroxylation, where the product has been released from the active site, the enzyme returns to its original state, with a water molecule returning to occupy the distal coordination position of the iron nucleus [11 - 15].

Cytochrome P450: Naming

Before we had a thorough understanding of this enzyme system, the CYP450 enzymes were named based on their catalytic activity toward a specific substrate, *e.g.*, aminopyrine N-demethylase now known as CYP2E1. Currently, all P450's are named by starting with "CYP" (**CY**tochrome **P**450, N1, L, N2 - the first number is the family (>40% homology), the letter is the subfamily (> 55% homology), and the second number is the isoform. The majority of drug metabolism is conducted by ~10 isoforms of the CYP1, CYP2 and CYP3 families in humans. The major human forms of CYP, quantitatively in the liver by the percentages of total CYP protein are: CYP3A4 – 28%, CYP2Cx – 20%, CYP1A2 – 12%, CYP2E1 – 6%, CYP2A6 – 4%, CYP2D6 – 4%. By number of drugs metabolized the percentages are: CYP3A4 – 35%, CYP2D6 – 20%, CYP2C8 and CYP2C9 – 17%, CYP2C18 and CYP2C19 - 8% CYP 1A1 and CYP1A2 -10%, CYP2E1 – 4%, CYP2B6 – 3%. The major functions of common CYP isozymes are listed in Table **1** [10].

Table 1. Common CYP isozymes with their functions [10].

CYP Family	Main Functions
CYP1	Xenobiotic metabolism
CYP2	Xenobiotic metabolism, Arachidonic acid metabolism
CYP3	Xenobiotic and steroid metabolism
CYP4	Fatty acid hydroxylation
CYP5	Thromboxane synthesis
CYP7	Cholesterol 7α-hydroxylation
CYP8	Prostacyclin synthesis
CYP11	Cholesterol side-chain cleavage, Steroid 11β –hydroxylation, Aldosterone synthesis
CYP17	Steroid 17α-hydroxylation
CYP19	Androgen aromatization
CYP21	Steroid 21-hydroxylation
CYP24	Steroid 24-hydroxylation
CYP26	Retinoic acid hydroxylation
CYP27	Steroid 27-hydroxylation
CYP46	Cholesterol 24-hydroxylation
CYP51	Sterol biosynthesis

Oxidoreductases

An oxidoreductase is an enzyme that catalyzes the electron transfer from reductant (*i.e.* the electron donor), to the oxidant (*i.e.* the electron acceptor) by utilizing NADP$^+$/NADPH or NAD$^+$/NADH as cofactors. Some of the most common oxidoreductases are: a) alcohol oxidoreductases (the CH-OH group of donors), b) the aldehyde or oxo oxidoreductases (C=O group of donors), c) the CH-CH oxidoreductases (CH-CH group of donors). One classic example of the oxidoreductases is *5-alpha reductase*, a CH-CH oxidoreductase, involved in steroid metabolism including bile acid biosynthesis, androgen and estrogen metabolism.

Transferases

A **transferase** is an enzyme that catalyzes the transfer of a functional group from a donor, often a coenzyme, to an acceptor, often a drug in the case of drug metabolism. Transferases are thus often termed as "*donor:acceptor group transferase.*" Some of the important transferases in drug metabolism are: a) methyltransferase (transfer one-carbon groups), b) acyltransferases (transfer acyl groups), c) sulfotransferases (transfer sulfur-containing groups), d) Glycosyltransferase (transfer glucuronic acid group), *e.g.*, **Uridine 5'-diphosph--glucuronosyltransferase** (UDP-glucuronosyltransferase, UGT). The UGT catalyzes glucuronidation (transfer of a glucuronic acid moiety to drug), a major phase II drug metabolism and is the most important pathway for the elimination of the top 200 drugs.

Hydrolases

Hydrolase catalyzes the hydrolysis of a wide range of chemical bonds and are named as *substrate hydrolase*, or more commonly as *substrate*ase. For example, an *ester*ase hydrolyzes ester bonds to corresponding carboxylic acids and alcohols. Some of the important hydrolases involved in drug metabolism are: esterases, phosphodiesterases, and phosphatase (ester bonds hydrolysis); Proteases/peptidases (peptide bond hydrolysis); amidases, ureases, lactamases (carbon-nitrogen bonds, other than peptide bonds hydrolysis).

Sites of Drug Metabolism

Smooth endoplasmic reticulum of liver cells is the major site of drug metabolism and is well organized with all enzyme systems. The epithelial cells of gastrointestinal tract, skin, kidney and lungs are considered minor sites of drug metabolism. A few examples of drug metabolism in the intestinal mucosa include: isoproterenol undergoes considerable sulfate conjugation in GI tract; levodopa, chlorpromazine and diethylstilbestrol are also reportedly metabolized in GI tract; esterases and lipases present in the intestine may be particularly important at carrying out hydrolysis of many ester prodrugs; bacterial flora present in the intestine and colon reduce many azo and nitro drugs (*e.g.*, sulfasalazine) [16].

First-pass effect

The first pass effect, or first-pass metabolism is the metabolism of drug during the absorption process, generally related to the liver and gut wall, thus greatly reducing the concentration of a drug before it reaches the systemic circulation [17]. The following drugs are metabolized extensively by first-pass effect: isoproterenol, lidocaine meperidine, morphine, pentazocine, propoxyphene, propranolol, nitroglycerin, and salicylamide.

Enterohepatic circulation

Intestinal β-glucuronidase can hydrolyze glucuronide conjugates of some drugs excreted in the bile, thereby liberating the free drug or its metabolite for possible reabsorption from the gastrointestinal tract into the systemic circulation, a phenomenon called *enterohepatic circulation or recycling*. This may result in multiple peaks in the plasma-concentration–time profile prolonging the apparent elimination half-life of the drugs. The common drugs undergoing enterohepatic circulation include: meloxicam, lorazepam, cyclosporin, norethisterone, piroxicam, ezetimibe, chloramphenicol, digitoxin, azithromycin, doxycycline, methotrexate, irbesartan, indomethacin, morphine, warfarin, ceftriaxone and imipramine [18].

Enzyme Induction & Inhibition

Many drugs have the ability to influence the activity of other co-administered drug by the process called enzyme induction or inhibition. The pharmacokinetic drug interactions depend upon: a) the isoform(s) required by the drug in question, b) the isoforms altered by concomitant therapy, c) the type of enzyme alteration (induction or inhibition). A few examples of *enzyme inducers* and *inhibitors* are listed in Table **2** [19, 20].

The enzyme induction is the process by which the rate of a specific enzyme's synthesis, such as CYP450, is increased due to the administration of *enzyme inducers*. As a consequence, co-administration of enzyme inducers may lead to drug-drug interactions. Thus knowledge of such enzyme inducers is clinically significant in evaluating certain unexpected pharmacologic and toxicological profiles of those drugs. As a result of enzyme induction, the co-administered drug, or the inducer drug itself, may be metabolized more rapidly to more potent, more toxic, or inactive metabolites leading to different pharmacological and/or safety parameters than expected under normal conditions. For example, cigarette smoking lowers the serum levels of theophylline, acetaminophen, and propoxyphene; and decreases the drowsiness from the benzodiazepines, diazepam and chlordiazepoxide, due to rapid metabolic elimination. Also, the duration of sedative effect of phenobarbital becomes shorter with repeated doses due to self-metabolic induction, known as auto-induction [19, 20].

Table 2. Enzyme inhibitors and inducers.

Enzyme Inducers	Drugs Affected
Cigarette (benzopyrene)	Benzodiazepines, acetaminophen, propoxyphene, theophylline, clozapine
Ethanol	Ethanol, barbiturates, phenytoin, warfarin
Barbiturates	Barbiturates, phenytoin, warfarin, corticosteroids, doxycycline, digitoxine
Inhibitor (enzyme)	**Drugs affected**
Cimetidine (hepatic microsomal)	Chlordiazepoxide, diazepam, phenytoin, antidepressants, theophylline, lidocaine
Sodium valproate (hepatic microsomal)	Acetaminophen, quinidine, testosterone, carbamazepine, phenytoin, phenobarbital, primidone
Chloramphenicol, Verapamil, diltiazem (MAO)	Theophylline, phenytoin, warfarin, corticosteroids, doxycycline, digitoxine, carbamazepine, cycloserine, carbamazepine
Erythromycin and clarithromycin, Bupropion, oral contraceptives, antifungals (CYP450)	Astemizole, cisapride and theophylline, Clozapine, metoprolol

The enzyme inhibition on the other hand is the process by which the rate of a specific metabolic enzyme synthesis, such as CYP450, is decreased due to the administration of certain drugs (*enzyme inhibitors*). The inhibition of CYP450 can be divided into three major categories: 1) reversible inhibition, 2) metabolite intermediate complexation of CYP450, 3) mechanism-based inactivation of CYP450. Reversible inhibition occurs when the inhibitor drug reversibly binds at the heme-iron active center of CYP450, the lipophilic site or both, and its action abolishes once the drug is discontinued. Cimetidine is a reversible inhibitor of hepatic microsomal CYP450 enzyme and thus affects the metabolism of diazepam, phenytoin and certain antidepressant drugs. When the metabolite of certain drugs forms stable covalent bond with the reduced ferrous heme intermediate, metabolite-intermediate complexation of CYP450 occurs. Macrolide antibiotics, erythromycin and clarithromycin, cause metabolite intermediate complexation of CYP450, and thus inhibit the metabolism of drugs astemizole, cisapride and theophylline. The mechanism based inhibitors generate reactive intermediates by CYP450 mediated oxidation of their functional groups, and irreversibly binding to the same enzyme (*suicidal inhibition*). For example, chloramphenicol forms reactive intermediate oxamyl chloride after oxidation by CYP450 (see later), which irreversibly binds the same enzyme resulting in toxic as well as enhanced activity of other co-administered drugs that are metabolized by this enzyme. As shown in Table **2**, several drugs, such as chloramphenicol, diltiazem and verapamil, also inhibit monoamine oxidase (MAO) [19, 20].

PATHWAYS OF METABOLISM

Drug metabolism has been assembled into phase I and II reactions. In phase I reactions, enzymes carry out oxidation, reduction, or hydrolytic reactions; while in phase II reactions the drug itself or its metabolites form conjugates with small polar endogenous substances. Phase I is often recognized as the functionalization reactions [21] where enzymatic reactions introduce functional groups, *e.g.*, -O-, -OH, -COOH, –NH2, or –SH groups. The addition of functional groups may not dramatically enhance aqueous solubility of the drug, but can significantly alter the biological properties of the drug. Phase I reactions alter pharmacologic activity and may produce biologically inactive or active or even toxic metabolites of the drug. On the other hand, phase II reactions result in metabolites with increased or sometimes decreased water solubility and increased molecular weight of the drug or its phase I metabolites.

Phase I Metabolic Reactions

A general summary of phase I reactions is presented in Table **3**. In general, Phase I reactions introduce or expose a functional group on the substrate molecule. Among the phase I reactions, oxidation is by far the most common metabolic reaction.

Oxidation

The mixed-function oxidases or monooxygenases are enzymes primarily involved in oxidation process. Mechanistically the oxidative process involves a series of steps, which are modulated by specific enzyme systems. Elemental oxygen exists as an unpaired diradical, which is relatively unreactive. It is postulated that reduction of oxygen molecules may develop superoxide radical anion, peroxide, hydroxyl radical, or oxygen atom (Fig. **6**). Any of those species can oxidize a susceptible substrate with the introduction of an oxygen atom. In the reductive process of oxygen, the electrons are supplied by NADPH [22 - 24].

Monooxygenase system can oxidize diverse number of oxidative transformation.

A large number of functional groups that are present in drugs can be oxidized by these enzymes. In an oxidation reaction drugs can gain oxygen, or lose a hydrogen, alkyl group or hetero atom. Oxidation can be followed by non-CYP450 reactions including conjugation or oxidation to ketones or aldehydes, with aldehydes being further oxidized to acids. Hydroxylation of the carbon α to heteroatoms often results in an unstable product which decomposes with cleavage of the carbon-heteroatom bond. This phenomenon is seen especially with N, O and S, which results in N-, S- or O-dealkylation. These form aldehyde (or ketone) plus an amine, hydroxyl or thio compounds. It is important to note that the carbon atom that is hydroxylated must have a replaceable hydrogen atom on it.

Table 3. General summary of phase I reactions.

Oxidative Routes
Aliphatic and Alicyclic Hydroxylation
Alkene and Alkyne Hydroxylation
Aromatic Hydroxylation
N-, O- and S-Dealkylations
Oxidative deamination
N- and S- oxidations
Oxidation of alcohols and aldehydes
Dehalogenation
Reductive Routes
Reduction of aldehyde and ketone
Azo and nitro reduction
Reduction of sulfur containing drugs
Hydrolytic Routes
Ester, amide, carbonate, carbamate, urea and hydrazine hydrolyses
Hydration of epoxides

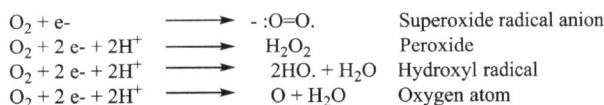

$$O_2 + e- \longrightarrow \; - :O=O. \qquad \text{Superoxide radical anion}$$
$$O_2 + 2\,e- + 2H^+ \longrightarrow H_2O_2 \qquad \text{Peroxide}$$
$$O_2 + 2\,e- + 2H^+ \longrightarrow 2HO. + H_2O \qquad \text{Hydroxyl radical}$$
$$O_2 + 2\,e- + 2H^+ \longrightarrow O + H_2O \qquad \text{Oxygen atom}$$

Fig. (6). Oxygen Activation.

Aliphatic and Alicyclic Hydroxylations

Oxidation is the addition of oxygen and/or the removal of hydrogen (dehydrogenation), while hydroxylation is the introduction of an OH group by oxidation. If a drug can be metabolized either by aliphatic hydroxylation or aromatic hydroxylation, aliphatic hydroxylation is usually predominant. The proposed mechanisms of hydroxylation and dehydrogenation of alkanes is shown in Fig. **7** [25].

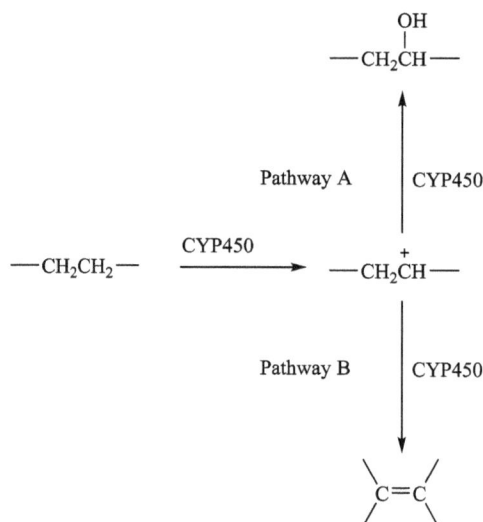

Fig. (7). Proposed mechanisms of alkane hydroxylation and dehydrogenation.

Although alkane groups are generally inert to metabolism, mixed-function oxidases can hydroxylate the β-position of an alkyl chain. Subsequent metabolism can result in formation of a carbonyl compound and also chain shortening.

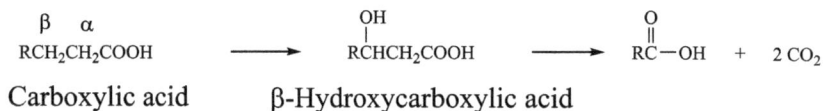

If an aromatic and aliphatic group is present on the same molecule, the following example illustrates the usual outcome.

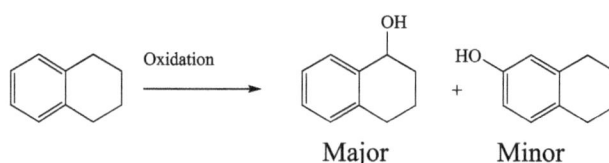

Alkyl side chains may undergo hydroxylation at the terminal carbon atom.

Oxidation at the terminal carbon is called ω oxidation, and oxidation at the carbon next to the last carbon is known as ω-1 oxidation, as shown in Fig. **8**.

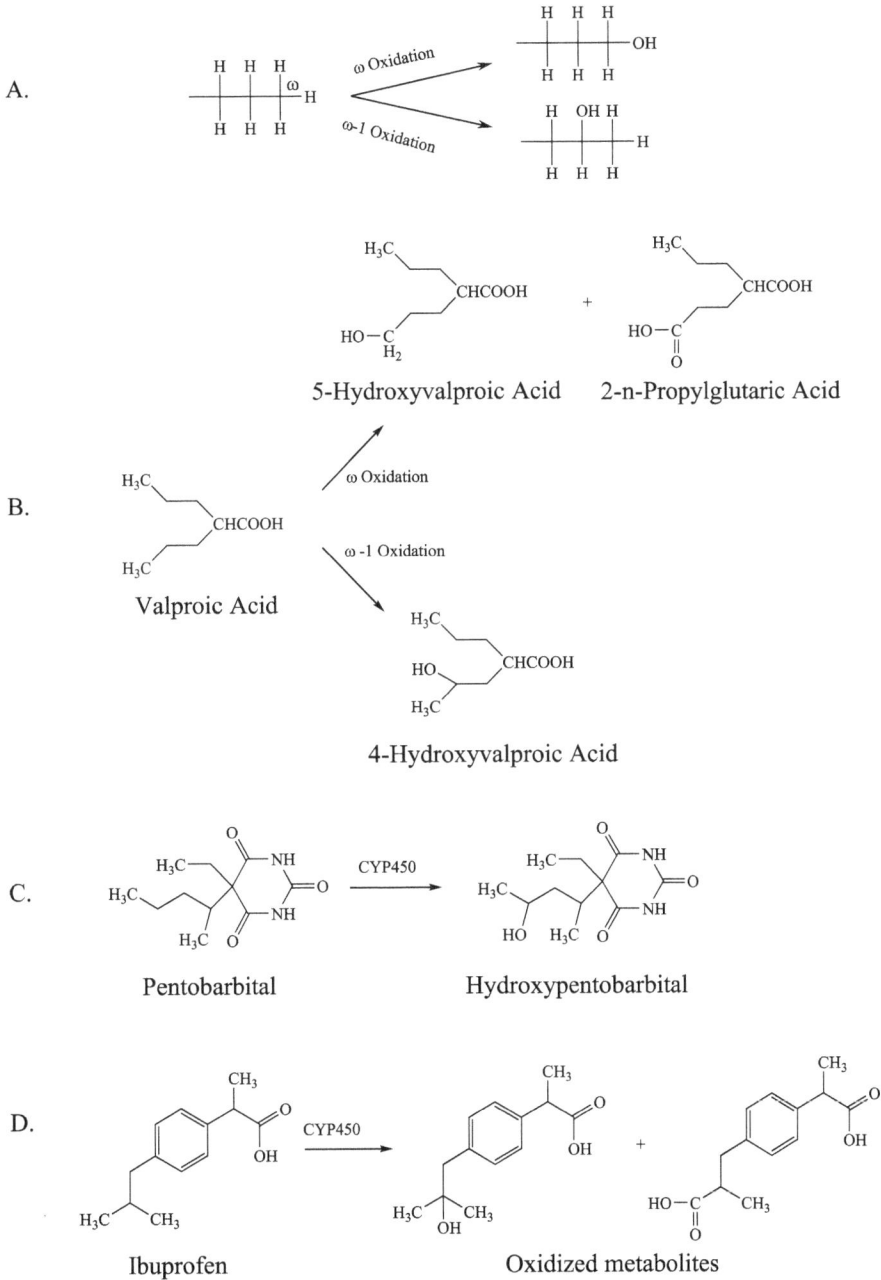

Fig. (8). The ω and ω-1 metabolisms. A. General scheme; B. Metabolism of valproic acid; C. Metabolism of pentobarbital; D. Metabolism of ibuprofen.

It has been reported that valproic acid, pentobarbital, secobarbital, ibuprofen, and many other drugs undergo ω and ω-1 oxidations developing hydroxyl metabolites (Fig. **8**) [26 - 31]. These types of oxidations of aliphatic chains generally need three or more carbon chain lengths and can take place in substrates with straight or branched alkyl chains. Many drug molecules contain cyclohexyl group (alicyclic or nonaromatic ring). The mixed function oxidase tends to hydroxylate at the 3 or 4 position of the ring, as shown in the case of acetohexamide. If position 4 is substituted, it becomes difficult to hydroxylate the molecules due to steric factors.

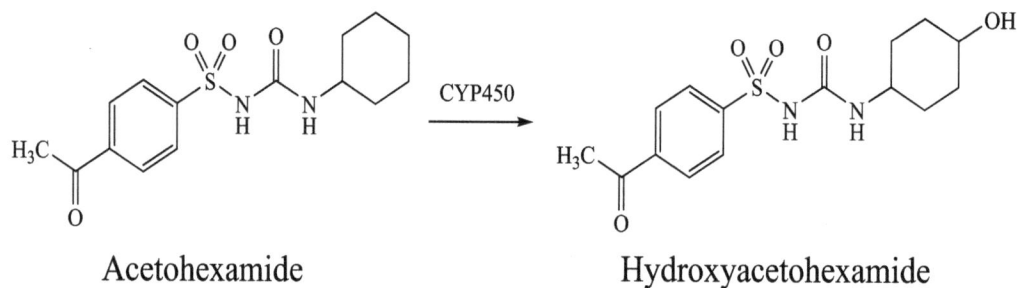

                 Acetohexamide   Hydroxyacetohexamide

Alkene and Alkyne Hydroxylation

Oxidation reactions of alkenes may give cyclic ethers in which both carbons of a double bond become bonded to the same oxygen atom. These products are called epoxides or oxiranes. Some epoxides are unstable but highly reactive and can covalently bind with hepatocytes, particularly with proteins or nucleic acids causing tissue necrosis or carcinogenicity. The reactive epoxides, however, are usually susceptible to enzymatic hydration by epoxide hydrolase to form trans-1,2-dihydrodiols (also called 1,2-diols or 1,2-dihydroxy compounds), as shown in Fig. **9**. Carbamazepine is metabolized to a relatively stable and active epoxide, which undergoes hydration before being excreted in the urine [32, 33].

Alkynes bear a carbon-carbon triple bond. Alkynes can be either terminal or internal; the terminal alkynes are mildly acidic (at the C-H bond). Electron rich alkynes undergo electrophilic reactions and reduction at rates faster than alkenes. The alkyne group undergoes predictable metabolism without adding much to the net size of the molecule. Most alkyne groups appear to be stable to metabolism. It has been suggested that some alkyne groups are oxidized to form very reactive oxirenes (Fig. **10A**). Both alkyne carbons may be attacked and two different reactive intermediates are developed. The reactive intermediates may alkylate protein side chain to form a protein adduct or heme alkylation resulting in enzyme inactivation (Fig. **10B**) [34, 35].

A.

Alkene   Epoxide   Trans dihydrodiol derivative

B.

Carbamazepine   Carbamazepine 10,11 epoxide   Carbamazepine trans 10,11 diol
(Active)   (Active & Toxic)   (Inactive)

C.

Diethylstilbestrol (DES)   DES-epoxide

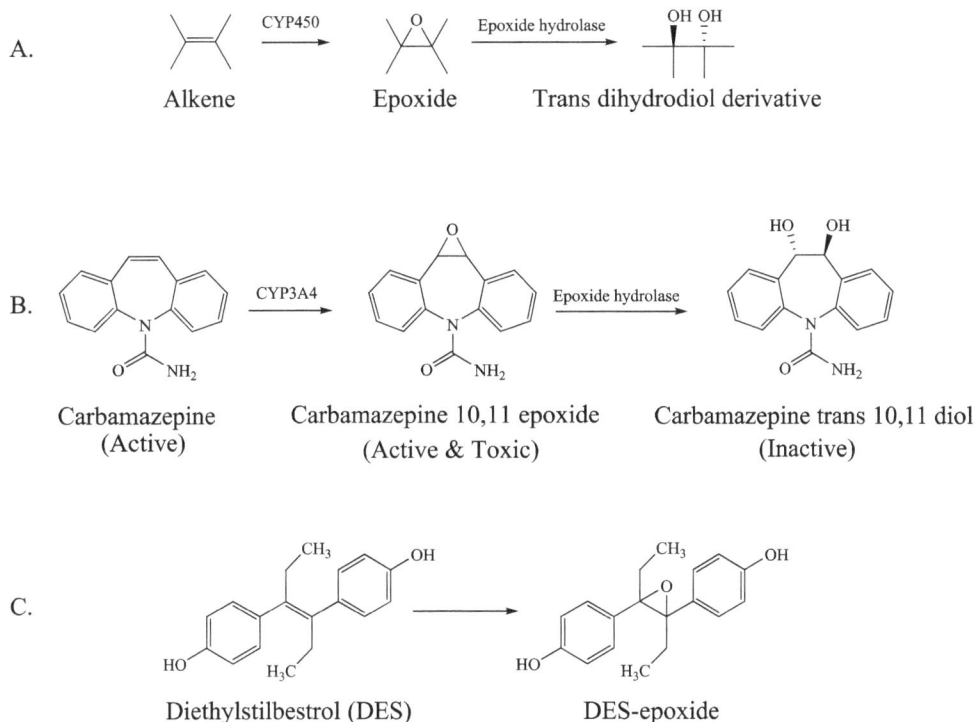

Fig. (9). Alkene hydroxylation. A. General scheme; B. Metabolism of carbamazepine; C. Metabolism of diethylstilbestrol.

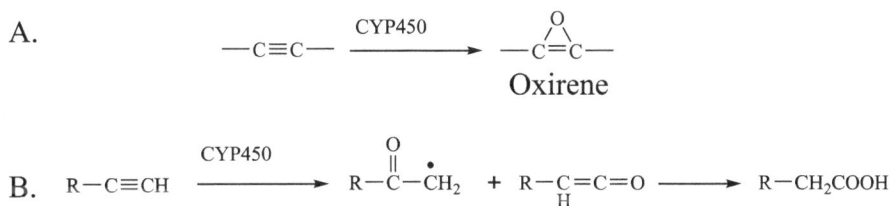

A.

Oxirene

B.

Fig. (10). Oxidation of alkynes.

Aromatic (Arene) Hydroxylation

Aromatic systems are flat structures with hexagonal bond angles (120 degrees). The majority of bioactive substances bear aromatic group. Oxidation of arenes to arenols *via* arene oxide, an epoxide intermediate, is the major route of metabolism of drugs with phenyl ring. The reaction occurs primarily at para position [36, 37]. Arene oxides are formed when a double bond in an aromatic ring undergoes epoxidation. Arene oxides are electrophilic and chemically reactive because of the presence of three membered epoxide ring (Fig. **11**).

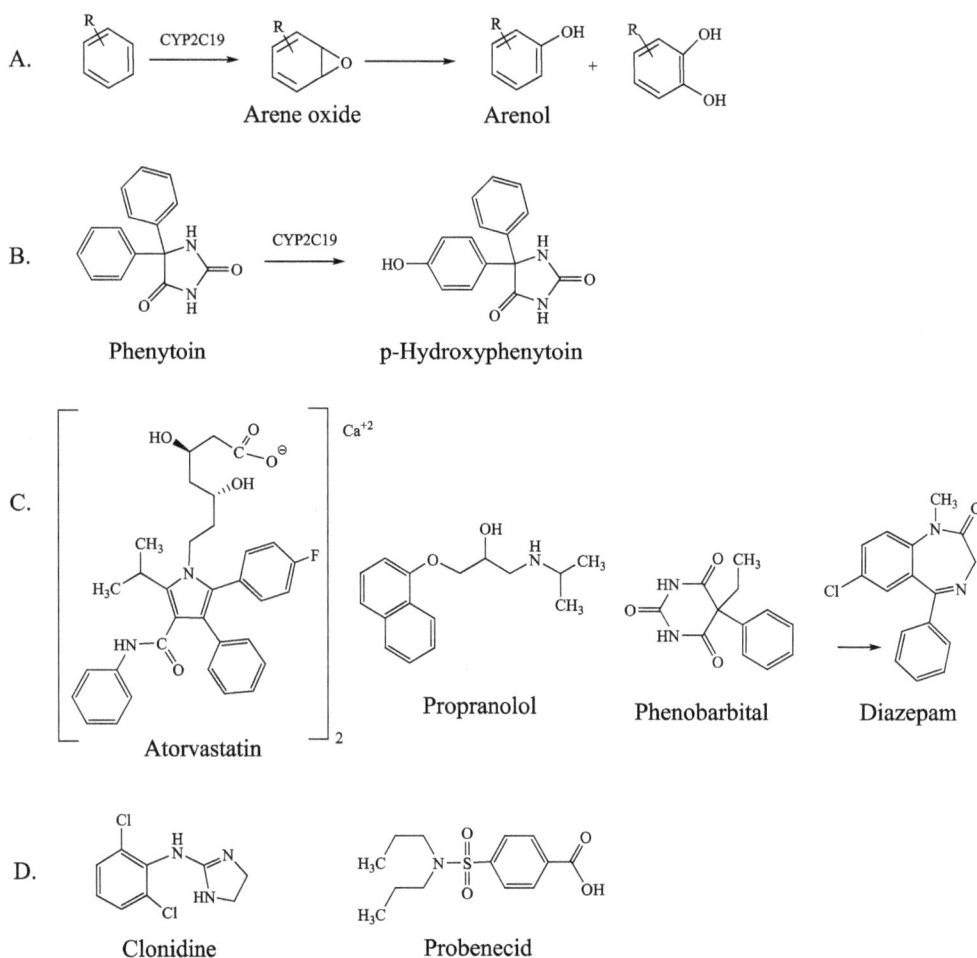

Fig. (11). Aromatic hydroxylation. A. General scheme; B. Aromatic hydroxylation of phenytoin; C. Few drug structures that undergo aromatic hydroxylation; D. Clonidine and probenecid are less likely to undergo aromatic hydroxylation.

Substituents attached to aromatic ring may influence the hydroxylation. Activated rings (with electron-rich substituents) are more susceptible to oxidation and they develop para- and ortho-hydroxylated products. For example, atorvastatin, phenytoin, propranolol, Phenobarbital, and many other important therapeutic agents undergo aromatic hydroxylation (Fig. **11**) [38 - 41]. Deactivated rings (with electron withdrawing groups, *e.g.*, -COOH, -SO$_2$NHR, -Cl, -N$^+$CH, -N$^+$R$_3$,) are generally less susceptible to and undergo meta-hydroxylation. For example, clonidine, an antihypertensive drug, exhibits very little aromatic hydroxylation [42]. Probenecid molecule contains carboxyl and sufamido groups, and as such, it is resistant to aromatic hydroxylation [43]. In addition, steric hindrance decreases the rate of oxidation of aromatic rings. In case of compounds like diazepam with multiple aromatic rings, hydroxylation takes place predominantly in the ring with higher electron density (Fig. **11C**) [44, 45].

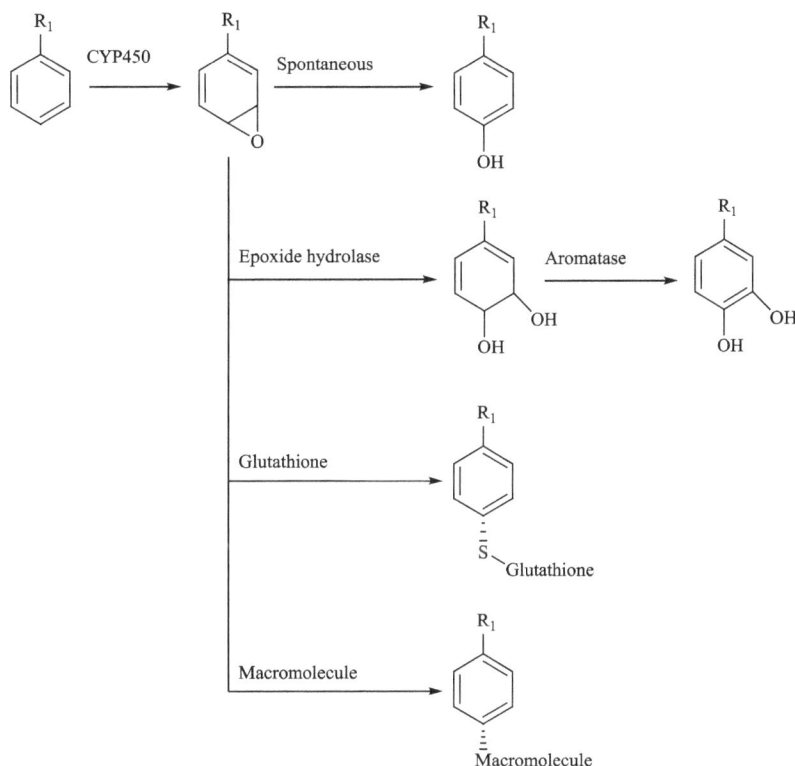

Fig. (12). Possible fates of arene oxide.

Fates of Arene Oxide: NIH Shift

Arene oxides are electrophilic and highly reactive in nature and undergoes several metabolic transformations (Fig. **12**). They can covalently bind with macromolecules or nucleophilic moieties in cells, *e.g.*, protein, DNA, and RNA resulting in tissue necrosis and cell death [46, 47]. Animal studies with bromobenzene indicate its toxic effects on the hepatic tissue. It has been reported that bromobenzene develops 4-bromobenzene oxide, a reactive intermediate, which readily conjugates with glutathione (GSH). At a higher level of exposure to bromobenzene, GSH is depleted and the reactive intermediate covalently binds to hepatic tissue causing necrosis [48]. This is the basis of toxicities of polycyclic aromatic hydrocarbons generated from auto and industrial emissions, cigarette smoke, and other combustion processes. Arene oxides, however, undergo spontaneous rearrangement to less toxic arenols or enzymatic hydration followed by conjugation with glutathione (GSH) before eliminated [49 - 53].

The spontaneous rearrangement of arene oxides to arenols is accompanied by intramolecular hybridization or group migration, usually 1,2-hydride shift (Fig. **13**) [54, 55]. This process was first elucidated by the National Institute of Health (NIH); hence, it is called "NIH shift." The migration of deuterium and hydrogen was reported in the aromatic hydroxylation of specifically deuterated, monosubstituted benzenes catalyzed by ammonia monooxygenase.

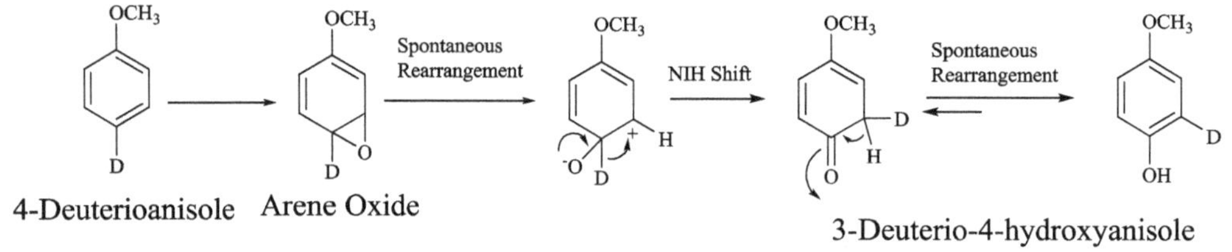
Spontaneous Rearrangement
NIH Shift
Spontaneous Rearrangement
4-Deuterioanisole
Arene Oxide
3-Deuterio-4-hydroxyanisole

Fig. (13). NIH Shift.

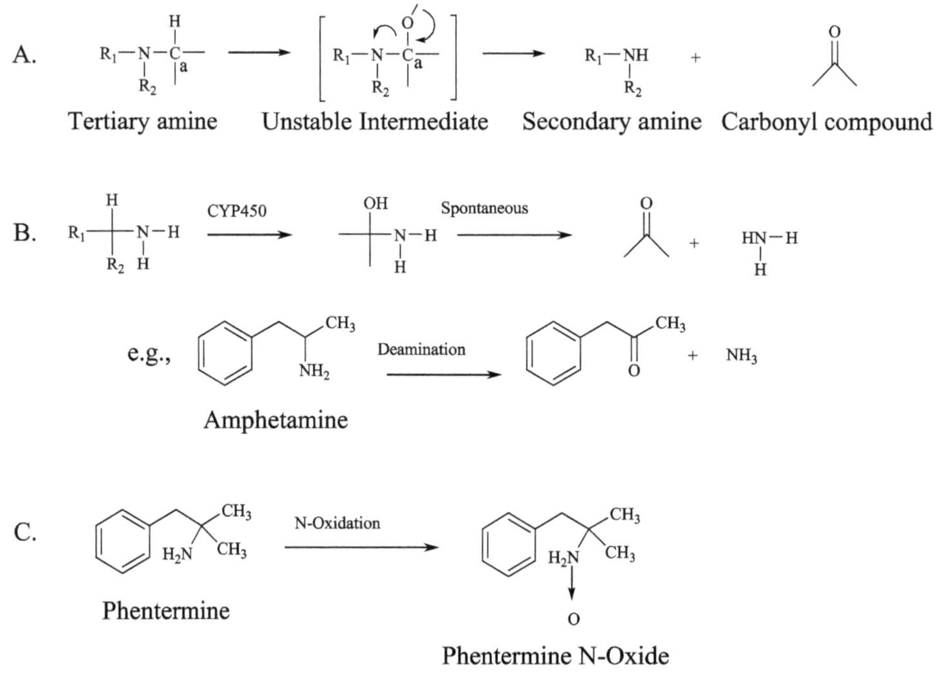
A.
Tertiary amine
Unstable Intermediate
Secondary amine
Carbonyl compound
B.
CYP450
Spontaneous
e.g.,
Deamination
Amphetamine
C.
N-Oxidation
Phentermine
Phentermine N-Oxide

Fig. (14). Oxidation Involving Carbon-Nitrogen Systems. A. *N*-Dealkylation; B. Deamination; C. *N*-Oxidation.

Oxidation Involving Carbon-Nitrogen Systems

All these reactions involve carbon-nitrogen systems. Thus, the drug molecules containing aliphatic, alicyclic, and aromatic amine or amide may undergo these types of metabolisms. A drug molecule may produce more than one metabolite, depending on its and its metabolic intermediate's susceptibility to the available enzyme systems. Oxidative *N*-dealkylation refers to the removal of alkyl groups from aliphatic or alicyclic amines. Dealkylation of secondary and tertiary amines yields primary and secondary amines respectively. *N*-dealkylation involves α-carbon hydroxylation that develops an unstable intermediate, which undergoes heterolytic degradation at C-N bond and develops an amine and a carbonyl compound (Fig. **14**).

Oxidative deamination and *N*-dealkylation differ only in the point of reference, *i.e.*, if the drug molecule contains R1 or R2 then it is termed as deamination reaction; and if the drug contains R3 or R4, it is called *N*-dealkylation as shown in Fig. **14**. Deamination takes place at α-substituted amines, as in amphetamine [56].

Monoamine oxidase (MAO) and diamine oxidase, mitochondrial flavin-containing enzymes, catalyze oxidative deamination reactions of primary and secondary amines where the N-atom is attached to α-carbon and both C and N atoms have at least one replaceable hydrogen atom. Secondary amines are metabolized by MAO if the substituent on the N-atom is a methyl group. Compounds with substitution at the α-carbon atoms are poor substrates of MAO. Two MAOs have been identified: MAO–A and MAO–B. Equal amounts are found in the liver, but the brain contains primarily MAO–B; MAO–A is found in the adrenergic nerve endings. MAO–A shows preference for serotonin, catecholamines, and other monoamines with phenolic aromatic rings and MAO–B prefers non–phenolic amines. β–Phenylisopropylamines such as amphetamine and ephedrine that contain a methyl substituent at the α-carbon are not metabolized by MAOs and are inhibitors of MAOs [57]. Xanthene oxidase is the other non-microsomal flavin-containing oxidase that oxidizes hypoxanthine and other purine bases to uric acid.

In general, *N*-oxidations of primary and secondary amines develop hydroxylamine, while tertiary amines and amides form *N*-oxides. For example, chlorpromazine yields chlorpromazine oxide upon *N*-oxidation [58]. In primary aliphatic amines, for example, amantadine [59] and phentermine [60], when α-carbon hydroxylation cannot take place *N*-oxidation can occur.

N-oxidation of primary aromatic amine forms N-hydroxylamine, which is further oxidized to nitroso product. For example aniline is N-oxidized to develop N-hydroxy product [61]. N-oxidation of heterocyclic nitrogen compounds is not as common as that of aromatic amines. It has been reported that metronidazole develops N-oxide metabolites upon N-oxidation of the heteroatom [62].

A. *N*-demethylation of imipiramine

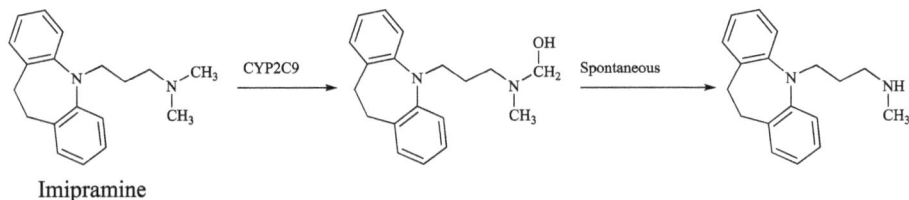

Imipramine

B. Other drug examples that undergo *N*-dealkylation

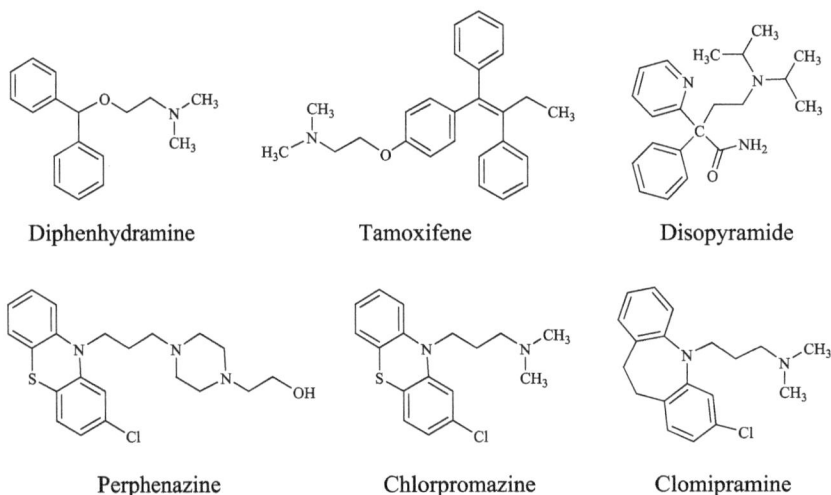

Diphenhydramine   Tamoxifene   Disopyramide

Perphenazine   Chlorpromazine   Clomipramine

Fig. (15). Tertiary amine drugs that undergo N-dealkylation.

Tertiary Amine drugs

Tertiary amines yield secondary amines and secondary amines yield primary amines, but the rate of dealkylation of the former is faster than the later. In general, least sterically hindered carbon (α) will be hydroxylated first. Hence, the more substituents on this C atom, the slower it proceeds. Furthermore, branching on the adjacent carbon slows it down, *i.e.* R1, R2 = H is fastest. Any group containing an α-H may be removed readily, *e.g.*, allyl, benzyl. Substituents without α-hydrogen, *e.g.*, *tert*-butyl are resistant to dealkylation. Quaternary carbon cannot be removed as it does not contain α-H. In addition, the more substituents on the nitrogen, the slower it proceeds due to steric hindrance. Similarly larger substituents slow down the dealkylation process (*e.g.* methyl vs. ethyl). On the other hand, smaller alkyl groups, *e.g.*, methyl, ethyl, isopropyl, n-propyl, n-butyl *etc.* favor dealkylation. There are many drugs that undergo oxidative N-dealkylation. For example, imipramine is demethylated to develop desipramine (Fig. **15**) [63]. Other drugs that undergo *N*-dealkylation include diphenhydramine [64], tamoxifen [65], disopyramide [66], perphenazine [67], chlorpromazine, and clomipramine [68].

Alicyclic 3° amines often generate lactams by α-C hydroxylation reaction as in nicotine and cyproheptadine (Fig. **16**) [69].

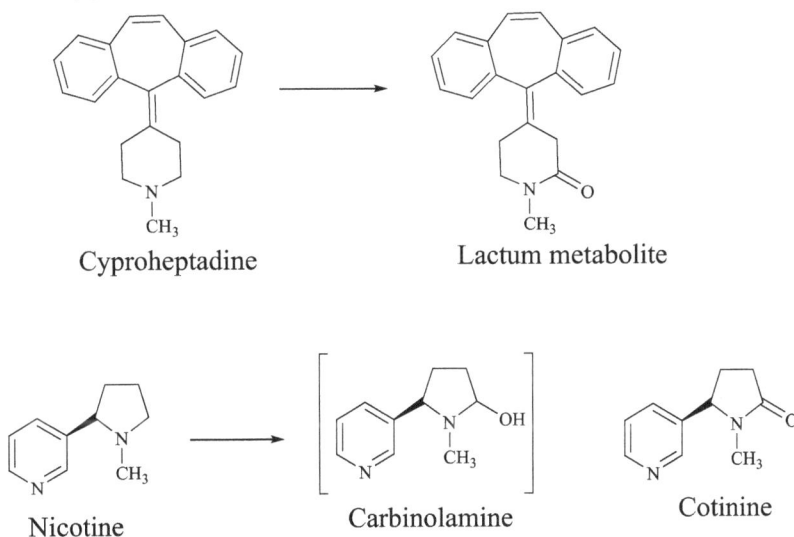

Cyproheptadine   Lactum metabolite

Nicotine   Carbinolamine   Cotinine

Fig. (16). Lactam formation by cyclic amines.

Secondary and Primary Amines

In general dealkylation in secondary amines occurs before deamination. The rate of deamination is easily influenced by steric factors both on the α-C and on the N. Hence, it is easier to deaminate a primary amine than a tertiary amine. Some 2° and 3° amines, however, can undergo deamination directly without dealkylation. 2° amines may undergo N-dealkylation, oxidative deamination, and N-oxidation reactions. For example, methamphetamine, a 2° amine drug, yields amphetamine, which undergoes deamination to develop phenylacetone (Fig. **17**) [70, 71].

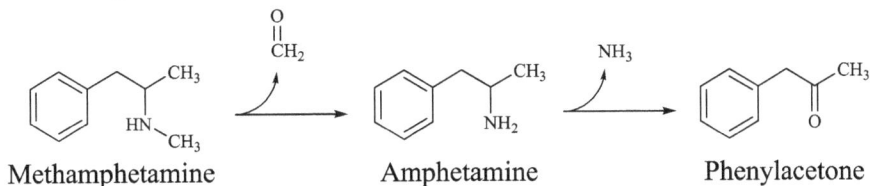

Methamphetamine   Amphetamine   Phenylacetone

Fig. (17). Oxidative deamination of methamphetamine.

Amides Oxidation

Amides may undergo oxidative α-hydroxylation and dealkylation or *N*-hydroxylation. The α-hydroxylation and dealkylation take place at the α-carbon, which is mechanistically similar to the dealkylation of amines. Alkyl group attached to the amide moiety of benzodiazepines are subjected to oxidative *N*-dealkylation. For example, diazepam develops desmethyldiazepam, an active metabolite, after *N*-demethylation (Fig. **18**) [72, 73].

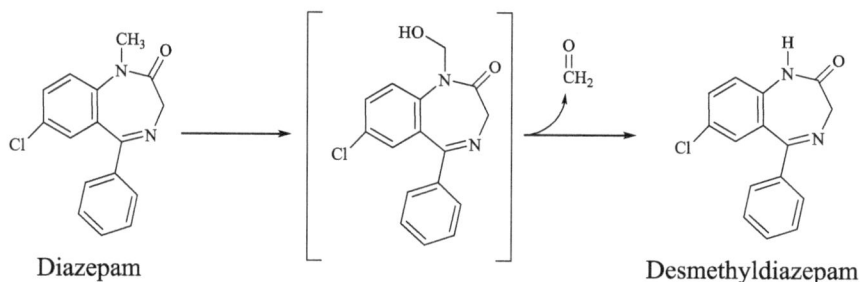

Diazepam　　　　　　　　　　　　　　　　　　Desmethyldiazepam

Fig. (18). *N*-Demethylation of diazepam.

N-Hydroxylation of aromatic amides is usually a minor route of their metabolism. Nevertheless, the *N*-hydroxylation sometimes develops active or toxic metabolites. The classical example in this context is acetaminophen metabolism (Fig. **19**). It has been suggested that acetaminophen undergoes *N*-hydroxylation reaction to form *N*-hydroxyacetaminophen. Spontaneous dehydration of this *N*-hydroxyamide produces *N*-acetylamidoquinone, a reactive metabolite (exact mechanism is not known yet) [74 - 77].

N-Acetylimidoquinone is a strong nucleophile (Fig. **19**). Usually the glutathione (GSH) present in the liver combines with it to form corresponding GSH conjugate. If GSH level is depleted by large doses of acetaminophen, the reactive metabolite covalently binds with liver macromolecules leading to cellular necrosis [78]. In adults, hepatotoxicity may occur after ingestion of a single dose of 150 to 250 mg/kg acetaminophen. Initially nausea, vomiting, anorexia, and abdominal pain occur. Clinical indication of hepatic damage is manifested in 2 to 4 days of ingestion of toxic doses. Initially plasma transaminases are elevated, and the concentration of bilirubin in plasma may be increased; in addition, the prothrombin time is prolonged. Severe liver damage occurs in patients with plasma conc. of acetaminophen greater than 300 mcg/ml at 4 hours and 45 mcg/ml at 12 hours after ingestion [79]. The principal antidote treatment involves administration of sulfhydryl compounds, which probably act, in part, by replenishing hepatic stores of glutathione. Oral *N*-acetylcystine (Fig. **19**) is an effective antidote, if the treatment begins within 24 hours of ingestion of acetaminophen [80 - 82].

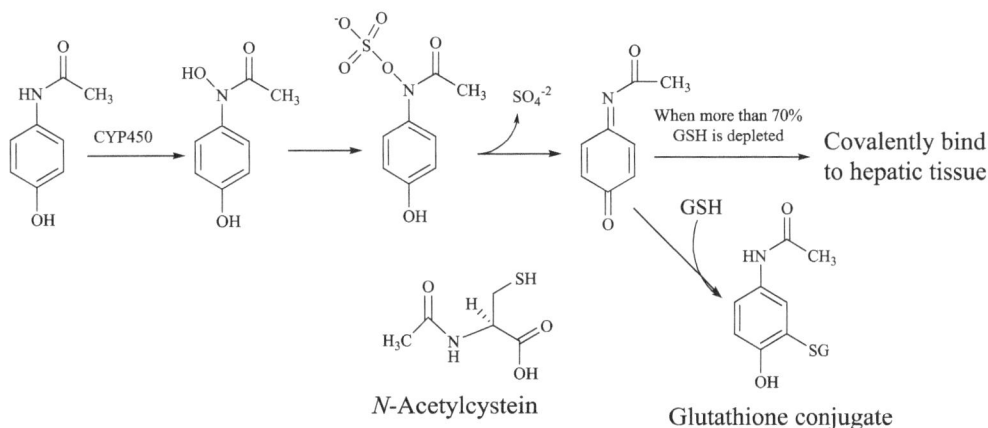

Fig. (19). Acetaminophen toxicity after *N*-hydroxylation, sulfate conjugation, desulfuration and then formation of the toxic *N*-acetylimidoquinone. *N*-Acetyl cysteine is an antidote for such toxicity.

Oxidation Involving Carbon-Oxygen Systems: O-Dealkylation

It is believed that *O*-dealklyation and *N*-dealkylation catalyzed by mixed-function oxidases occur by similar mechanism (Fig. **20**). Both involve an initial α-carbon hydroxylation forming an intermediate, which undergoes spontaneous cleavage of carbon-oxygen bond producing alcohols, phenols, aldehydes, and ketones. One exception that appears to be a form of *O*-dealkylation is the oxidation of ethanol by CYP2E1 [83]. In this case, R3 is hydrogen instead of carbon to form the terminal alcohol rather than ether (Fig. **20**). The enzyme involved is CYP2E1 and has been historically referred to as the microsomal ethanol oxidizing system (MEOS).

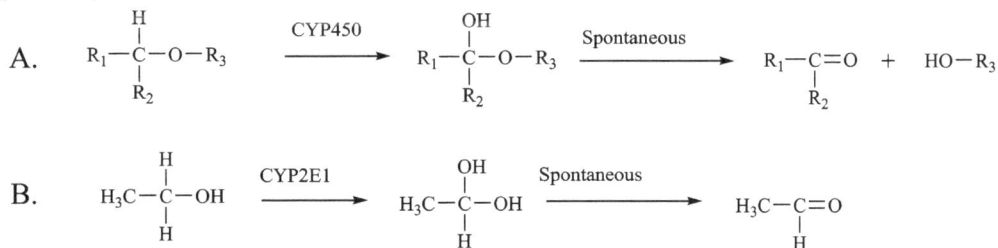

Fig. (20). *O*-Dealkylation. A. General scheme; B. Oxidation of ethanol by CYP2E1.

Oxidative dealkylation appears to be the main rout of metabolism of ether containing drugs (Fig. **21**). Small alkyl groups attached to oxygen are preferentially removed over the larger groups. For example, codeine is *O*-demethylated to morphine [84]. Few examples of drugs that undergo this kind of reaction include indomethacin [85], metoprolol [86], prazosin [87], and trimethoprim [88, 89].

A. *O*-Demethylation of trimethoprim

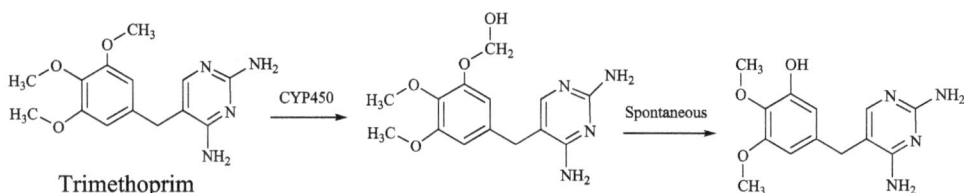

B. *O*-Demethylation of codeine

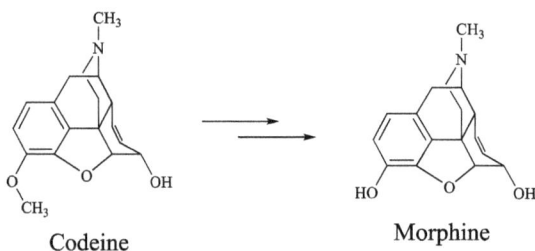

C. Other drugs that undergo *O*-dealkylation

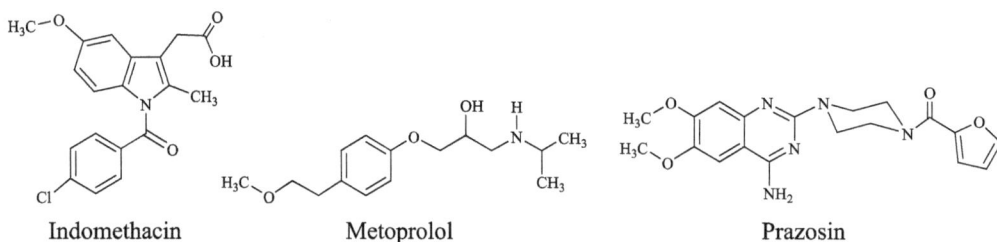

Indomethacin   Metoprolol   Prazosin

Fig. (21). *O*-Dmethylation of some common drugs. A. Demethylation of trimethoprim; B. Demethylation of codeine; C. Other drug structures that undergo demethylation.

Oxidation of Carbon-Sulfur Systems: S-Dealkylation, Desulfuration, and S-Oxidation

S-deakylation involves an α-carbon hydroxylation (Fig. **22**) similar to *N*- and *O*-dealkylation. Aliphatic and aromatic methyl thioethers undergo *S*-dealkylation to thiol and carbonyl compound. For example, 6-methylpurine yields 6-mercaptopurine upon *S*-dealkylation [90].

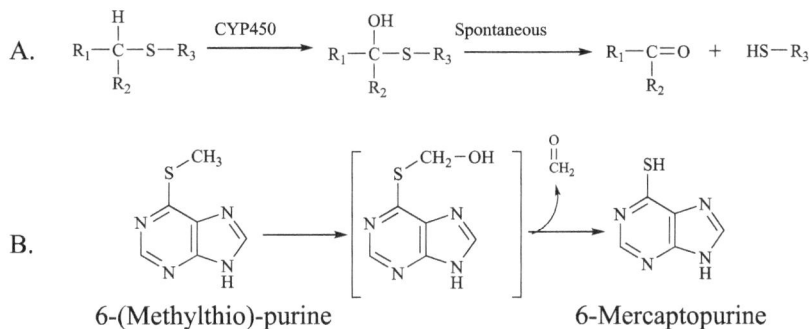

Fig. (22). *S*-Dealkylation. A. Mechanism of *S*-dealkylatio; B. *S*-Dealkylation of methylthiopurine.

Desulfuration refers to the conversion of carbon-sulfur or phosphorus-sulfur double bonds to carbon-oxygen or phosphorus-oxygen double bonds. For example, thiopental undergoes oxidative desulfuration to develop pentobarbital [91]. Desulfuration of parathion yields paraxenon, active metabolite with anticholinesterase activity [92]. Similarly, *N,N′,N″*-triethylenethiophosphoramide (thio-TEPA) is desulfurated to its active metabolite, TEPA (Fig. **23**) [93].

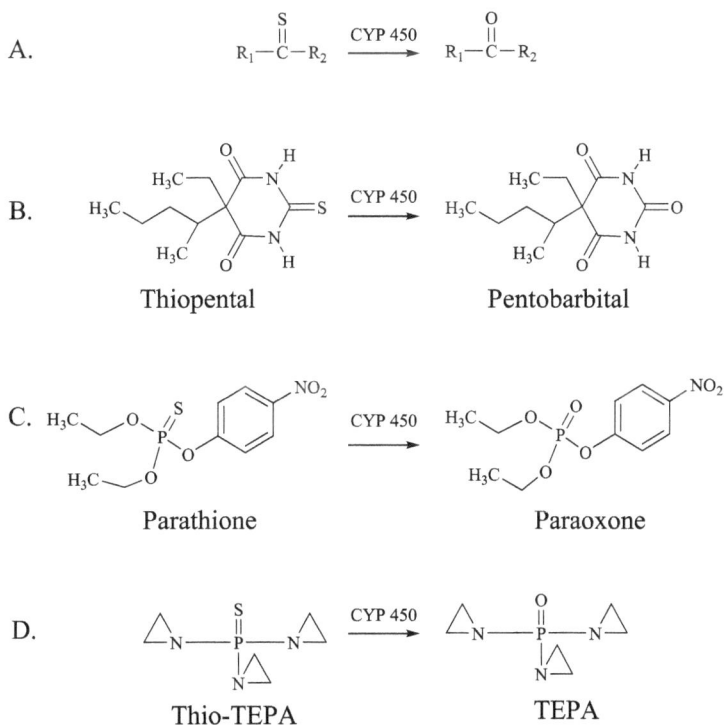

Fig. (23). Desulfuration reaction. A. General scheme; B. Desulfuration of thiopental; C. Desulfuration of parathion; D. Desulfuration of thiotepa.

In addition to *S*-dealkylation, sulfides may also undergo *S*-oxidation to form sulfoxides, which is further oxidized to sulfones (Fig. **24**). Thioridazine undergoes *S*-oxidation at 2-methylthio group developing two major active metabolites sulphoxide and mesoridazine, which are further oxidized to sulphone, sulforidazine [94].

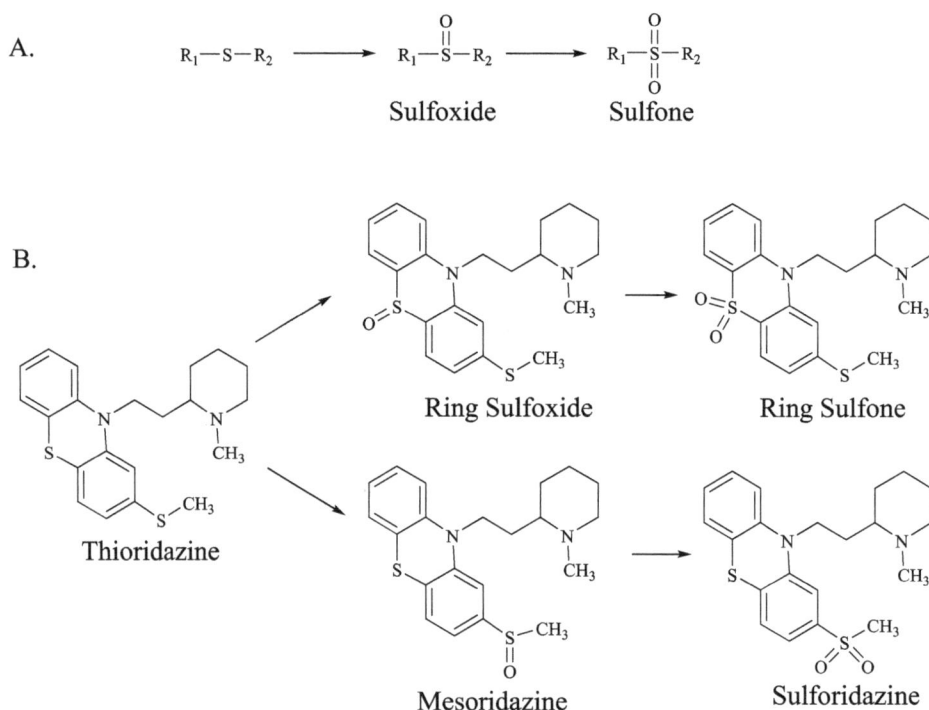

A.

B.

Thioridazine

Ring Sulfoxide

Ring Sulfone

Mesoridazine

Sulforidazine

Fig. (24). S-Oxidation. A. General scheme; B. Thioridazine metabolism.

Dehalogenation

The presence of a halogen on a molecule imparts a dipole owing to its electron withdrawing (electronegativity) characteristics. Among the common halogens, fluorine is the most electronegative and iodine the least. The carbon directly attached to the halogen experiences the greatest effect of the halogen and is considered to be partially electron deficient. This is the reason that substitution reactions occur at this carbon. In addition to the dipole features, halogens add lipophilicity to a drug molecule. In general, halogens cannot form hydrogen bond and this drawback renders them more lipophilic.

Although halogenated drugs are not readily metabolized, many halogenated hydrocarbons are metabolized by different dehalogenation reactions (Fig. **25**). Many drugs, pesticides, industrial solvents, and flame retardants contain halogen, which may develop toxic metabolites after dehalogenation [95]. Many halogenated hydrocarbons undergo oxidative dehydrohalogenation biotransformation [96]. The oxidation generates reactive intermediates, acyl or carbonyl halides, which can covalently bind with cellular proteins causing toxicity or react with water to form carboxylic acid. For example, oxidative dehalogenation of chloramphenicol yields a reactive oxamyl chloride that can acylate microsomal proteins or react with water to form oxamic acid (Fig. **25**) [97, 98].

Halothane undergoes hydroxylation reaction by CYP450 to form a carbinol intermediate (Fig. **25**), which undergoes dehalogenation to generate trifluoroacetyl chloride, a strong electrophilic species. Trifluoroacetetyl chloride can covalently bind with microsomal proteins resulting in tissue necrosis [99, 100]. It has been suggested that chloroform is metabolized by oxidative dehalogenation to yield phosgene, which is responsible for hepato- and nephro-toxicities [101 - 103].

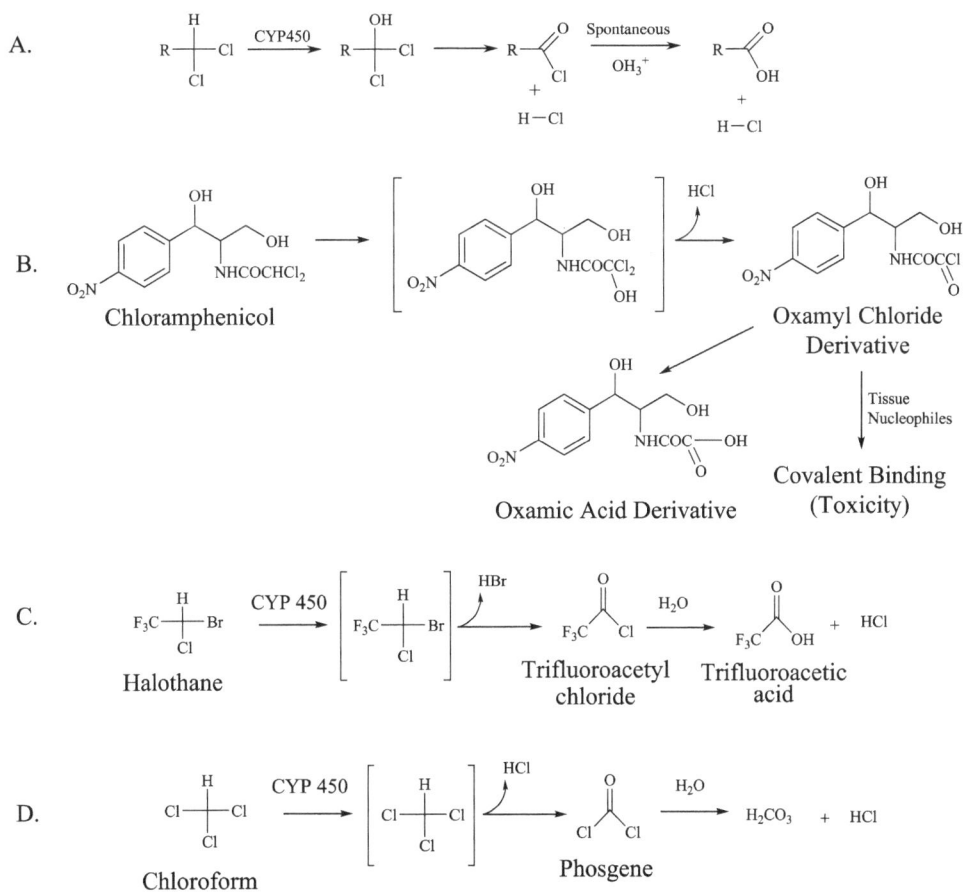

Fig. (25). Oxidative dehalogenations. A. General scheme; B. Chloramphenicol metabolism; C. Halothane metabolism; D. Chloroform metabolism.

Oxidation of Alcohols and Aldehydes

One of the important physicochemical properties of alcohols is their ability to form hydrogen bond. The unequal sharing of electrons between the oxygen and the hydrogen in alcohol molecules renders this property. Thus, alcohol-containing drugs can hydrogen bond to water, receptors, enzymes or other hydrogen bond donors and acceptors. When two hydroxy groups are present on a molecule it is called a diol; for three it is triol; and then polyol.

In phenols (Ph-OH) hydroxyl groups are directly attached to the benzene nucleus. Phenols also form hydrogen bond, but cause a localized acidity when hydrogen bonding occurs. The aromatic portion of a phenol is lipid and the hydroxyl group is a good hydrogen bonding group; hence, phenols tend to have two properties at odds. Phenol is a unique class of alcohols and has a pKa near 10 (weakly acidic). When two hydroxy groups are situated ortho or para on a benzene ring they are called ortho and para hydroquinones, respectively.

The alcohol functional group is relatively stable but it is particularly susceptible to oxidation (primary and secondary alcohols) and dehydration (especially tertiary alcohols). Primary alcohols are usually oxidized to aldehydes at a relatively rapid rate. Aldehydes are further metabolized to carboxylic acids. On the other hand, oxidation of secondary alcohols to ketones occurs at a slower rate. Ketones are often rapidly reduced back to alcohol. Thus, the oxidation of secondary alcohol is not as important as that of primary alcohols. Tertiary alcohols are resistant to oxidative metabolism.

Oxidation of alcohol (Fig. **26**) is catalyzed predominantly by alcohol dehydrogenase (ADH), an NAD-specific enzyme, available in the mitochondrial and soluble fraction of tissue homogenates. The minor microsomal ethanol oxidizing system (MEOS) pathway contributes in the metabolism when there is a high blood alcohol levels [104, 105]. Thus, ADH is considered as non-microsomal enzyme. ADH has broad specificity, catalyzing various alcohols and steroids and catalyzing the oxidation of fatty acids.

Ethanol is metabolized by ADH to acetaldehyde (Fig. **26**), which is further oxidized to acetic acid. Thereafter enzymatic reaction converts acetic acid to acetyl-CoA, which directly enters into the *citric acid cycle*. Continued and excess exposure to ethanol results in hepatic injury that involves ethanol metabolism *via* CYP2E1 and consequent oxidant stress. In addition, direct effects of ethanol on membrane proteins, independent of ethanol metabolism, are responsible for liver damage [106, 107].

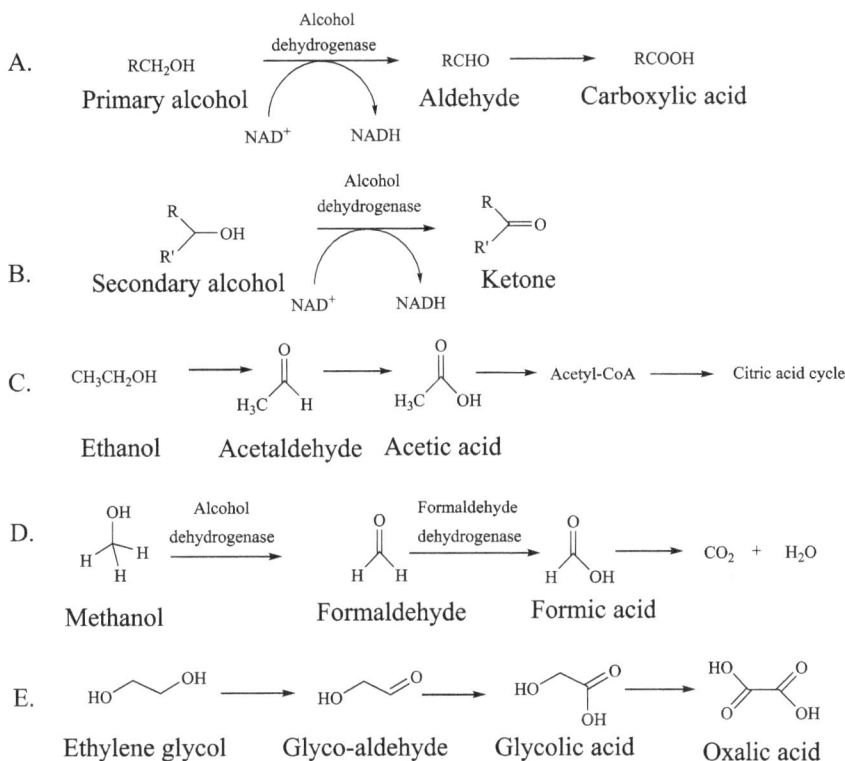

Fig. (26). Oxidation of alcohols and aldehydes. A. Oxidation of primary alcohol; B. Oxidation of secondary alcohol; C. Ethanol oxidation; D. Methanol oxidation; E. Ethylene glycol oxidation.

Methanol is oxidized by ADH to formaldehyde, which is further oxidized to formic acid. Formic acid is further oxidized to carbon dioxide and water (Fig. **26**). The rate of elimination of methanol is relatively slow compared to that of ethanol. Excessive accumulation of formic acid is responsible for toxicities of methanol that include metabolic acidosis and blindness [108].

Ethylene glycol, an organic solvent, widely used as an automotive antifreeze and a precursor to polymers. It is an odorless, colorless, sweet-tasting liquid. Ethylene glycol is toxic, and ingestion can result in death. Oxidation of ethylene glycol by ADH yields glyco-aldehyde, which is further oxidized to glycolic acid. Glycolic acid is responsible for metabolic acidosis. Glycolic acid is metabolized by various pathways, but the major one is the formation of oxalic acid which rapidly forms complex with calcium and precipitates as calcium-oxalate in various tissues and in the urine. It has been reported that calcium oxalate crystals are responsible for the membrane damage and cell death in normal human and calcium oxalate accumulation in the kidney is responsible for the renal toxicity associated with ethylene glycol exposure. Ethylene glycol toxicity is complex and not fully understood, but is mainly due to the severe metabolic acidosis caused by glycolic acid and to the calcium oxalate precipitation [109, 110]. 4-Methylpyrazole (Fomepizole), an alcohol dehydrogenase inhibitor, is recommended as first-line antidote in methanol and ethylene glycol toxicities.

Reductive Reactions

Bioreduction of carbonyl compounds (C=O, *e.g.*, aldehyde and ketone) generates alcohol (aldehyde → 1° alcohol; ketone → 2° alcohol), while Nitro and azo compounds yield amino derivatives. Reduction of N-oxides to their corresponding 3° amines and reduction of sulfoxides to sulfides are less frequent. Reductive cleavage of disulfide (-S-S-) linkages and reduction of C=C are minor pathways in drug metabolism. Reductive dehalogenation is a minor reaction primarily differs from oxidative dehalogenation in that the adjacent carbon does not need to have replaceable hydrogen and generally removes one halogen from a group of two or three. Table **4** presents reductive reactions of some functional groups.

Table 4. Classes of substrates for reductive reactions.

Functional group	Product
$\underset{R \quad R_1}{\overset{O}{\|\|}}$ (ketone/aldehyde)	$\underset{R \quad R_1}{\overset{OH}{\|}}$
$\underset{R}{\overset{O}{\|\|}}N \quad \underset{R}{\overset{O}{\|\|}}N=O$	$\underset{R}{\overset{OH}{\|}}NH$
$\underset{R}{\overset{OH}{\|}}NH$	$R-NH_2$
$R-N=N-R_1$	$R-NH_2 \ + \ H_2N-R_1$
$R-X$	$R^+ \ + \ X^-$

Reduction of Aldehydes and Ketones

Many drugs contain C=O functional group, especially the ketones. In addition, aldehydes and ketones often arise from deamination. A number of soluble enzymes found in liver and other tissues called aldo-keto oxidoreductases carry out bioreduction of aldehydes and ketones. Those enzymes require NADPH as coenzyme. Ketones tend to be converted to alcohols which can then be eliminated by glucuronidation or other processes. Aldehydes can also be converted to alcohols, but can undergo additional pathway of oxidation to carboxylic acids.

Reduction of ketones often leads to the creation of an asymmetric center and thus two stereoisomeric alcohols are possible. Biotransformation of α, β-unsaturated ketones found in steroidal drugs results in reduction of ketones and C=C groups (Fig. **27**).

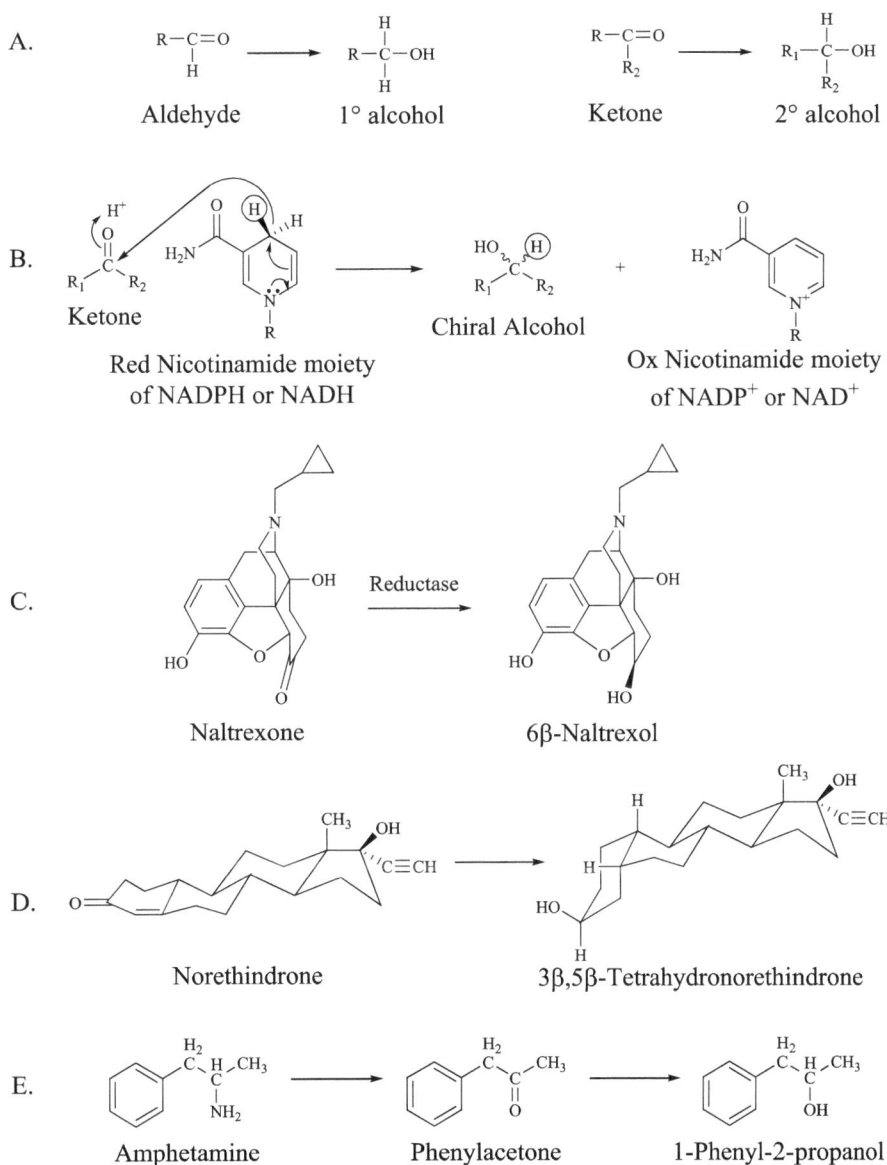

Fig. (27). Reduction of aldehyde and ketones. A. General scheme; B. General mechanism of NADH or NADPH catalysis; C. Reduction of naltrexone; D. Reduction of norethindrone; E. Reduction of oxidized metabolite of amphetamine.

Naltrexone, an opioid, is reported to be reduced to 6β-naltrexol in human [111]. Levonorgestrel and norethindrone undergo extensive reduction of the alpha, beta-unsaturated ketone in ring A [112]. Oxidative deamination of amphetamine develops phenylacetone (a non-aromatic ketone), which is reduced to 1-phenyl-2-propanol (Fig. **27**) [113, 114].

Nitro and Azo Reduction

Aromatic nitro reduction catalyzed by CYP450 in presence of NADPH requires anaerobic conditions. Bacterial nitro and azo reductases in the gastrointestinal tract also catalyzes nitro and azo groups' reduction. The reduction of nitro group to nitroso group is the rate-limiting step. For example, Clonazepam, an anticonvulsant, undergo nitro reduction to its corresponding amine (Fig. **28**) [115].

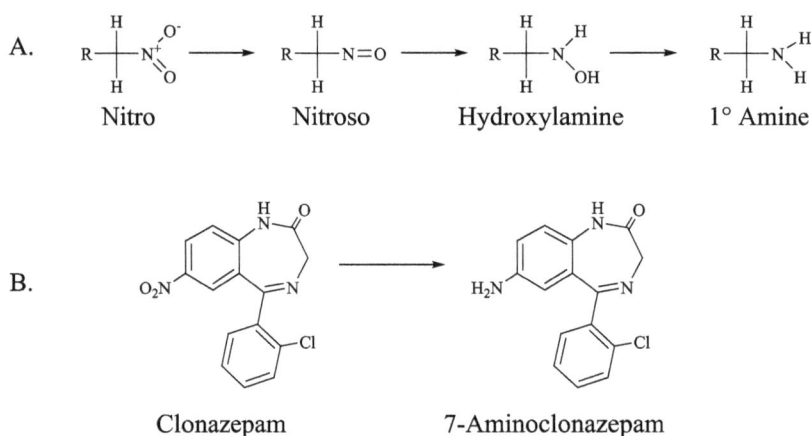

Fig. (**28**). Nitro reductions. A. General scheme; B. Reduction of clonazepam.

Reduction of azo group (RN=NR) is catalyzed by cytochrome P450 and by NADPH-cytochrome P450 reductase under anaerobic conditions. Bacterial azo reductase in gastrointestinal tract mediates this type of reaction. For example, prontosil, sulfasalazine, nitrofurazone are reduced to primary amines by azoreductase. Olsalazine is converted to 5-amino salicylic acid by intestinal bacterial azo reductase (Fig. **29**) [116].

Azido Reduction

Azides may be converted to amines by cytochrome P450 isoenzymes and NADPH-cytochrome P450 reductase (Fig. **30**). The anti-HIV drug 3'-azido--'-deoxythymidine (Zidovudine or AZT) undergoes reductive reaction to a toxic metabolite 3'-amino-3'-deoxythymidine (AMT) [117].

Fig. (29). Azoreduction. A. General scheme; B. Reduction of prontosil; C. Other drugs that undergo similar reduction.

Fig. (30). Azido reduction scheme. Zidovudine undergoes similar reductive metabolism.

Reduction of Sulfur Containing Compounds

Disulfide and sulfoxide, for example disulfiram (Antabuse®), undergo reductive reactions while sulfones do not as shown in the schemes in Fig. **31** [118].

Sulfoxide reduction (cannot reduce a sulfone)

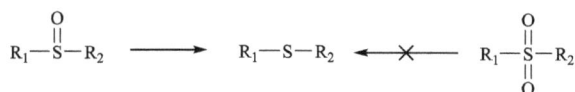

Disulfide reduction

$$R_1-S-S-R_2 \longrightarrow R_1-SH \quad + \quad HS-R_2$$

Disulfiram reductive

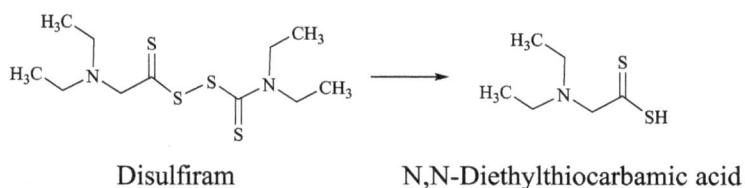

Disulfiram   N,N-Diethylthiocarbamic acid

Fig. (31). Reduction of sulfur containing compounds.

Reductive Dehalogenation

Under anaerobic conditions, halothane, an anesthetic agent, undergoes reductive dehalogenation to 2-chloro-1,1-difluoroethene by CYP450 [119]. The intermediate product, 1-chloro-2,2,2-trifluoroethyle radical, however, is either further reduced to 2-chloro-1,1,1-trifluoroethene or covalently bind with hepatocytes resulting in hepatitis (Fig. **32**). It is notable that halothane may also undergo oxidative dehalogenation by a different mechanism.

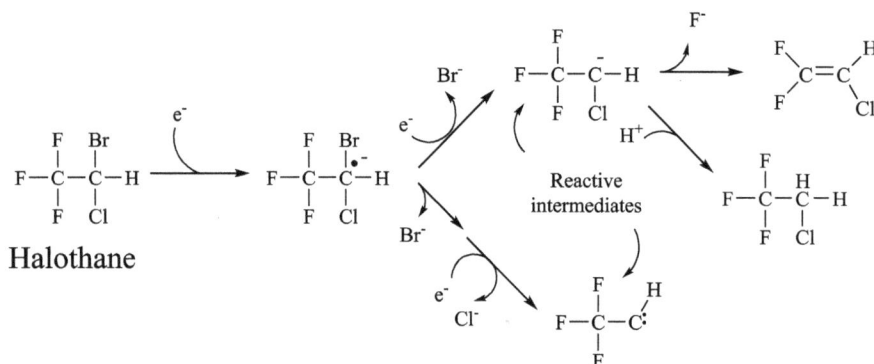

Halothane

Fig. (32). Reductive dehalogenation of halothane.

Hydrolytic Reactions

Esters, amides and their isosteres undergo hydrolytic degradation leading to formation of corresponding carboxylic acid, alcohols, and amines. The schemes of hydrolysis of various groups are presented in Fig. **33**. Hydrolysis of those compounds is often catalyzed by various carboxylases, *e.g.*, carboxylesterase, aryl carboxylesterase cholinesterases. The degree of susceptibility to enzymatic hydrolysis depends on the presence of substituents in the structure of a drug. In general amides are more stable to enzymatic hydrolysis than are esters (Fig. **33**). Thus, hydrolysis of procainamide is slow relative to that of procaine (Fig. **34**). Steric hindrance in an ester slows its hydrolysis.

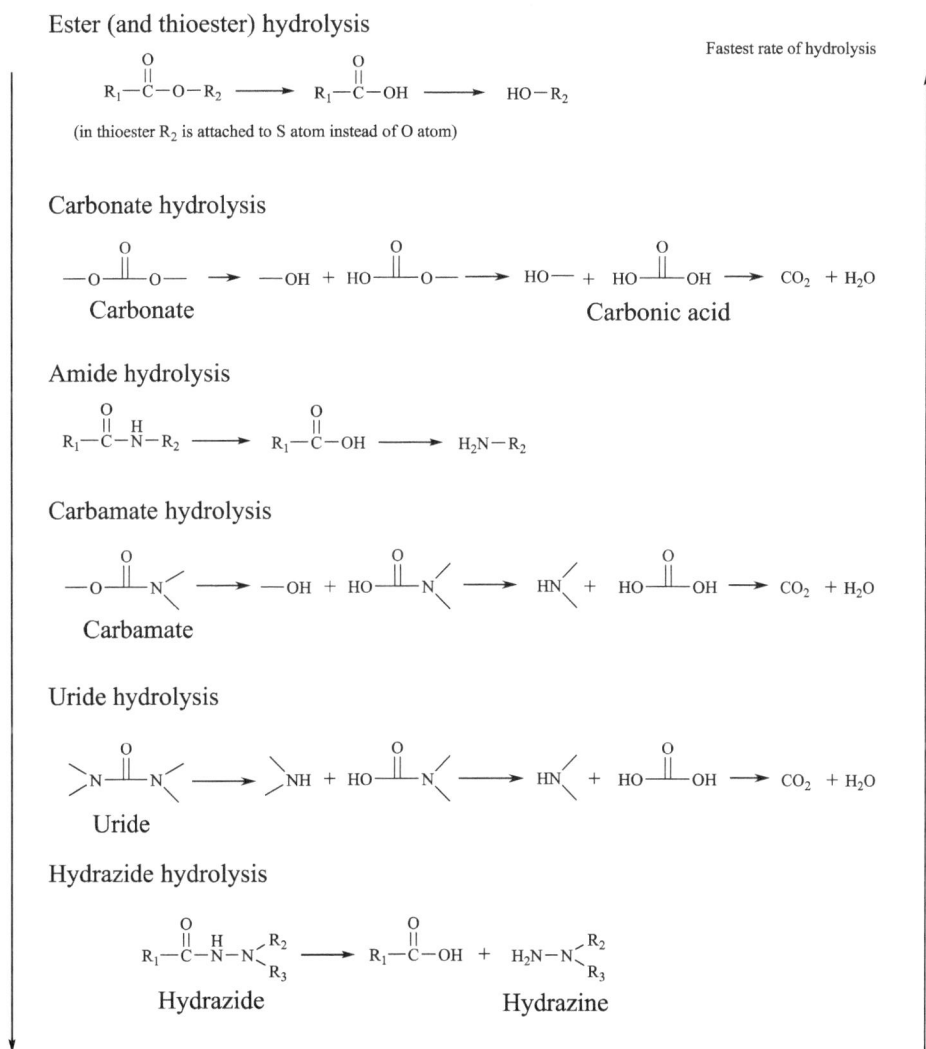

Ester (and thioester) hydrolysis

$$R_1-\overset{\overset{\text{O}}{\|}}{C}-O-R_2 \longrightarrow R_1-\overset{\overset{\text{O}}{\|}}{C}-OH \longrightarrow HO-R_2$$

Fastest rate of hydrolysis

(in thioester R_2 is attached to S atom instead of O atom)

Carbonate hydrolysis

$$-O-\overset{\overset{\text{O}}{\|}}{C}-O- \longrightarrow -OH + HO-\overset{\overset{\text{O}}{\|}}{C}-O- \longrightarrow HO- + HO-\overset{\overset{\text{O}}{\|}}{C}-OH \longrightarrow CO_2 + H_2O$$

Carbonate   Carbonic acid

Amide hydrolysis

$$R_1-\overset{\overset{\text{O}}{\|}}{C}-\overset{\overset{\text{H}}{}}{N}-R_2 \longrightarrow R_1-\overset{\overset{\text{O}}{\|}}{C}-OH \longrightarrow H_2N-R_2$$

Carbamate hydrolysis

$$-O-\overset{\overset{\text{O}}{\|}}{C}-N< \longrightarrow -OH + HO-\overset{\overset{\text{O}}{\|}}{C}-N< \longrightarrow HN< + HO-\overset{\overset{\text{O}}{\|}}{C}-OH \longrightarrow CO_2 + H_2O$$

Carbamate

Uride hydrolysis

$$>N-\overset{\overset{\text{O}}{\|}}{C}-N< \longrightarrow >NH + HO-\overset{\overset{\text{O}}{\|}}{C}-N< \longrightarrow HN< + HO-\overset{\overset{\text{O}}{\|}}{C}-OH \longrightarrow CO_2 + H_2O$$

Uride

Hydrazide hydrolysis

$$R_1-\overset{\overset{\text{O}}{\|}}{C}-\overset{\overset{\text{H}}{}}{N}-N<^{R_2}_{R_3} \longrightarrow R_1-\overset{\overset{\text{O}}{\|}}{C}-OH + H_2N-N<^{R_2}_{R_3}$$

Hydrazide   Hydrazine

Fig. (33). Hydrolytic cleavage mechanism of different functional groups showing the relative rates as well.

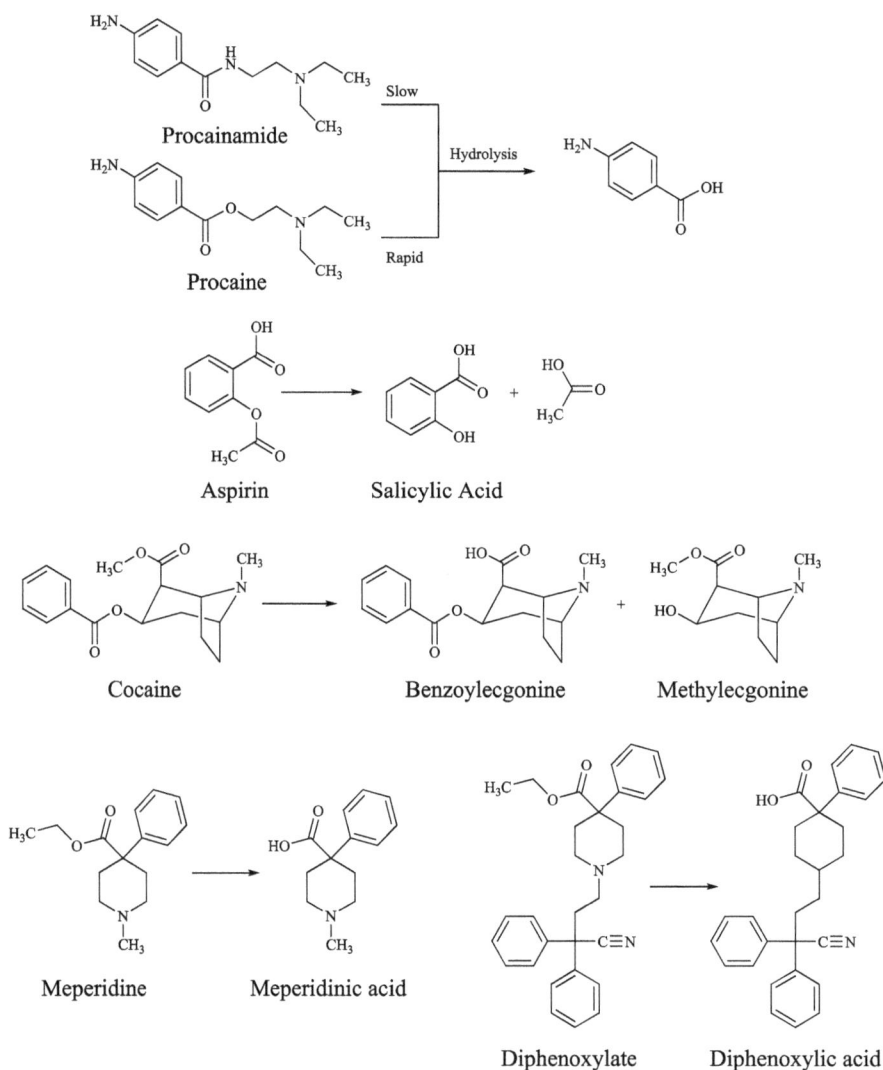

Fig. (34). Hydrolytic metabolism of procaine, procainamide, aspirin, cocaine, meperidine and diphenoxylate to their corresponding acids.

The hydrolytic degradation of several drugs are shown in Fig. **34**. Most mammalian tissues are capable of hydrolytic degradation of xenobiotics, but intestine, liver, blood, and kidney are the major sites of hydrolysis of a drug as there are a number of nonspecific esterases and amidases are available in those tissues. Thus, a drug may be hydrolyzed at different sites. For example, acetylsalicylic acid is hydrolyzed to salicylate by carboxylesterases in the liver. About 60% of an oral dose reaches the systemic circulation where it is hydrolyzed by cholinesterases and aryleserases available in plasma and red blood cell, respectively [120].

In conclusion, the phase-I biotransformation reaction: i) introduces a new functional group into the drug molecule, ii) modifies an existing functional group, or iii) exposes a functional group on the drug molecule. Certain phase-I metabolites, hereafter, are excreted and/or undergo phase-II or conjugation reactions.

Phase II Metabolic Reactions

Phase II metabolism is terminology coined by RT Williams [121], commonly referred to as conjugation reaction, which is the attachment of small, polar endogenous molecules such as glucuronic acid, sulfonate and amino acids to Phase I metabolites or parent drugs. In general, Phase II metabolites are more water-soluble and easily excretable. This process often diminishes pharmacological activity and thus toxicity by trapping highly electrophilic molecules with endogenous nucleophiles such as glutathione thus preventing damage to important macromolecules (DNA, RNA, proteins). Thus, it is commonly regarded as detoxifying pathway although the product is not always less toxic or inactive. A few examples of conjugated products that are pharmacologically active and/or toxic include: morphine-6-glucuronide (pharmacologically active), N-acetyl procainamide (pharmacologically active), glutathione conjugates of haloalkenes (nephrotoxic), and acyl glucuronides (reactive, binding covalently to proteins) [122]. Conjugation reactions are usually of two general types: 1) Type 1, in which an appropriate transferase enzyme activates the transferring group (glucuronate, sulfonate, methyl, acetyl *etc.*) in an activated coenzyme form that combines with substrate to yield the conjugated product. 2) Type 2, in which the drug substrate is activated to combine with an endogenous group such as amino acid to yield a conjugated product. A distinguishing feature of both type of Phase 2 reactions is that it involves metabolite activation by a high–energy intermediate with the aid of an appropriate transferase enzyme [123]. However, in glutathione conjugation neither the metabolite nor the endogenous species is activated.

A few clinically significant conjugation reactions that increase polarity and thus water solubility are: 1) glucuronic acid conjugation, 2) sulfonate conjugation, 3) amino acid conjugation and 4) glutathione conjugation. Other Phase 2 reactions that increase lipophilicity (LWPC) but serve to terminate pharmacological activity are: 1) acetylation, 2) methylation, and 3) fatty acid and cholesterol conjugations. Fig. (**35**) shows the common endogenous compounds that are attached to the drug molecules and/or their metabolites to get conjugated products.

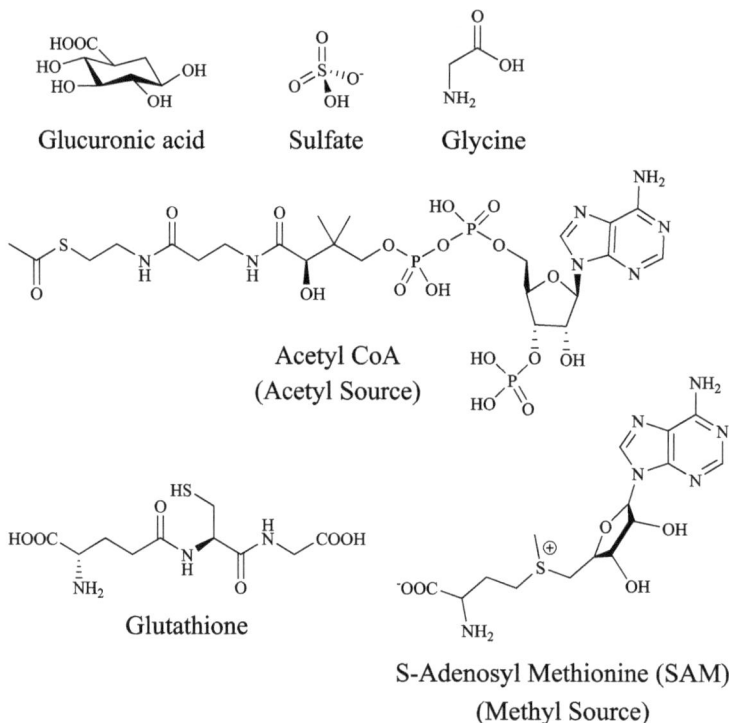

| Glucuronic acid | Sulfate | Glycine |

Acetyl CoA
(Acetyl Source)

Glutathione

S-Adenosyl Methionine (SAM)
(Methyl Source)

Fig. (35). Structure of common endogenous compounds involved in conjugation reactions.

Glucuronic Acid Conjugation

Glucuronic acid conjugation, or *glucuronidation* is the most common conjugation pathway accounting for metabolism of majority of clinically used drugs [124]. This conjugation pathway is clinically highly significant primarily because of the readily available supply of D–glucuronic acid derived from D–glucose. Secondly, the glucuronyl moiety greatly increases water solubility with its easily ionizable carboxylic acid (pKa 3.2) and three hydroxyl groups. A large number of functional groups can combine with glucuronic acid.

Formation of β–glucuronides involves several steps (Fig. **36**): 1) synthesis of activated coenzyme form, uridine diphosphate glucuronic acid (UDPG), synthesized from α-D-Glucose phosphate; 2) oxidation to UDPGA (the actual coenzyme); and 3) transfer of the glucuronyl group to the substrate by microsomal enzymes called UDP–glucuronyl transferases found primarily in the liver but also occur in many other tissues, including kidney, intestine, skin, lung, and brain [125].

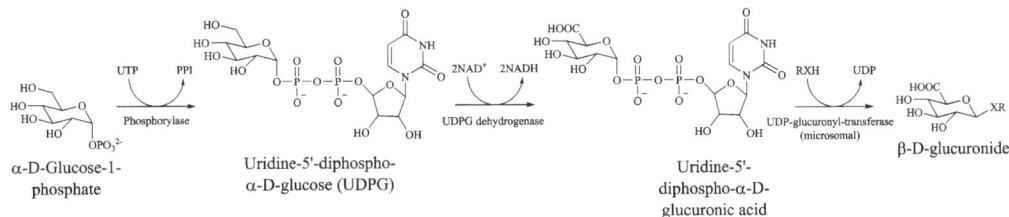

Fig. (36). Formation of glucuronide conjugates.

The activated UDP-glucuronic acid contains D-glucuronic acid in the α-configuration at the anomeric center, but glucuronate conjugates are β-glycosides. This means that inversion of stereochemistry is involved in the glucuronidation. Glucuronides are highly hydrophilic and water soluble. UDP-glucuronosyltransferase is closely associated with CYP450 so that Phase I products of drugs are efficiently conjugated [126].

Four general classes of glucuronides are possible, namely: *O-, N-, S-, and C*-glucuronides (Fig. **37**) [127 - 131]. Phenolic and alcoholic hydroxyls (secondary and tertiary only) and the carboxyl group of aryl or arylaliphatic acids are the most common functional groups undergoing glucuronidation. The carboxyl group can be attached directly or removed by up to three atoms from the aromatic system. Occurring less frequently is glucuronidation of other hydroxyl groups such as enols, *N*–hydroxylamides, N–hydroxylamines. These are known as *O*–glucuronides because the glucuronyl is attached to an Oxygen atom. *N*–glucuronides with aromatic amines, aliphatic amines, amides and sulfonamides occur occasionally, but it is a minor pathway for nitrogen containing drugs. *S*–glucuronides are known and *C*–glucuronides have also been isolated [126].

Glucuronide conjugates are primarily excreted in the urine. However, as the molecular weight (MW) exceeds 500, biliary excretions may become an important route of elimination [132]. Biliary excretion is patterned on the excretion of bile acid conjugates and therefore, is geared primarily to the excretion of high molecular weight anions. Drugs excreted in the bile usually have a MW threshold of 500 to 600 and a polar group which may be charged (anionic) or uncharged. Compounds with MW between 300 and 500 are excreted in both the urine and bile. Polar substance with MW below 300 are usually excreted in the urine only [133]. Products of biliary excretion are deposited in the duodenum and may be excreted in the feces or reabsorbed by the intestines through enterohepatic recycling after hydrolysis by β–glucuronidases present in the intestines [18]. Most glucuronides are hydrolyzed by β–glucuronidases; however, *C*-glucuronides (*e.g.* ethchlorvynol, phenylbutazone) and some quaternary amine glucuronides are stable to β-glucuronidase [134, 135].

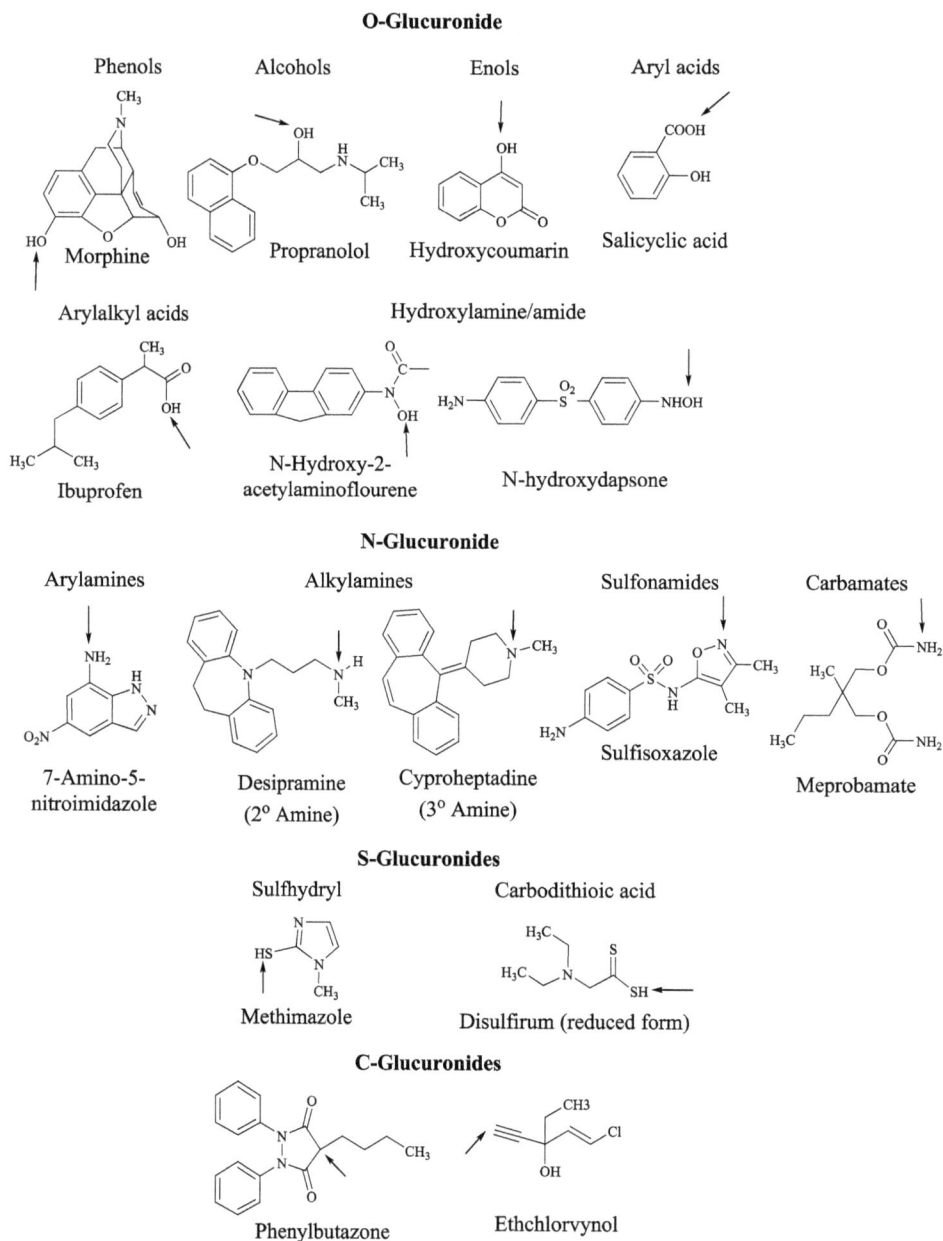

Fig. (37). Examples of different functional groups forming glucuronides. Arrows represent the site of glucuronidation.

In neonates and children, glucuronidation processes are often not fully developed. In such patients, drugs (*e.g.*, chloramphenicol) and endogenous compounds (*e.g.*, bilirubin) which are normally glucuronidated may accumulate and cause serious toxicity. Inability of infants to metabolize chloramphenicol leads to "gray baby" syndrome characterized by ashen appearance developed soon after therapy starts. It may be fatal to 40% of the neonates due to circulatory collapse [126].

Sulfonate Conjugation

The addition of an endogenous sulfonate group on a hydroxyl function of a drug molecule is considered as sulfonation. The amount of available sulfate from where active sulfonate donor is formed is limited and a significant portion of the sulfate pool is utilized by the body to conjugate endogenous compounds. This limits sulfonate conjugation. The sulfonate conjugation process involves three-enzyme catalyzed activation of inorganic sulfate to the coenzyme 3″-phosphoadenosine-5′-phosphosulfate (PAPS) (Fig. **38**). Subsequent transfer of the sulfonate group from PAPS to the accepting substrate is catalyzed by various soluble (cytosolic, not membrane bound) sulfotransferases or sulfokinases present in high levels in the liver, notable in intestine, common throughout the body.

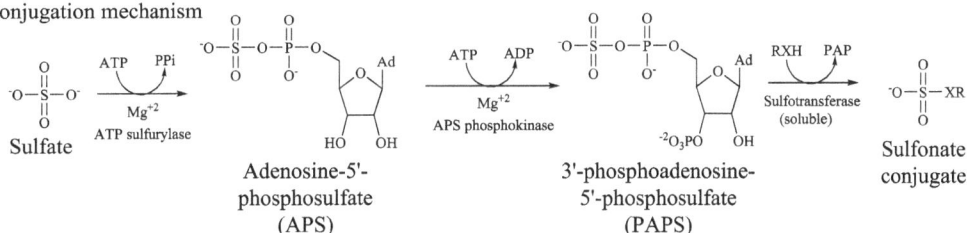

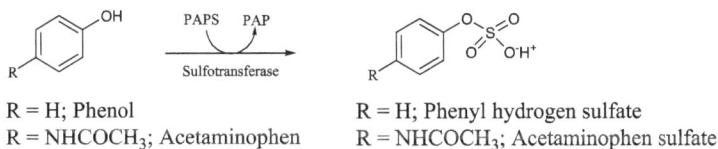

R = H; Phenol
R = NHCOCH$_3$; Acetaminophen

R = H; Phenyl hydrogen sulfate
R = NHCOCH$_3$; Acetaminophen sulfate

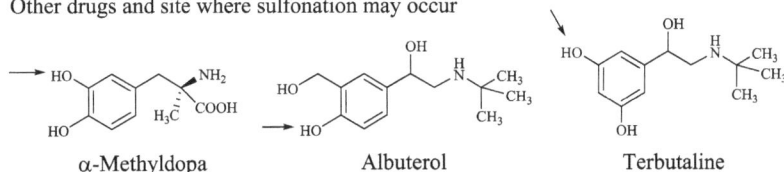

α-Methyldopa   Albuterol   Terbutaline

Fig. (38). Formation of sulfonate conjugate with examples.

Glucuronidation of phenols is frequently a competing reaction and may predominate for some drugs. Because sulfonation is usually high affinity, low capacity pathway while glucuronidation is low affinity, high capacity, sulfonate conjugation predominates in low substrate concentrations (low dose), and glucuronidation predominates in high substrate concentrations (high dose) [136]. Interestingly, in infants and children (ages 3 to 9) due to underdeveloped glucuronidation system, sulfonate conjugation predominates.

Sulfonate conjugation occurs primarily with phenols and occasionally with alcohols, aromatic amines and N–hydroxy compounds (Fig. **38**). Because sulfate esters are strong acids with pKa < 1, at pH 7.4 they are essentially 100% ionized leading to water–soluble metabolites that are usually excreted through urine. The larger conjugates of molecules can be excreted in the bile. Sulfonate conjugates are usually inactive, but minoxidil sulfate is an active metabolite that stimulates hair growth [137].

Sulfonation of hydroxylated aromatic amines sometimes lead to reactive intermediates and toxicity [138]. For example, phenacetin and acetaminophen, two hydroxylated aromatic amines, are metabolized in several pathways. However, the toxicity of these drugs is attributed to the formation of a reactive electrophilic metabolite, N-acetyl-p-benzoquinone imine (NAPQI), according to the mechanism shown in Fig. **39** [139]. This electrophilic metabolite (NAPQI; Fig. **39**) can bind covalently with the critical cellular macromolecules. This toxic metabolite is detoxified by binding to glutathione (GSH) by conjugation between the ortho carbon of the amino group of p-aminophenol and the SH group of GSH. This eventually leads to the excretion of the cysteine and N-acetyl cysteine (mercapturic acid) conjugates of acetaminophen in urine. Thus, about 5-10% of the dose of acetaminophen/phenacetin found in the form of thioether metabolites in the urine is the reflection of its toxic reactions [140].

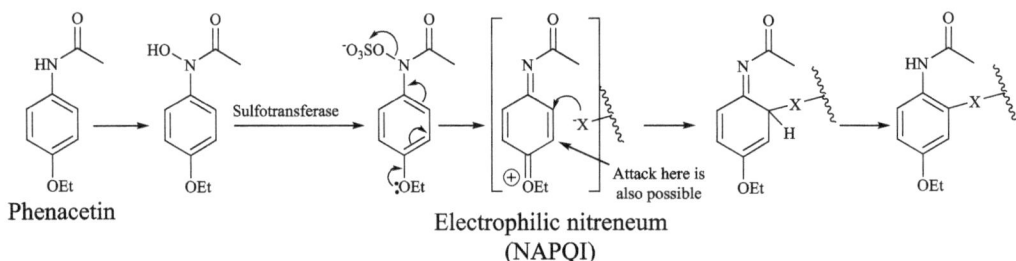

Fig. (39). Mechanism of phenacetin toxicity through sulfonate conjugation of hydroxylated aromatic amines pathway.

Amino Acid Conjugation

Carboxylic acid containing parent drugs and/or metabolites may undergo amino acid conjugation. Glycine conjugate of benzoic acid, hippuric acid, is the first mammalian drug metabolite isolated (Fig. **40**). Aromatic, arylacetic, and heterocyclic carboxylic acids usually undergo amino acid conjugation forming amide bond. Glycine conjugation is the most common although taurine, arginine, asparagine, histidine, lysine, glutamate, aspartate, alanine, and serine conjugation may also occur [126]. The quantity of amino acid conjugates formed is quite small because of the limited availability of amino acids in the body and competition with glucuronidation.

Mechanism of amino acid conjugation

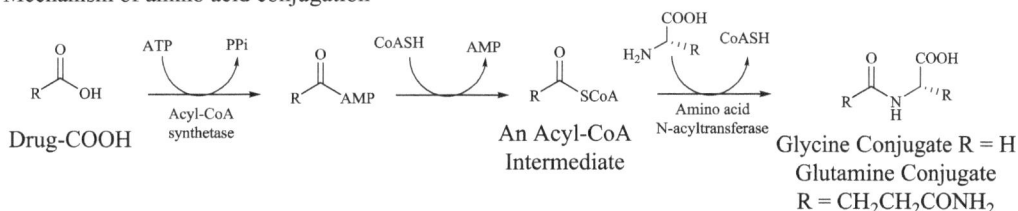

Drug-COOH   An Acyl-CoA Intermediate   Glycine Conjugate R = H
Glutamine Conjugate
R = CH$_2$CH$_2$CONH$_2$

Glycine conjugation of parent drug

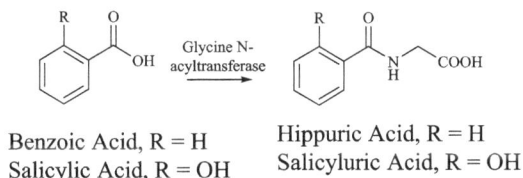

Benzoic Acid, R = H   Hippuric Acid, R = H
Salicylic Acid, R = OH   Salicyluric Acid, R = OH

Glycine conjugation of brompheniramine metabolite

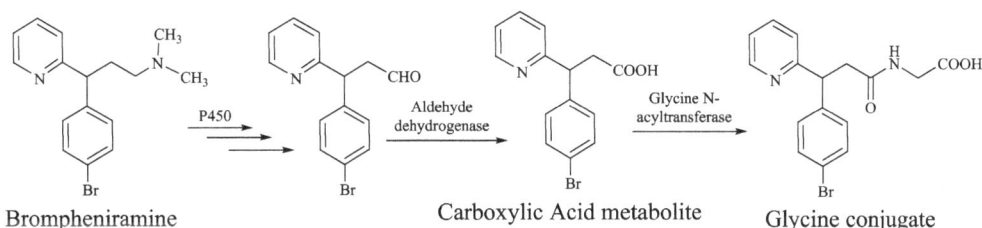

Brompheniramine   Carboxylic Acid metabolite   Glycine conjugate

Fig. (40). Mechanism of amino acid conjugation with examples.

Amino acid conjugation is a three-step process occurring in the mitochondria of liver and kidney cells. The carboxylic acid substrate is activated with ATP and Coenzyme A to form an acyl Coenzyme A complex. N–Acyltransferases catalyze the acetylation reaction of this complex with the amine functional group of amino acid to form amide (Fig. **40**). The polar and water soluble amino acid conjugates are mainly excreted in urine.

Glutathione Conjugation

Glutathione (GSH/GSSG; Fig. **41**) is a tripeptide (Glu-Cys-Gly) – found virtually in all mammalian tissues. Its thiol group functions as a nucleophilic scavenger of harmful electrophilic parent drugs or their metabolites. Thus, GSH conjugation is an important pathway to detoxify the toxic electrophiles to protect vital cellular constituents by virtue of its nucleophilic sulfhydryl group. The sulfhydryl group reacts with electron deficient compounds to form S–substituted glutathione adducts. A family of cytoplasmic GSH S-transferases catalyzes these conjugation reactions without the need for an activated coenzyme of GSH or the substrate. However, the substrates should be highly electrophilic for such conjugation reactions [141]. The conjugation may occur with or without the influence of enzyme glutathione transferases, which is found in most tissues.

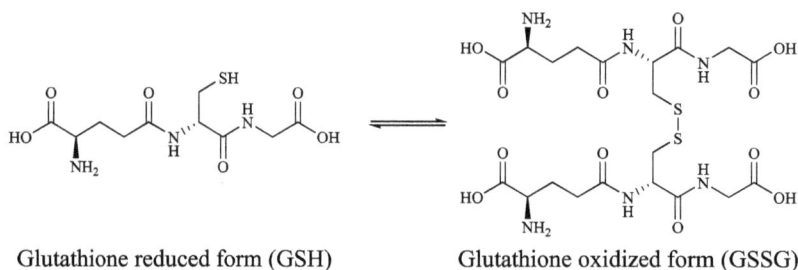

Glutathione reduced form (GSH)   Glutathione oxidized form (GSSG)

Fig. (41). Structures of oxidized and reduced forms of glutathione.

Compounds deactivated by glutathione are of four categories. 1) electrophilic carbons having leaving groups such as halogens, sulfates, and nitro groups on aliphatic carbons (not aromatic), 2) small rings (epoxides, β–lactam), 3) β carbon of an α,β–unsaturated carbonyl compound, and highly reactive products of Phase 1 metabolism. Thus, drugs containing aliphatic and arylalkyl halides, arene oxides, sulfates, sulfonates, nitrates, epoxides and olefins are electrophilic enough to undergo GSH conjugation reactions. Mechanistically there are three types of GSH conjugation reactions: 1) S_N2 reaction, 2) S_NAr reaction, and 3) Michael addition (Fig. **42**).

S_N2 Reaction

As shown in Fig. **42A**, in this type of reaction, a nucleophilic leaving group is replaced by GS⁻ functional group to form a conjugate. Methanesulfonate acts as a leaving group in case of busulfan, and nitro ($-NO_2$) group acts as a leaving group in nitroglycerin.

S_NAr Reaction

This reaction mechanism is very similar to S_N2 mechanism; however, the leaving group is attached to an aromatic ring. Azathioprine is a good example of drugs undergoing such mechanism (Fig. **42B**).

A. S_N2 Mechanism:

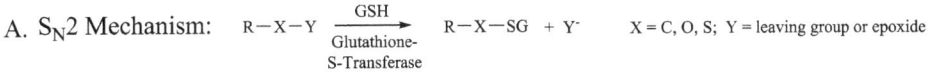

$X = C, O, S;$ Y = leaving group or epoxide

Examples

1.

Busulfan

2.

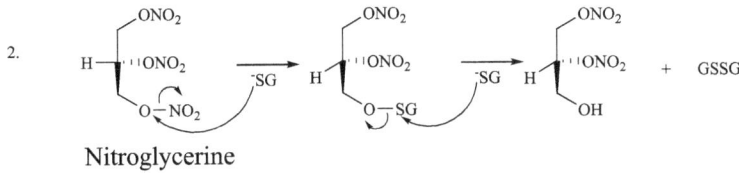

Nitroglycerine

B. S_NAr Mechanism:

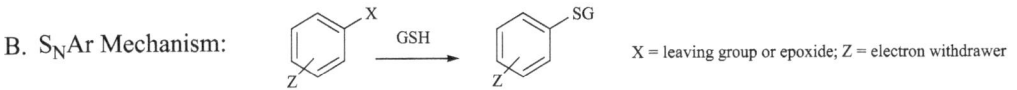

X = leaving group or epoxide; Z = electron withdrawer

Example

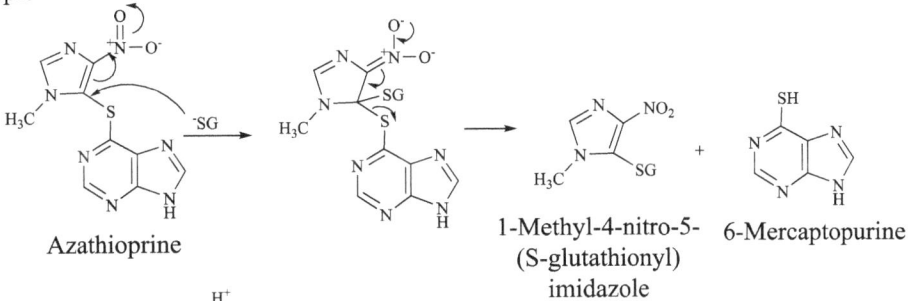

Azathioprine   1-Methyl-4-nitro-5-   6-Mercaptopurine
(S-glutathionyl)
imidazole

C. Michael Addition:

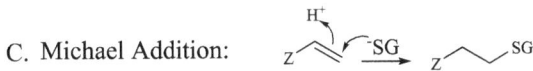

Example:

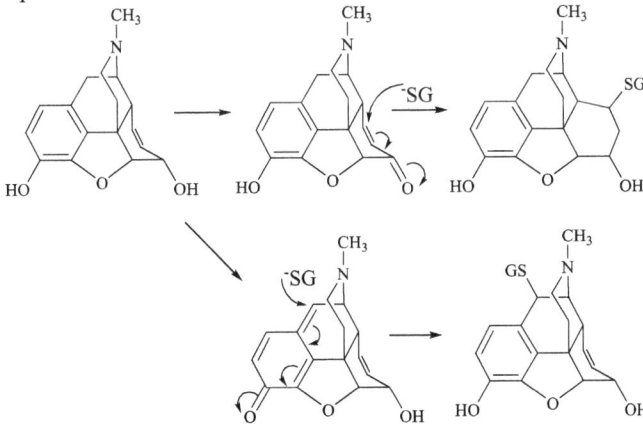

Fig. (42). Mechanism of glutathione conjugation with examples. A) S_N2 mechanism, B) S_NAr mechanism, C) Michael addition.

Michael Addition

Michael addition is the addition of the nucleophilic GS⁻ on the one end of the olefinic double bond and addition of proton on the other end of the double bond making an addition product as shown in Fig. **42C.**

The glutathione conjugates may be excreted intact through the biliary tract into the feces, or further break down into cysteine conjugates in bile or kidneys, which are then presumably reabsorbed, and N-acetylated to form the mercapturic acid and re-excreted [142]. Overall, the mercapturic acid conjugates are formed by sequential cleavage of glutamic acid and glycine with the action of peptidases, γ-glutamyl transpeptidase and cysteinyl glycinase (aminopeptidase M), respectively, and then N-acetylation (Fig. **43**).

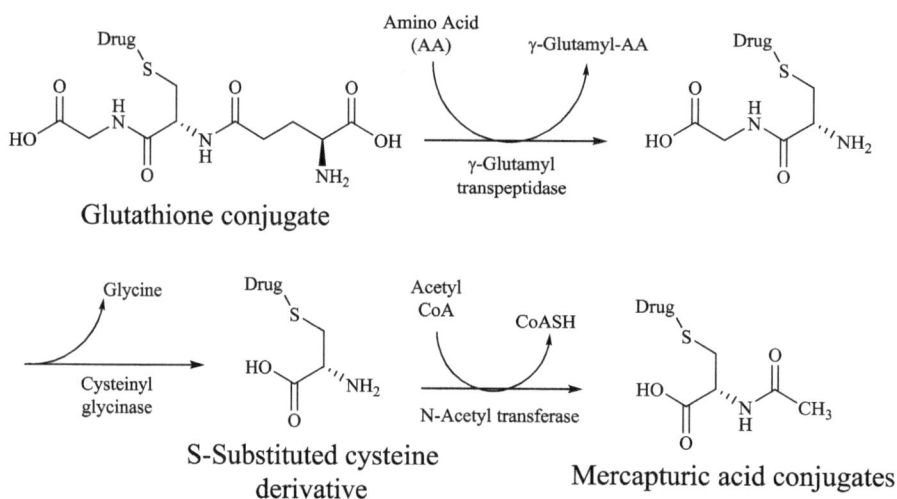

Fig. (**43**). Conversion of glutathione conjugates to mercapturic acid (N-acetyl cysteine) conjugates.

As shown in the previous section under sulfonate conjugation, acetaminophen may produce reactive electrophile at overdoses. This species is deactivated by glutathione, but the stores of glutathione are limited and can be exhausted. When this occurs, the intermediate reacts with important macromolecules, resulting in cell death, which leads to irreversible damage to the liver. This in turn can cause death. A drug used in acetaminophen overdoses is Mucomyst (acetylcysteine), which works in a similar mechanism to glutathione and produce mercapturic acid conjugate.

Acetyl Conjugation

Acetylation constitutes an important metabolic route for drugs containing primary amino groups, including primary aromatic amines, sulfonamides, hydrazines, hydrazides. Aromatic amines are especially susceptible. The amide metabolites are generally nontoxic and inactive. Since water solubility is decreased, it appears the function of acetylation is to terminate pharmacological activity. However, *N*-acetyl procainamide is as potent as the parent antiarrhythmic drug procainamide (Procanbid), and *N*-acetyl isoniazid is more toxic than its parent drug.

Acetylation is a two-step, covalent catalytic process. The acetyl group is supplied by Acetylcoenzyme A and transfer is mediated by transacylases present in hepatic cells. Because the substrate are amines, the transferases are more specifically known as *N*-acetyl transferase (NAT). The endogenous acetate ion is activated by the action of Coenzyme A, forming Acetyl Coenzyme A, an activated thio ester. This coenzyme is analogous to the acyl Coenzyme A encountered in amino acid conjugation. However, the R group is a methyl. The acetyl transferase enzymes are similar and like in amino acid conjugation, the product formed is an amide (Fig. **44**) [126].

Acetylation of primary amine

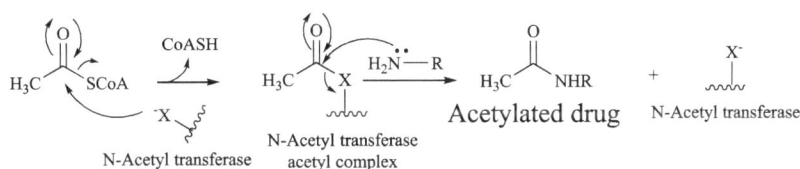

Examples

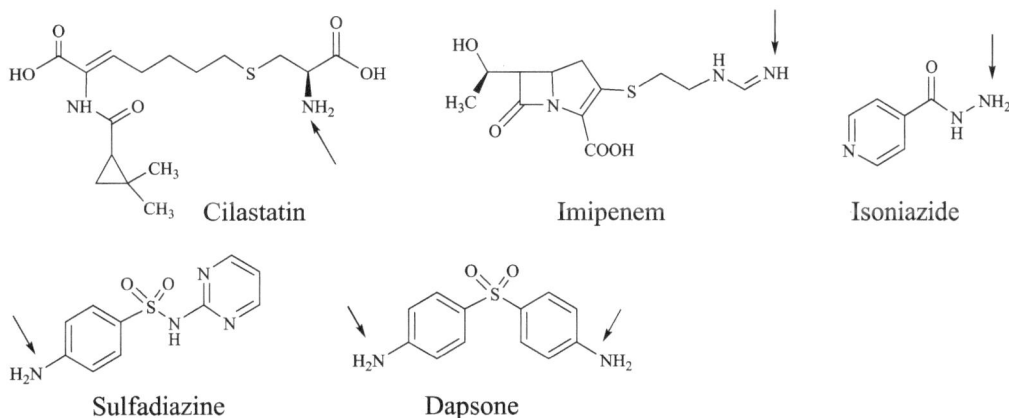

Cilastatin   Imipenem   Isoniazide

Sulfadiazine   Dapsone

Fig. (44). Mechanism of acetyl conjugation with examples. Arrows on the drug structures show the site of acetylation.

Sulfonamides are metabolized by *N*–acetylation at the *N*–4 position. These metabolites are less water soluble than the parents and have the potential of crystallizing out in the renal tubules (crystalluria), causing kidney damage.

Individuals are classified as either *slow* or *fast acetylators*. This variation is genetic. *Slow acetylators* are prone to develop adverse reactions while *fast acetylators* may show inadequate response to standard doses. This phenomenon known as genetic polymorphism abounds in acetylation due to differences in NAT type 2 (NAT2) activity. Polymorphism in NAT1 activity has minor effect on drug metabolism. Multiple NAT2 alleles (NAT2*5, *6, *7, and *14) have substantially decreased acetylation activity and are common in populations of Caucasians and African descent. In these groups, most individuals carry at least one copy of a *slow acetylator* allele, and less than 10% are homozygous for the wild type (*fast acetylator*) trait. The ratio of NAT2 activity is 7 in Caucasians to 18 in the Chinese population. Thus, in general Chinese are considered fast acetylators, while Caucasians are *slow acetylators* [143].

Fatty acid ester formation

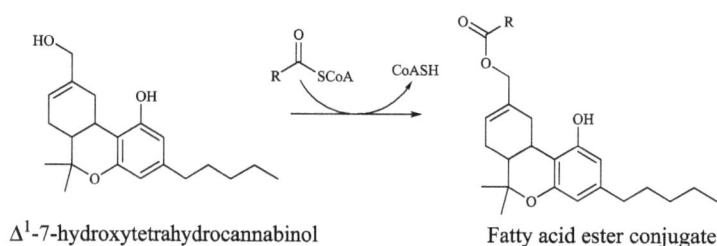

Δ^1-7-hydroxytetrahydrocannabinol   Fatty acid ester conjugate

Cholesterol esterification of prednimustine

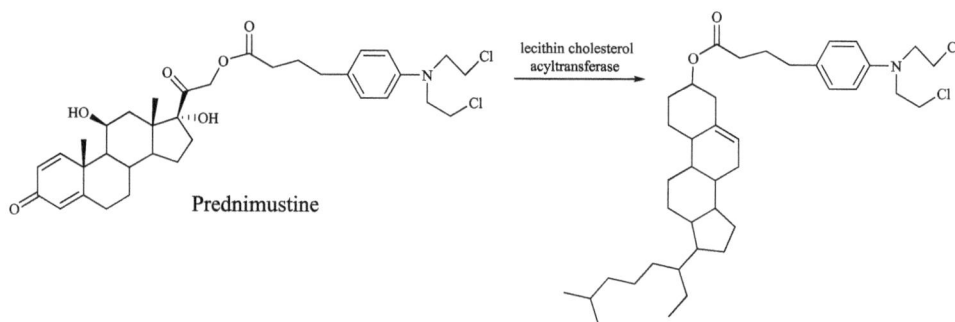

Prednimustine

Fig. (45). Examples of fatty acid and cholesterol conjugation.

Fatty Acid and Cholesterol Conjugation

Hydroxyl-containing drugs can undergo conjugation with a wide range of endogenous fatty acids, such as saturated acids (from C10 to C16) and unsaturated acids (oleic and linoleic acids). Fatty acid ester metabolites are highly lipophilic due to large carbon chain attached while masking the polar hydroxyl group of the drug, or its metabolite. For example, Δ^1-7-hydroxytetrahydrocannabinol is metabolized in this pathway to produce fatty acid ester (Fig. **45**), which is accumulated in the adipose tissues instead of excreting out of the body [126]. The fatty acid ester may form with the action of fatty acyl coenzyme A in a similar mechanism to amino acid and acetyl conjugation, discussed in the previous sections. Cholesterol (an alcohol) ester metabolites have been detected for drugs containing either an ester or a carboxylic acid. This also produce highly lipophilic metabolites, which are deposited into adipose or liver tissues. Thus, cholesterol conjugate of chlorambucil was detected after transesterification of prednimustin, which is catalyzed by lecithin cholesterol acyltransferase (Fig. **45**) [126].

Methyl Conjugation

Methylation (or methyl conjugation) is very important in the biosynthesis and metabolism of endogenous compounds (neurotransmitters), but methyl conjugation is a minor pathway in drug metabolism. Methyl conjugation decreases water solubility except where it forms a quaternary ammonium ion, when it increases polarity. Most methylated products tend to be pharmacologically inactive, with a few exceptions.

Methyl conjugation is catalyzed by a variety of methyl transferases, such as catechol *O*-methyl transferase (COMT), phenol-*O*-methyltransferase, *N*-methyl transferase and S-methyltransferase, in two-step process (Fig. **46**). These transferases transfer methyl group from coenzyme *S*–adenosylmethionine (SAM) to the drug molecules or their metabolites (RXH, where X = O, N, or S).

The major substrate for *N*–methyltransferases are aliphatic primary amines; for *O*–methyltransferases are phenolic hydroxyls (competing reaction with glucuronidation); and any thiol for S–methyltransferases. Most methyltransferases are found in the liver and show a high degree of structural specificity that limits methyl conjugations. Only if drugs resemble the endogenous substrates are they methylated. For example, COMT carries out *O*–methyl conjugations of important neurotransmitters and terminates their activity by methylating the C–3 hydroxyl. COMT also methylates drugs which contain the catechol system, such as isoproterenol at the same position [144].

N–Methylation of drugs occurs only to a limited extent as most *N*–methyltransferases are highly specific. For example, Phenylethanolamine *N*–methyltransferase which methylates endogenous and exogenous phenylethanolamines but not phenylethylamines. Histamine *N*–methyltransferase is specific for histamine. Indoleethylamine *N*–methyltransferase is nonspecific and will methylate a variety of endogenous amines and drugs. S–methylation of drugs is catalyzed by a nonspecific *S*–methyltransferase.

S-AdenosylL-methionine

S-adenosyl-L-homocysteine

Drug Examples:

Isopreterenol   6-Mercaptopurine   Morphine   Phenylpropanolamine

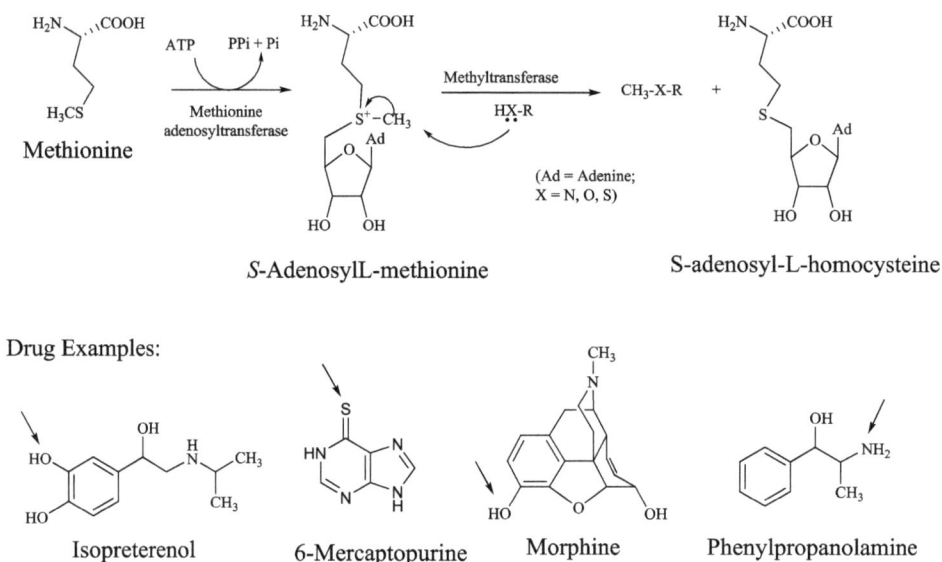

Fig. (46). Mechanism of methyl conjugation reaction with drug examples.

STEREOCHEMISTRY AND DRUG METABOLISM

The metabolism and pharmacokinetics of chiral drugs are frequently influenced by their stereochemistry [145, 146]. The different stereoisomers may be metabolized preferentially by different enzymes or by the same enzyme at different rates due to the distinct interactions between chiral drugs and the receptors, enzymes, and transport proteins. Following are a few examples that show such differences in drug metabolism based on their stereochemistry.

Itraconazol (ITZ) Metabolism

Only (2R,4S)-ITZ have been shown to metabolize by CYP3A4 to hydroxy-, keto-, and N-desalkyl-ITZ and no metabolism was detected for (2S,4R)-ITZ (Fig. **47**). Thus, after a single dose of ITZ, the (2S,4R)-ITZ produced higher plasma concentration than the corresponding diastereomeric pair, which produced keto- and N-desalkyl-ITZ metabolites in plasma (Fig. **46**) [147].

Etodolac Metabolism

It had been shown that S-etodolac is more rapidly metabolized than R-etodolac in human, and S-etodolac is preferentially glucuronidated by UGT1A9 while R-etodolac is preferentially hydroxylated by CYP2C9 (Fig. **48**) [148].

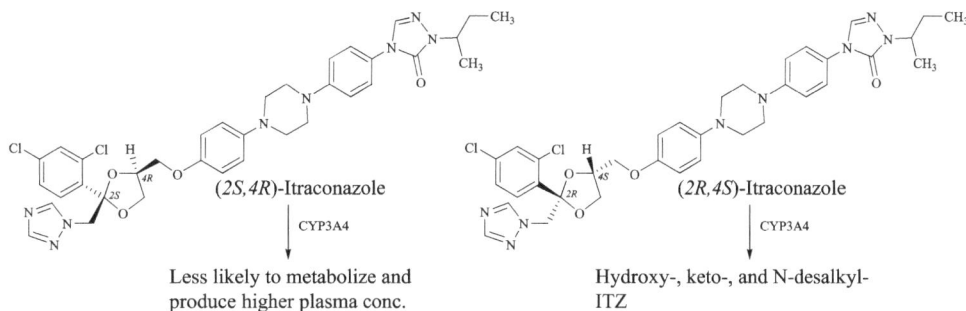

Fig. (47). Stereoselective metabolism of itraconazole.

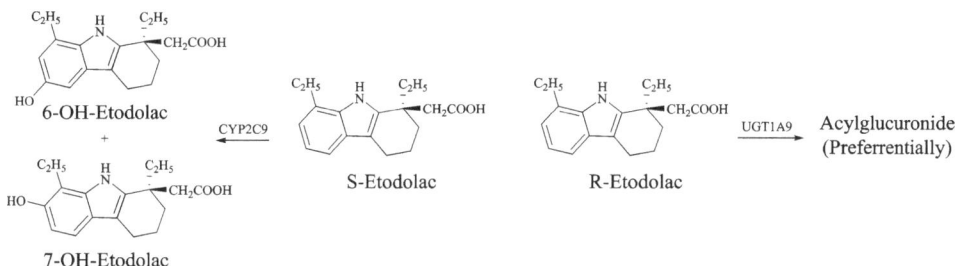

Fig. (48). Stereoselective metabolic pathway of etodolac in man.

Etomidate Metabolism

The hypnotic or general anesthetic drug etomidate (Amidate) is stereoselectively metabolized in the liver. The *R*-(+)-isomer is more rapidly hydrolyzed to the major metabolite *R*-(+)-1-(1-phenylethyl)-1H-imidazole-5-carboxylic acid that accounts for 80% of its urinary excretion, but *S*-(-)-isomer is more rapidly hydroxylated and then dealkylated (Fig. **49**) [149].

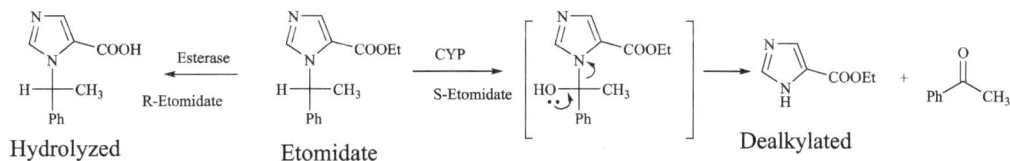

Fig. (49). Stereoselective metabolism of etomidate.

Warfarin Metabolism

The metabolism of warfarin was examined to determine the effect of chirality. It was shown that *R*-warfarin was oxidized to 6-hydroxywarfarin and was reduced to (*R*,*S*)-warfarin alcohol, while *S*-warfarin was oxidized to both 6- and 7-hydroxy warfarin and was reduced to the (*S*,*S*)-warfarin alcohol (Fig. **50**) [150].

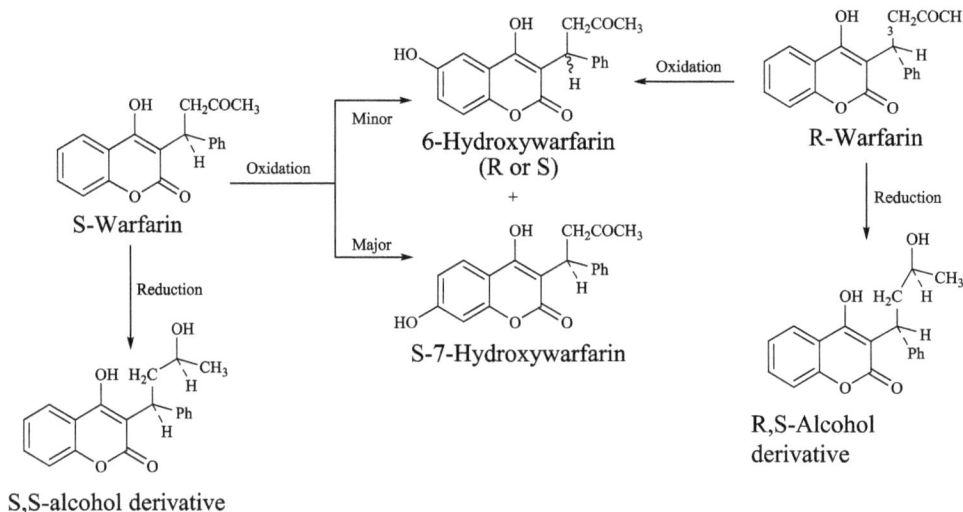

Fig. (50). Stereoselective metabolism of warfarin.

Propranolol Metabolism

The glucuronidation of the β-adrenergic blocker propranolol is catalyzed by UGT1A9 and UGT1A10 displaying opposite stereoselectivity [151]. The biologically active S-propranolol is preferentially glucuronidated by UGT1A9, whereas R-enantiomer is preferentially conjugated by UGT1A10. Thus, the former is glucuronidated predominantly in the human liver microsomes that express UGT1A9, which also shows extensive first pass metabolism. On the other hand the latter is predominantly glucuronidated into the human intestine microsomes that express UGT1A10.

PHARMACOLOGICALLY ACTIVE METABOLITES

Both Phase I and Phase II drug metabolism are important mechanism for drug clearance; however, these also lead to active metabolites. Over hundreds of drug examples are available that lead to active metabolites by both phase I and Phase II mechanisms. Some of the examples are discussed below with their significance in efficacy and safety. These are divided into several category: 1) Drugs with active metabolites that dominate the activity (*e.g.*, tamoxifen, and thioridazine); 2) Drugs with active metabolites that contribute comparably to the parent (*e.g.*, metoprolol and morphine); 3) Drugs with metabolites that possess target potency but contribute minimally towards *in vivo* effect (*e.g..*, diazepam, temazepam, and oxazepam); 4) Drugs with metabolites possessing activity at related targets (*e.g.*, doxepine); and 5) Drugs that generate active metabolites but assessment of *in vivo* contribution is ambiguous (*e.g.*, atorvastatin) [152].

The Active Metabolites of Tamoxifen and Thioridazine

Tamoxifen, a selective estrogen receptor modulator, after sequential hydroxylation and N-demethylation catalyzed by CYP2D6 and CYP3A, respectively (Fig. **51**), produce the secondary metabolite, endoxifen, which is in fact is clinically significant active metabolite. The action of CYP2D6 generates 4-hydroxytamoxifen, which is also pharmacologically active. It is suggested that 4-hydroxytamoxifen and endoxifen are two orders of magnitude more potent than their respective nonhydroxylated counterparts.

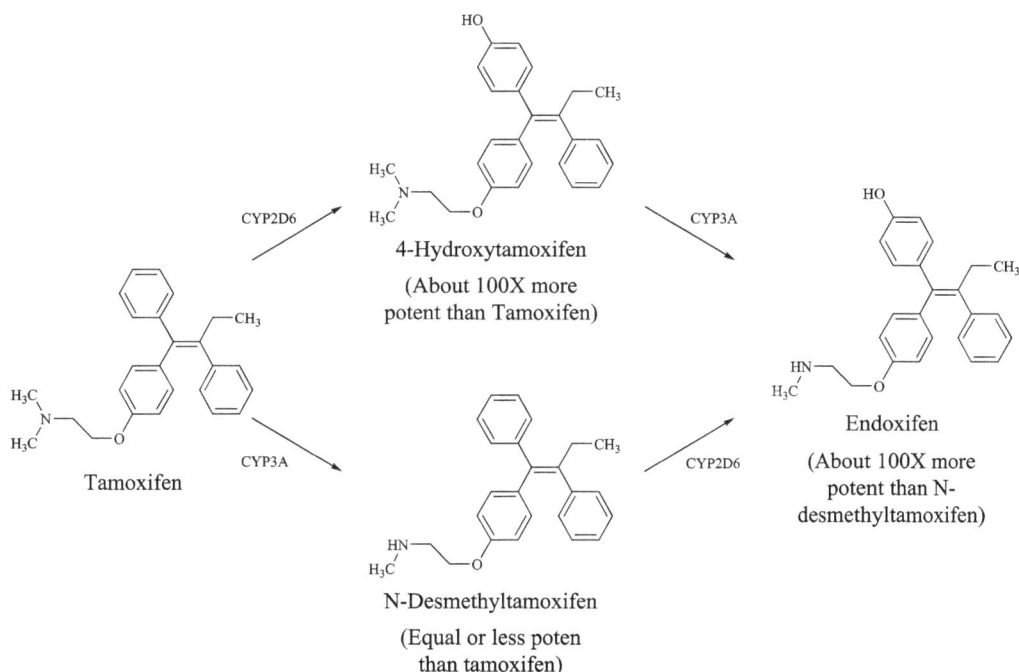

Fig. (51). Active metabolites of tamoxifen that are clinically more significant.

The classic antipsychotic, thioridazine, undergoes *S*-oxidation to form mesoridazine (sulfoxide) and sulforidazine (sulfone) (Fig. **52**). Both the metabolites are somewhat more potent dopamine receptor antagonists compared to parent drug (thioridazine). Although mesoridazine and sulforidazine are about 6-fold and 18-fold less exposed, respectively in the blood after administration of thioridazine due to vastly less protein bound nature of the metabolites, they are highly clinically significant. In fact, mesoridazine is clinically used as a separate entity.

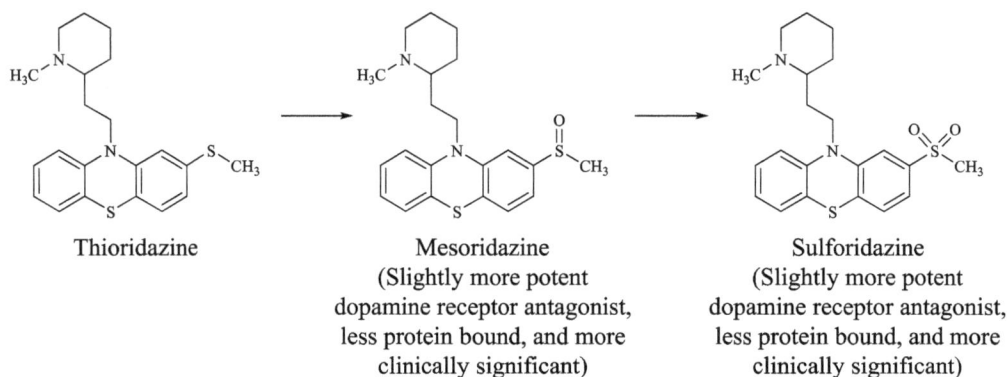

Thioridazine

Mesoridazine
(Slightly more potent
dopamine receptor antagonist,
less protein bound, and more
clinically significant)

Sulforidazine
(Slightly more potent
dopamine receptor antagonist,
less protein bound, and more
clinically significant)

Fig. (52). Active metabolites of thioridazine that are clinically more significant.

The Active Metabolites of Metoprolol and Morphine

Metoprolol is a β-blocker that undergoes extensive metabolism to mostly inactive metabolites with the exception of CYP2D6 mediated α-hydroxymetoprolol metabolite (Fig. **53**), which is about 17-fold more potent β-adrenergic blocker than parent metoprolol. It has been suggested that the α-hydroxymetoprolol possibly contribute up to 100% of the activity relative to metoprolol.

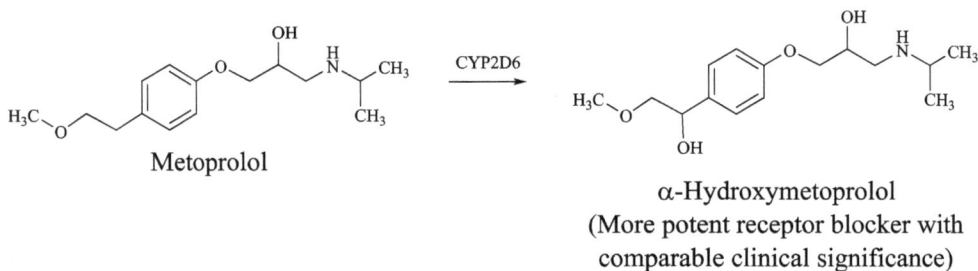

Metoprolol

α-Hydroxymetoprolol
(More potent receptor blocker with
comparable clinical significance)

Fig. (53). Active metabolites of metoprolol with comparable clinical significance.

Morphine is metabolized by glucuronide conjugation pathway by the action of UGT2B7 to 3-and 6-glucuronides (Fig. **54**). The 3-glucuronide does not possess μ-opioid receptor affinity; however, morphine-6-glucuronide is nearly equivalent μ-opioid receptor agonist as the parent drug. It has been shown that morphine--glucuronide has good, but delayed analgesic efficacy.

Diazepam, Temazepam, and Oxazepam

Among several oxidative metabolites of diazepam, a benzodiazepine receptor agonist, three are clinically used drugs (Fig. **55**). Nordiazepam (N-demethylated; major metabolite), temazepam (a hydroxyl metabolite) and oxazepam (the corresponding hydroxyl metabolite of nordiazepam) have lower affinity for the benzodiazepine receptor (4- to 7-fold and about 8- to 50-fold lower circulating concentrations than the parent drug. Thus, it is unlikely that any of these metabolites contribute substantially to the effects of diazepam. However, they are clinically useful when administered as a separate drug.

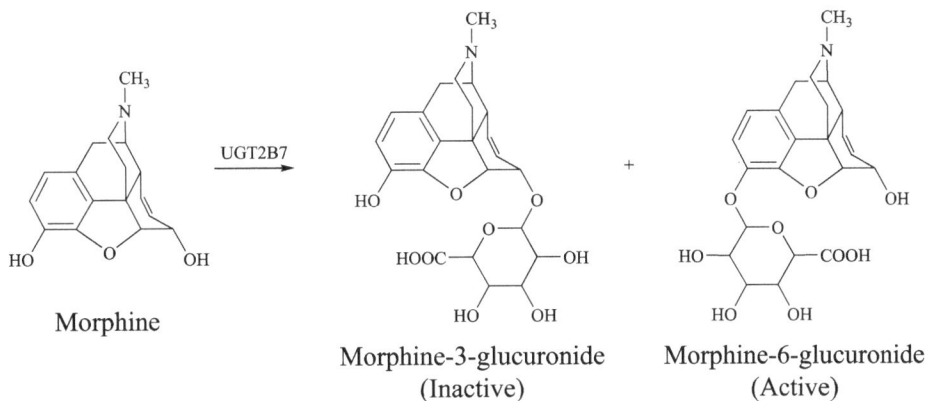

Morphine

Morphine-3-glucuronide
(Inactive)

Morphine-6-glucuronide
(Active)

Fig. (54). Active metabolite of morphine.

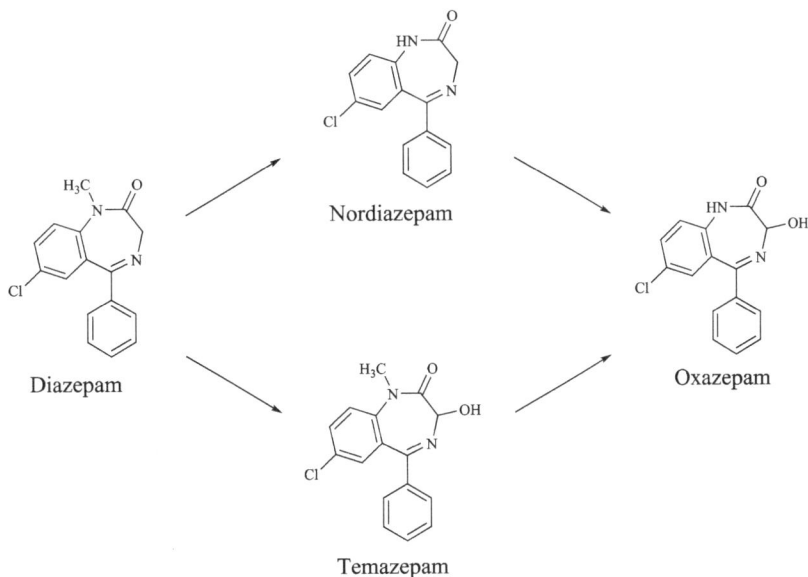

Nordiazepam

Diazepam

Temazepam

Oxazepam

Fig. (55). Target active metabolites of diazepam with little to no *in vivo* contribution, but are clinically used drugs.

Active Metabolite of Doxepine

Doxepin is metabolized to *N*-desmethyl product that has very different pharmacological activities than doxepine itself. At lower doses, doxepin is clinically indicated as an antihistamine that readily penetrates the brain and thus causes sedation. *N*-Desmethyldoxepin (Fig. **56**) is 7-fold more potent norepinephrine reuptake inhibitor (NRI) than doxepine and thus produce antidepressant activity similar to tricyclic antidepressants. Thus, doxepine is indicated as antidepressant at high doses.

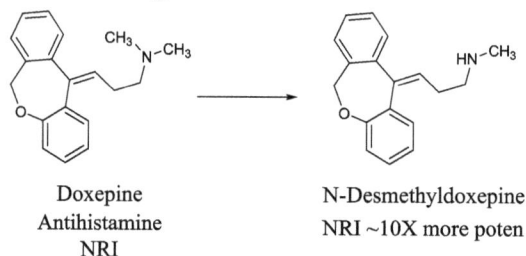

Doxepine
Antihistamine
NRI

N-Desmethyldoxepine
NRI ~10X more poten

Fig. (56). Active metabolite of doxepine possessing activity at a related target.

The Active Metabolites of Atorvastatin

Atorvastatin is a lipid lowering agent act by inhibiting HMGCoA reductase. It is claimed to form two active metabolites, 2- and 4-hydroxyatorvastatin, by the action of CYP3A (Fig. **57**). However, actual inhibition potencies of the metabolites for the target enzyme are not available in the scientific literature. Although the product label for Lipitor claims that 70% of the activity is attributable to metabolites the assessment of its clinical contribution in the pharmacological effect of atorvastatin is ambiguous.

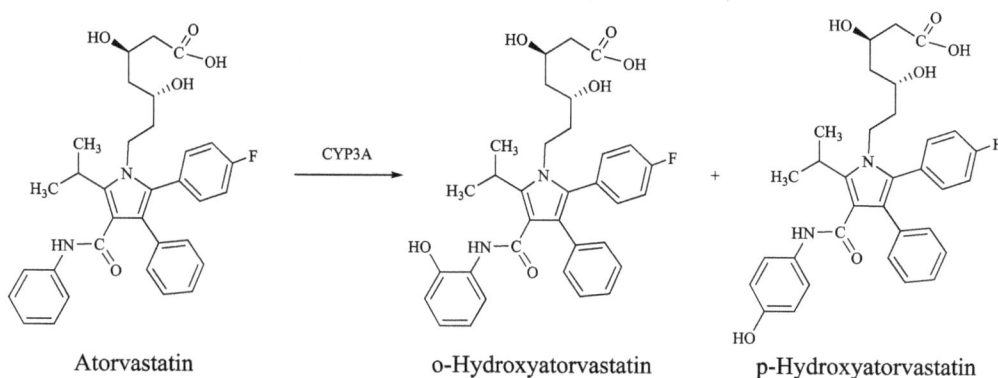

Atorvastatin

o-Hydroxyatorvastatin

p-Hydroxyatorvastatin

Fig. (57). Active metabolite of atorvastatin with ambiguous assessment of *in vivo* contribution.

PRODRUGS, SOFT DRUGS AND ANTEDRUGS

Introduction

Albert coined the term ***prodrug*** to refer a pharmacologically inactive compound that is transformed by the mammalian system into an active substance by either chemical or metabolic means [153]. As defined by The International Union of Pure and Applied Chemistry (IUPAC) – "prodrug is any compound that undergoes biotransformation before exhibiting its pharmacological effects. Prodrugs can thus be viewed as drugs containing specialized non-toxic protective groups used in a transient manner to alter or to eliminate undesirable properties in the parent molecule" [154].

In a prodrug, the modification in drug structure is done with a *promoiety* to improve the physicochemical, biopharmaceutical or pharmacokinetic properties of the parent drugs. Generally, the inactive *promoiety* undergoes only one to two chemical or enzymatic transformation to yield the active parent drug. About 5–7% of drugs approved worldwide can be classified as prodrugs.

Hard Drugs are compounds designed to contain the structural characteristics necessary for pharmacological activity but in a form that is not susceptible to metabolic or chemical transformation, have high lipid solubility and thus accumulation, or high water solubility thus quick excretion.

Soft Drug concept is introduced by Bodor in 1977 [155], and is defined by IUPAC as "a compound that is degraded *in vivo* to predictable nontoxic and inactive metabolites, after having achieved its therapeutic role" Indeed soft drugs are a group of modified compounds that are also designed to deliver the drugs into the brain (the chemical delivery system).

Antedrugs overlap soft drugs to some extent in concept but are strictly designed for local use of drugs and undergo rapid metabolic inactivation once in the systemic circulation [156]. The prodrug and antedrug concepts can be summarized as follows (Fig. **58**) [157]:

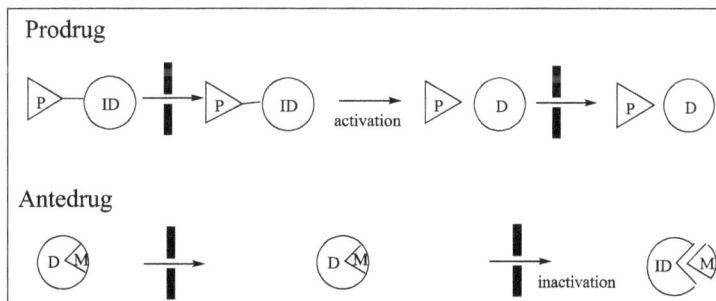

Fig. (58). The prodrug and antedrug concepts. In a prodrug, the potency of a drug is diminished *via* the attachment of promoiety (P), which renders it essentially inactive. Once administered and the promoiety is removed by enzymatic and chemical reaction, the drug is activated. In an antedrug, the active drug, which contains a positive modifier (M), is administered locally. Upon leaving the target tissue, the modifier is removed by facile biotransformation, rendering the drug inactive, thereby minimizing untoward systemic effects. D, Active drug; ID, Inactive drug.

Prodrugs: Prodrugs can be **carrier-linked prodrugs** where the active drugs have been attached through a metabolically labile promoiety as discussed above. **Bioprecursor prodrugs** contain no promoiety but rather rely on metabolism to introduce the functionality necessary to create an active species. For examples, prontosil is reduced to sulfanilamide antibacterial; cyclophosphamide is metabolized to phosphoramide mustard. Many of the antineoplastic and antiviral drugs fall under this category.

This chapter will focus on the carrier mediated prodrugs. Advantages and/or purposes of a carrier mediated prodrug can be summarized as follows [158, 159]:

1. Improved lipophilicity or permeability
2. Improved aqueous solubility
3. Improved parenteral administration
4. Exploit carrier-mediated absorption
5. Improved ophthalmic and dermal delivery
6. Enable site-selective drug delivery
7. Increased stability and thus prolonged action
8. Sustained release
9. Reduce toxicity
10. Improve patient acceptability

Common Functional Groups in Prodrugs

Drugs with carboxylic, hydroxyl, amine, phosphate/phosphonate and carbonyl groups are most commonly converted into prodrugs by modification of these groups to esters, carbonates, carbamates, amides, phosphates and oximes. The commonly occurring drug functional groups and the corresponding prodrugs are summarized in Fig. **59**.

Prodrugs with Increased Lipophilicity

Enalapril: Enalapril is a prodrug of enalaprilat where the promoiety is released as ethanol after ester hydrolysis due to the action of esterases, resulting in active drug enalaprilat (Fig. **60**). The purpose of this prodrug formation is to improve its *lipophilicity or permeability* and thus *oral bioavailability*. Similar to enalapril other ACE inhibitors, *e.g.*, quinapril, benazepril, moexipril, ramipril and trandolapril are also used as ethyl ester prodrugs for oral administration, and all undergo similar metabolic activation.

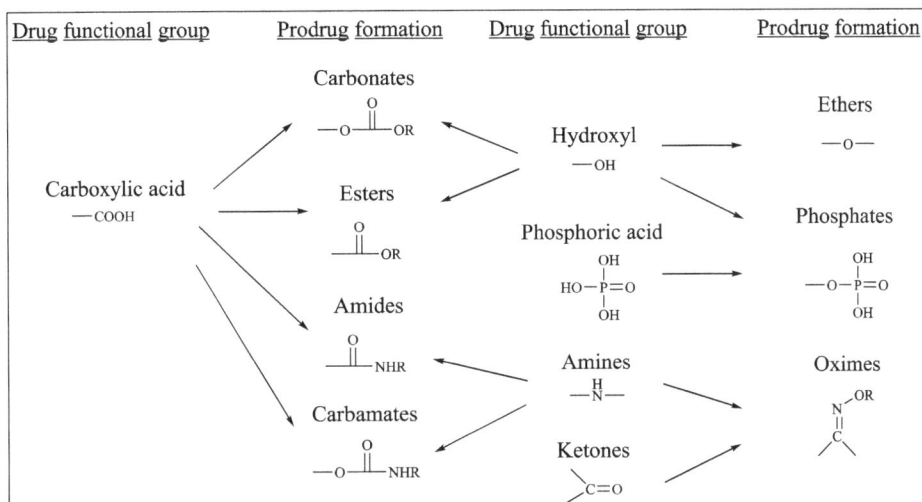

Fig. (59). Common carrier-linked prodrugs.

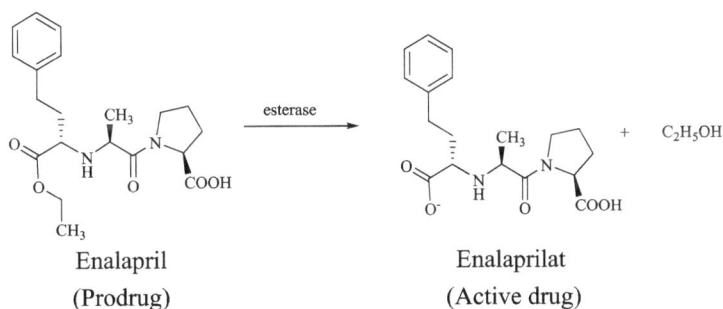

Fig. (60). Activation of enalapril by esterase.

Dipivefrin is a prodrug requiring similar metabolic activation by esterases in cornea, conjunctiva, and aqueous humor after *ophthalmic application* to the active hydroxyl drug epinephrine (Fig. **61**). It is 600-fold more *lipophilic* and is able to permeate the human cornea 17-times faster than epinephrine and thus achieve higher intraocular concentration [160, 161].

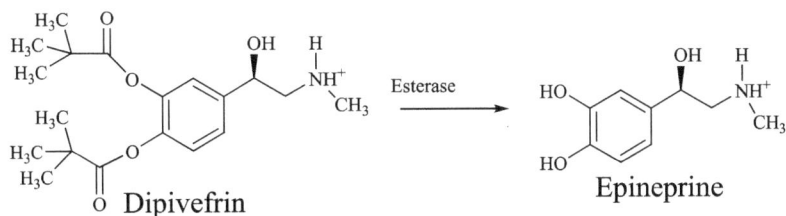

Fig. (61). Metabolic activation of dipivefrin.

Chloramphenicol's *bitter taste is masked* by making palmitate ester prodrug, which once swallowed, is quickly hydrolyzed by esterases to the active *hydroxyl* drug chloramphenicol [159]. **Carbenicillin indanyl ester** is activated by esterase hydrolysis in blood plasma to carbenicillin, a *carboxylic acid* drug. The *lipophilic* indanyl ester furnishes improved *oral bioavailability*.

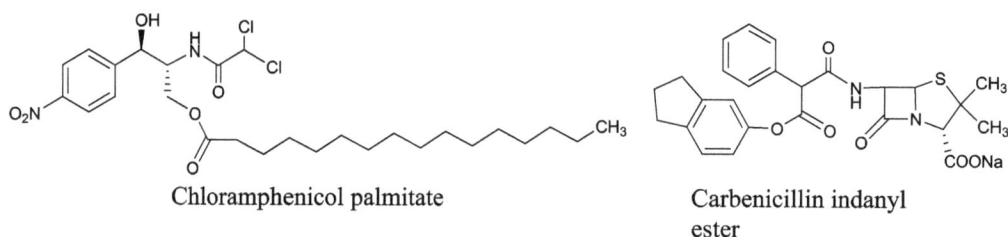

Chloramphenicol palmitate

Carbenicillin indanyl
ester

Some carboxylic esters are not hydrolyzed easily *in vivo* by esterases where *double ester approach* is used. For example, **pivampicillin** is such a prodrug of ampicillin, which is first hydrolyzed by esterase and then spontaneously to ampicillin (Fig. **62**). It serves the same purposes of increasing *lipophilicity* and thus *oral bioavailability*. By this approach, the oral bioavailability of ampicillin (32-55%) is increased considerably (to 87-94%) [158].

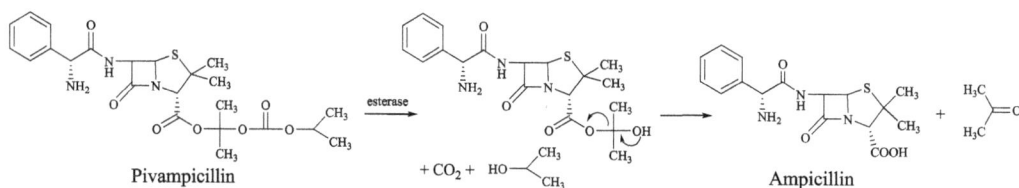

Pivampicillin

Ampicillin

Fig. (62). Metabolic activation of pivampicillin, a double ester prodrug of ampicillin.

Famciclovir, an antiviral agent, is an ethylester prodrug of penciclovir. It is activated by esterases and oxidation to penciclovir (Fig. **63**). The *oral bioavailability* of 4% for penciclovir has been increased to 75% for famciclovir [158].

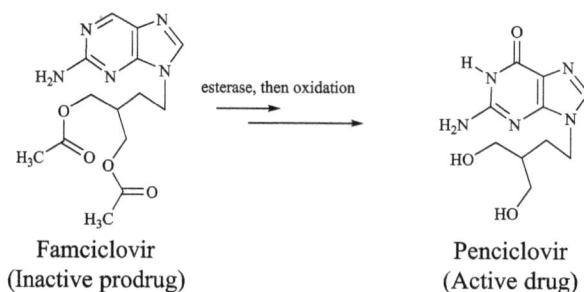

Famciclovir
(Inactive prodrug)

Penciclovir
(Active drug)

Fig. (63). Metabolic activation of famciclovir to its active form penciclovir.

Prodrugs with Increased Hydrophilicity as well as Parenteral Applicability

Sulindac, an NSAID, is an oxide prodrug of sulindac sulfide (most appropriately classified as a *bioprecursor* prodrug). It provides ~ 100-fold increase in *aqueous solubility* and is reduced to the active sulfide form after oral absorption (Fig. **64**) [158].

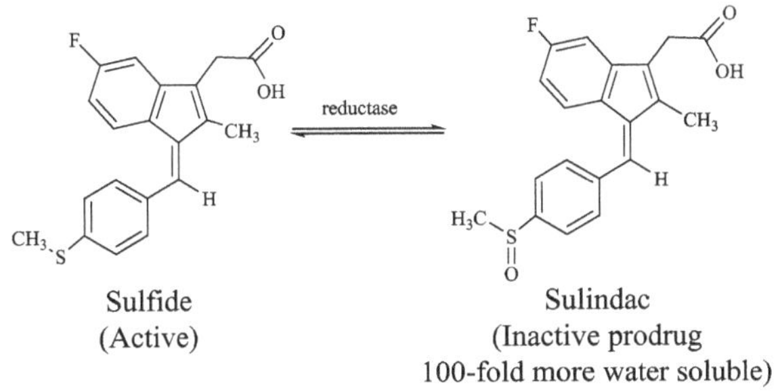

Sulfide
(Active)

Sulindac
(Inactive prodrug
100-fold more water soluble)

Fig. (64). Reductive activation of sulindac.

Fosamprenavir (antiviral) is a phosphate ester prodrug of amprenavir, which is activated by alkaline phosphatases (Fig. **65**). Phosphate prodrug exhibits 10-fold increased *aqueous solubility* as well as patient compliance when given in liquid dosages. Estramustine phosphate (anticancer), prednisolone phosphate, and fludarabine phosphate are other examples of phosphate ester prodrugs. Estramustine phosphate is marketed both as injectable and oral formulations for the treatment of prostate carcinoma. Prednisolone phosphate is a glucocorticoid that enabled the development of a liquid formulation, and thus, improved children's compliance to prednisolone treatment. The antiviral fludarabine phosphate prodrug is marketed as both parenteral and oral dosage forms [158].

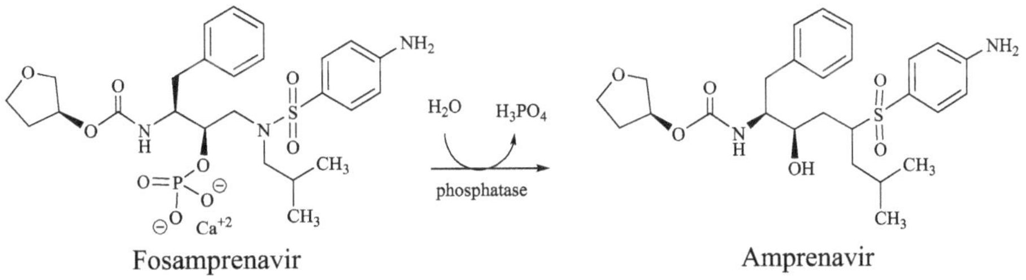

Fosamprenavir

Amprenavir

Fig. (65). Activation of fosamprenavir by alkaline phosphatases.

Fosphenytoin (anticonvulsant) is a phosphonooxymethyl amine of phenytoin, which is rapidly converted to phenytoin by alkaline phosphatases. This prodrug increased aqueous solubility from 20–25 µg per ml of phenytoin to 140 mg per ml of fosphenytoin and is thus suitable for intravenous administration.

Phosphonooxymethyl ether of propofol and propofol phosphate are similar prodrugs for intravenous administration of general anesthetic propofol with significantly increased the aqueous solubility.

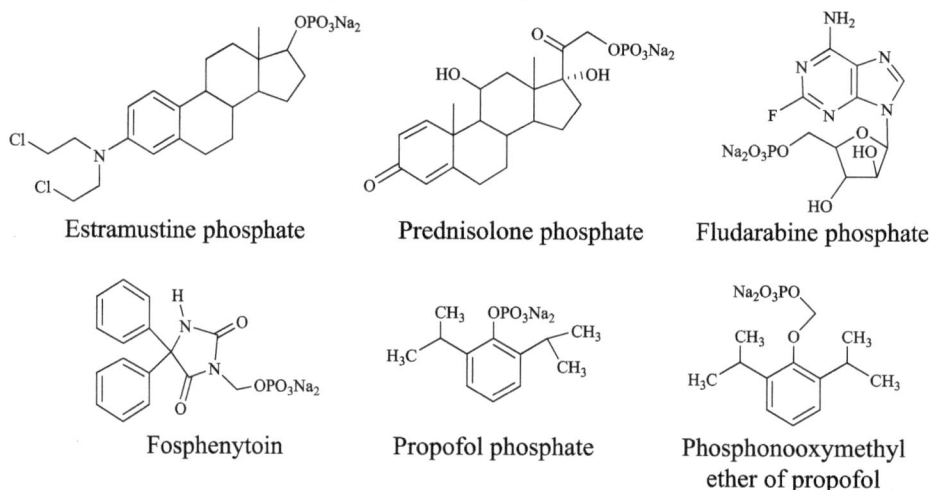

Estramustine phosphate   Prednisolone phosphate   Fludarabine phosphate

Fosphenytoin   Propofol phosphate   Phosphonooxymethyl ether of propofol

Irinotecan (anticancer) dipiperidino carbamate of camptothecin has increased *aqueous solubility* from 2–3 μg per ml (in water) of camptothecin to 20 mg per ml (at pH 3–4) of irinotecan. The prodrug is activated to camptothecin (Fig. **66**) primarily in liver and to a minor extent in tumors by carboxylesterases. Both irinotecan and camptothecin exist in a pH-dependent equilibrium between open and closed lactone forms; only the closed lactone form of camptothecin is pharmacologically active.

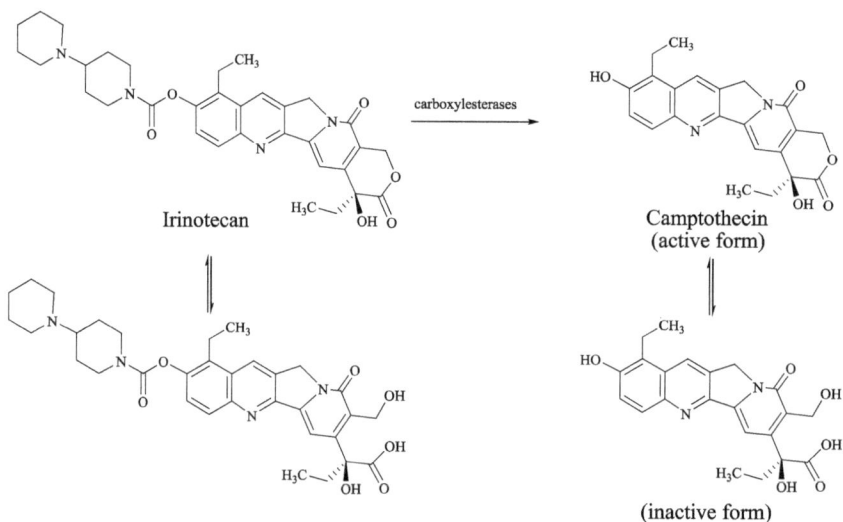

Irinotecan   carboxylesterases →   Camptothecin (active form)

(inactive form)

Fig. (66). Metabolic transformation of irinotecan.

Prodrugs with Improved Site-Specific Delivery

The site specific *delivery* of drugs is an important way of increasing drug's *therapeutic index*. The knowledge of prodrug and drug metabolism is used to concentrate drugs at its target site, thus minimizing systemic toxicity.

Levodopa (anti-Parkinson drug) crosses the blood–brain barrier (BBB) and enters the brain by using L-amino acid transport 1 (LAT1). It is then decarboxylated to dopamine by aromatic amino-acid decarboxylase (Fig. **67**) [162,163].

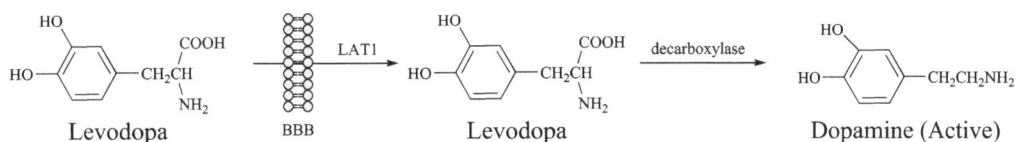

Fig. (67). Brain delivery and activation of levodopa.

The anti-viral drugs **valacyclovir** and **valganciclovir** are the valyl esters of acyclovir and ganciclovir, respectively. Valacyclovir is activated by valacyclovir hydrolase (valacyclovirase) and ganciclovir is activated by intestinal and hepatic esterases. Both of these drugs are transported by human peptide transporter 1 (hPEPT1) and thus improves the oral bioavailability of valacyclovir to 54% (acyclovir's bioavailability is 12-20%) and valganciclovir's bioavailability to 61% (gancicloir's bioavailability is 6%) (Fig. **68**) [158].

XP13512 is an isobutanoyloxyethoxy carbamate of **gabapentin** activated by esterases to use in the treatment of restless leg syndrome and neuropathic pain. The prodrug is transported by both monocarboxylic acid transporter 1 (MCT1) and sodium-dependent vitamin transporter (SMVT) and thus increases the oral bioavailability from 25% (gabapentin) to 84% (XP13512) in monkeys (Fig. **68**) [158].

Other Prodrugs

Bambuterol, a long acting beta-adrenoceptor agonist (LABA), is a bisdimethylcarbamate prodrug of terbutaline. It is designed to have improved metabolic stability against first-pass metabolism on its aromatic hydroxyl functional groups and therefore prolonged duration of action in the lungs. It has to undergo series of hydrolysis and oxidation reactions to the active terbutaline form (Fig. **69**) [158].

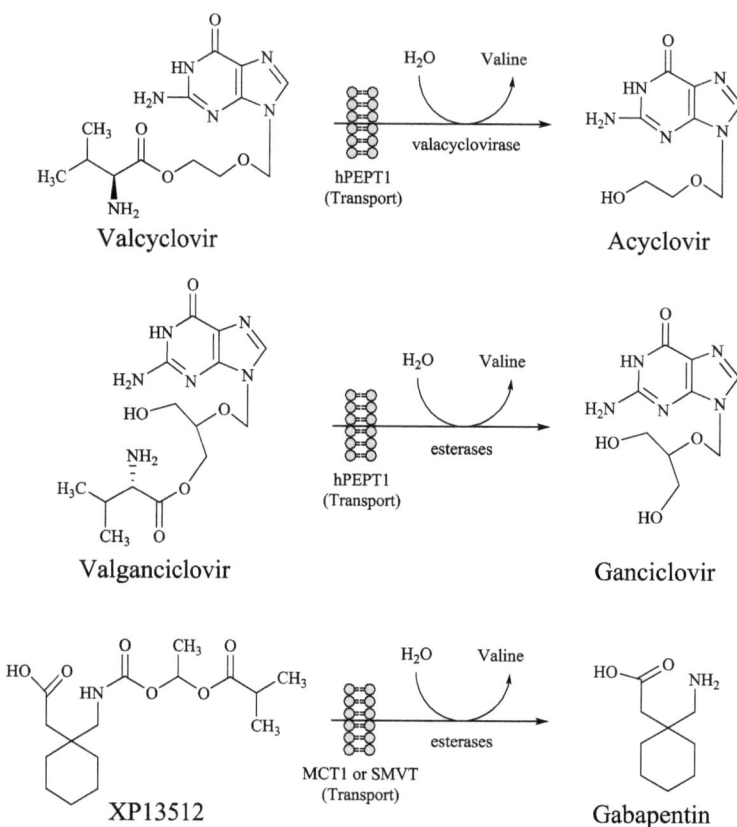

Fig. (68). Prodrugs with improved membrane transport and bioavailability.

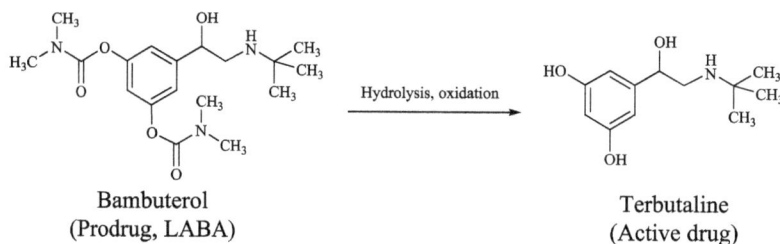

Bambuterol
(Prodrug, LABA)

Terbutaline
(Active drug)

Fig. (69). Bambuterol as a long acting prodrug of terbutaline.

Cyclophosphamide is a successful orally administered anticancer prodrug with reduced toxicity. It is free of toxicity *via* alkylating activity to the gut wall but becomes a toxic alkylating agent after being metabolized. It is possible that the high level of phosphoramidase enzyme present in some tumor cells leads to a greater concentration of the active alkylating species in these cells resulting in selectivity of action (Fig. **70**) [164].

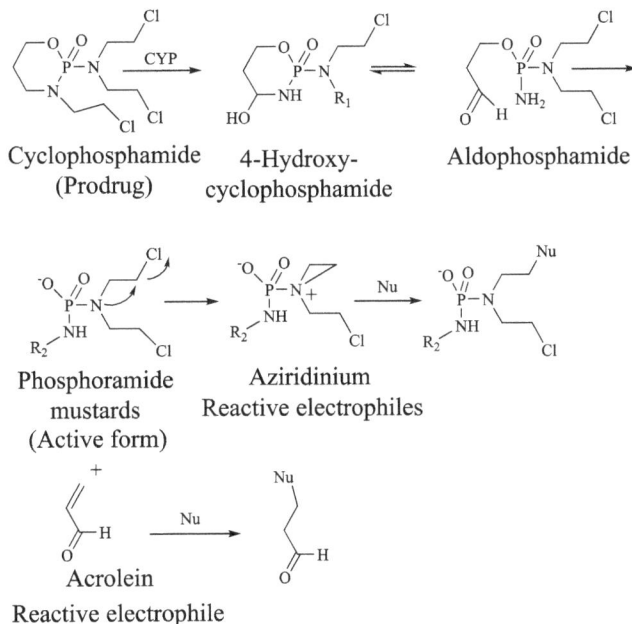

Fig. (70). Cyclophosphamide as the site specific prodrug with reduced toxicity.

The Soft Drugs and Antedrugs

As discussed earlier, both soft drugs and antedrugs overlap with regards to chemical features. However, the antedrugs are intended especially to act locally and not systemically unlike the soft drugs which are intended for a variety of purposes. Some important advantages from soft drugs and antedrugs are listed below:

1. Localization of the drug effects
2. Elimination of toxic metabolites, increasing the therapeutic index
3. Avoidance of pharmacologically active metabolites that can lead to long-term effects
4. Elimination of drug interactions resulting from metabolite inhibition of enzymes
5. Simplification of PK problems caused by multiple active species

Esmolol, a cardioselective β-blocker commonly used in surgical patients to prevent tachycardia by slow intravenous injection, is an example of a systemically used soft drug. The ester-methyl side chain is rapidly hydrolyzed (half-life ~9 minutes) by plasma esterases to an inactive carboxylic acid (Fig. **71**) [165, 166].

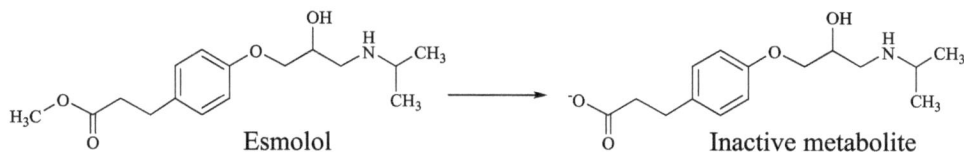

Fig. (71). Esmolol as a short acting softdrug.

Loteprednol etabonate is the first corticosteroid synthesized based on soft drug approach (Fig. **72**). The drug is also deactivated in systemic circulation by esterase hydrolysis to a carboxylic acid form and intended to work topically in the eye. Thus it is also an antedrug and lacks serious systemic side effects when used in inflammatory and allergy-related ophthalmic disorders as well as asthma, rhinitis, colitis, and dermatological problems [157].

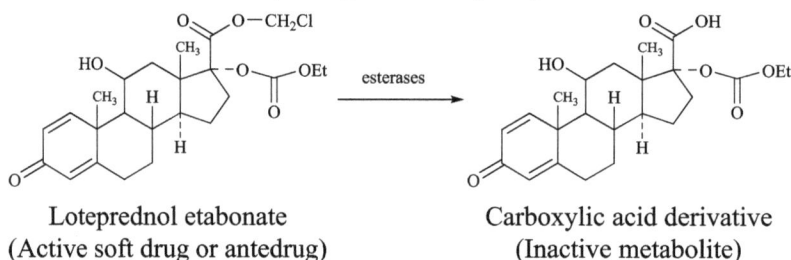

Loteprednol etabonate
(Active soft drug or antedrug)

Carboxylic acid derivative
(Inactive metabolite)

Fig. (72). Inactivation of soft drug, or antedrug loteprednol.

Fluticasone propionate, an anti-inflammatory steroid approved for use in asthma and allergic rhinitis, is believed to work according to the same soft drug or antedrug principle (Fig. **73**). It is inactivated *in vivo* by hydrolytic or oxidative cleavage of the thiol ester bond to the carboxylic acid derivative. This, in principle allows reduced systemic exposure of the glucocorticoid and thus reduces systemic toxicity [157].

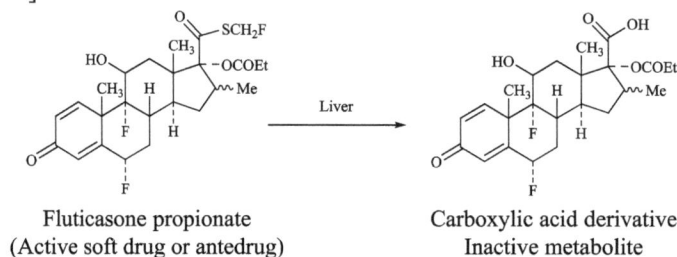

Fluticasone propionate
(Active soft drug or antedrug)

Carboxylic acid derivative
Inactive metabolite

Fig. (73). Fluticasone propionate as an antedrug.

FACTORS AFFECTING DRUG METABOLISM

Drug metabolism is affected by multiple factors including the following:

- Genetic factors
- Due to differences in the expression of metabolizing enzymes
- Genetic polymorphism
- Physiologic factors
- Age
- Sex
- Pregnancy
- Nutritional status
- Pharmacodynamic factors
- Dose frequency
- Route of administration
- Protein binding.
- Environmental and other factors
- Toxic chemicals such as CO and pesticides may induce and inhibit metabolic enzymes (see enzyme induction and inhibition)
- Smoking
- Dietary factors

A few of the most important factors are described in detail below.

Influence of Age on Drug Metabolism

A few notable physiologic changes in elderly including low plasma clearance, blood flow, glomerular filtration, hepatic enzyme activity and plasma protein bonding profoundly impair both normal metabolism, especially the phase II metabolism, as well as first-pass and elimination of many drugs. For example the first pass metabolism of diazepam, theophylline, morphine, propranolol and amitriptyline are considerably low in elderly. During the first part of gestation, human fetus is only capable of metabolize by CYP 3A, which is present in placenta. In general, fetus can perform negligible Phase-II metabolism due to underdeveloped enzyme system and are thus at high risk of toxicity by pennant's metabolites. In about 1 to 2 months after birth, most of the metabolic enzymes are to adult levels, except UDP glucuronide transferase which takes longer. This is why infants are unable to glucuronide chloramphenicol leading to gray-baby syndrome. Neonatal hyperbilirubinemia is also the result of their inability to glucuronide bilirubin.

Influence of Sex on Drug Metabolism

The difference in drug metabolism in male and female varies in different species and animals. In human, nicotine, aspirin diazepam, are known to be metabolized more rapidly in male compared to female possibly due to more anabolic action of steroid.

Genetic Polymorphism

Genetic polymorphism can be defined as the genetic differences in the natural expression of enzyme isoforms. The genetic polymorphisms of metabolizing enzymes, transporter proteins, drug target receptors/ and ion channels translate into altered drug pharmacokinetics as well as pharmacodynamics. This has led to the study of pharmacogenomics and personalized medicine as a new emphasis in drug therapy. Genetic factors have profound effect on drug metabolism in in human. For example, rapid acetylation of drugs (such as isoniazid) is observed in 90% of Asians and Eskimos due to expression of more acetyl *N*-transferase. They are called rapid acetylators. Egyptians and Mediterranean are on the other hand called slow acetylators, due to low level of acetyl *N*-transferase expression.

Clinically significant polymorphism has also been seen in oxidizing enzymes and many other conjugation enzymes. For example, codeine is metabolized to active narcotic analgesic morphine by the action of CYP2D6, which shows genetic polymorphism leading to differential clinical outcomes.

Effects of Smoking on Drug Metabolism

Chemicals in cigarette smoke such as polycyclic aromatic hydrocarbons can induce certain cytochrome P450 enzymes. As a result, drug metabolism by those enzymes are changed; for example theophylline clearance is higher in smokers than nonsmokers regardless of age [167]. Smoking induces CYP1A2 activity, which significantly increases metabolism of olanzapine, amitriptyline, clozapine, duloxetine, fluvoxamine, haloperidol, imipramine, ondansetron, propranolol, theophylline, warfarin resulting in diminished effects of these drug [168, 169]. Smoking is also known to induces CYP2B6 thus increases the metabolism of Cyclophosphamide, bupropion and clopidogrel [169].

Effects of Grapefruit Juice on Drug Metabolism

It has been reported that various natural flavonoids present in grapefruit juice are potent inhibitors of certain cytochrome P450 isoenzymes including CYP3A4, CYP2C9, and CYP2D6 [170]. Inhibition of CYP 3A4 results in an increase in Cmax and area under the concentration time curve (AUC) of drugs with a high first-pass metabolism of felodipine and amiodarone [171].

CASE STUDIES

Case 1. After oral administration of a new drug "X", plasma concentrations vs. time profile yielded the following curve. Based on this information, explain the origin of second peak.

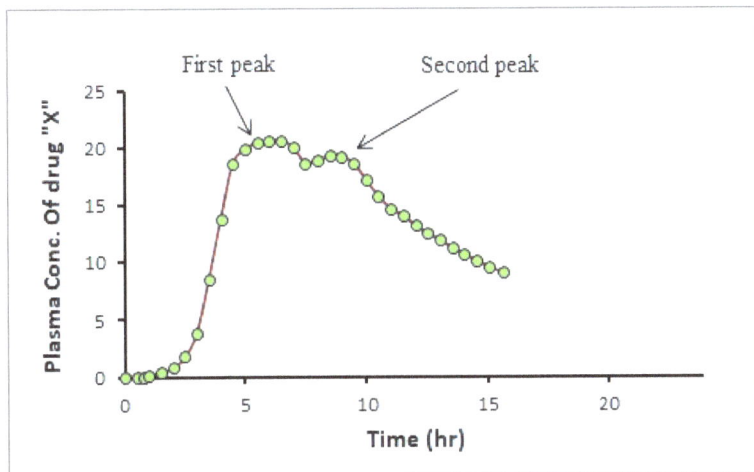

Answer: Drug "X" is most likely excreted unchanged *via* liver. Excreted drug is reabsorbed from the intestine, due to enterohepatic recirculation, which contributes to the second peak. This is known as double-peak phenomena. Drugs like Alprazolam exhibit this type of phenomena. Alprazolam has only one relatively metabolically labile functional group, a methyl on the triazole ring. This methyl group gets metabolized by CYP enzyme but it takes time, thus giving a high bioavailability. Also, because it is relatively a larger lipophilic molecule it is prone to excrete *via* liver into intestine from where it gets reabsorbed [172].

Case 2. A six years old child weighing 56 pounds was presented to ER with consumption of about 45 mL of antifreeze (Ethylene glycol). Based on metabolism scheme of ethylene glycol, provide a detail explanation of consequences if the patient is not treated. The structure of ethylene glycol is HO-CH_2-CH_2-OH.

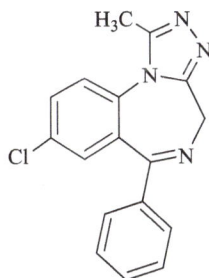

Alprazolam

Answer: Ethylene glycol (EG) is a solvent used in many products, such as antifreeze fluid, de-icing solutions, carpet, and fabric cleaners. It has a sweet taste, which leads accidental ingestion. It is rapidly absorbed from the stomach and small intestine, and is quickly distributed throughout the body. The ingested amount of EG required to produce toxicity in animals is approximately 1.0 to 1.5 mL per kg, or 100 mL in an adult. Toxicity results from the depressant effects of ethylene glycol on the central nervous system (CNS) while metabolic acidosis and renal failure are caused by its metabolites. Oxidative reactions convert EG to glycoaldehyde, and then to glycolic acid (Fig. **74**), which is the major cause of metabolic acidosis. Glycolic acid is further oxidized to oxalic acid. Oxalic acid does not contribute to the metabolic acidosis, but it is deposited as calcium oxalate crystals in many tissues, which can result in kidney failure and death.

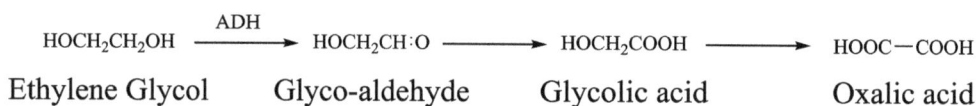

$$HOCH_2CH_2OH \xrightarrow{ADH} HOCH_2CH{:}O \longrightarrow HOCH_2COOH \longrightarrow HOOC-COOH$$

Ethylene Glycol   Glyco-aldehyde   Glycolic acid   Oxalic acid

Fig. (74). Metabolism of ethylene glycol.

The clinical syndrome of EG intoxication has traditionally been divided into three stages: progressive involvement of the CNS, the cardiopulmonary systems, and the kidneys. EG produces CNS depression similar to that of ethanol. Symptoms of ethylene glycol toxicity include confusion, ataxia, hallucinations, slurred speech, and coma. Symptoms are most severe six to 12 hours after ingestion, when the acidic metabolites of ethylene glycol are at their maximal concentration [173].

Case 3. Based on metabolic pathway of methanol (Fig. **75**), explain why it is not safe to drink methanol. (Structure of methanol is CH_3OH)

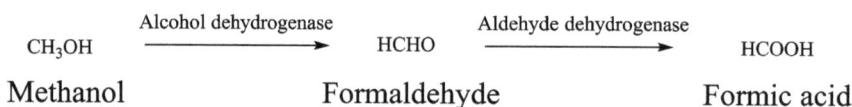

$$CH_3OH \xrightarrow{\text{Alcohol dehydrogenase}} HCHO \xrightarrow{\text{Aldehyde dehydrogenase}} HCOOH$$

Methanol   Formaldehyde   Formic acid

Fig. (75). Metabolism of methanol.

Answer: Methanol poisoning is manifested by nausea, vomiting, abdominal pain, and mild central nervous system depression. Methanol is oxidized by alcohol dehydrogenase to formaldehyde, which is oxidized to formic acid by formaldehyde dehydrogenase. In cases of methanol poisoning, formic acid accumulates. The acidosis, results from methanol poisoning, is caused by formic acid accumulation. It has been reported that there is a direct correlation between the formic acid concentration and morbidity and mortality. Formic acid inhibits cytochrome oxidase and is the prime cause of ocular toxicity ranging from altered visual fields to complete blindness [174].

Case 4. A patient calls you and informs that when he opens his aspirin bottle, the tablet smells like vinegar like. From medicinal chemistry stand point, explain the root cause of vinegary smell from the aspirin bottle.

Answer: Ester and amide linkages in many drugs undergo hydrolytic reaction. Aspirin (acetyl salicylic acid), in presence of moisture from the environment, undergoes hydrolytic degradation developing acetic acid and salicylic acid. Vinegar is a dilute solution of acetic acid. Thus, the open bottle of uncoated aspirin tablets smells like vinegar.

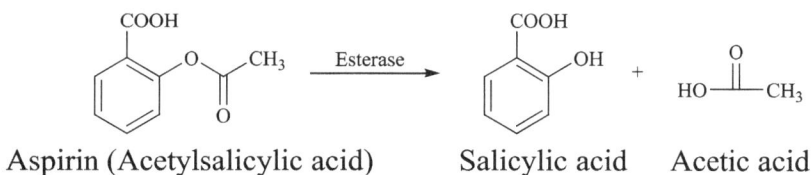

Aspirin (Acetylsalicylic acid)   Salicylic acid   Acetic acid

The metabolism of ester and amide linkages in many drugs is catalyzed by hydrolytic enzymes present in various tissues like liver, kidney, intestine, and in plasma. The metabolic products formed (carboxylic acids, alcohols, phenols, and amines) generally are polar and functionally more susceptible to conjugation and excretion than the parent drugs.

Case 5. A patient diagnosed with latent tuberculosis infection presents a prescription for isoniazid 300 mg daily for nine months. When you ask the patient if his doctor has suggested him to take any other medicine, he says "No." Based on the metabolic pathways of isoniazid, what should you do and why?

Answer: Patient should be advised to take pyridoxine (Vitamin B_6) to prevent peripheral neuropathy.

The acetylation pattern of many drugs, *e.g.*, isoniazide, hydralazine, procainamide *etc.* in human population displays a bimodal character in which the drug is conjugated either rapidly or slowly with acetyl-CoA. This phenomenon is termed *acetylation polymorphism*. Individuals are classified as having either slow or rapid acetylator phenotypes. This variation in acetylation is genetic and is caused mainly by differences in N-acetyltransferease (NAT) activity. A high proportion of Eskimos and Asians are rapid acetylators, whereas Egyptians and some Western Europian groups are mainly slow acetylators. For some drugs, slow acetylators seem more likely to develop adverse reactions, whereas rapid acetylators are more likely to show an inadequate therapeutic response.

The plasma half-life of isoniazide in rapid acetylators is about 45-80 minutes; in slow acetylators the half-life ranges from 140 to 200 minutes. Slow acetylators tend to accumulate higher plasma conc. of isoniazide than the rapid acetylators. Accordingly slow acetylators are more prone to develop peripheral neuritis and drug-induced systemic lupus erythematosus syndrome than the rapid acetylators. However, patients who are rapid acetylators appear to be more likely to develop isoniazide associated hepatitis. The liver toxicity presumably arises from initial hydrolysis of N-acetylated metabolite N-acetylisoniazide to acetylhydrazine (Fig. **76**). The latter metabolite is further converted (by cytochrome P-450 enzyme systems) to chemically reactive acylating intermediates that covalently bind to hepatic tissue, causing necrosis [175].

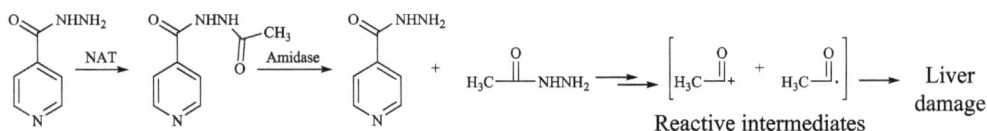

Fig. (**76**). Metabolic intoxication mechanism of isoniazid.

Case 6: CF, a 29-year-old Caucasian schizophrenic woman, was admitted to a psychiatric unit for acute psychotic attack. She is on oral contraceptives (OC) for long time and does not smoke any tobacco product. She was also diagnosed with oral candidiasis, for which she was prescribed simultaneously with fluconazole and miconazole gel for about 1 week after about 3 weeks treatment with antipsychotic drug clozapine. After about a month of clozapine treatment, she was showing good response to her psychotic symptoms. However, after 3 weeks, her clozapine plasma level had increased to toxic level showing symptoms of eosinophilia, *e.g.*, nausea, vomiting, palpitations, echocardiographic abnormalities and pericardial effusion. The symptoms were resolved after discontinuation of OC and interruption of clozapine [176].

Question: Why CF experienced the clozapine toxicity, and why had it resolved after discontinuation of OC? Explain clearly with the metabolic pathways of clozapine.

Answer: Clozapine is extensively metabolized by many CYP450 enzymes including CYP3A4 and CYP1A2 forming pharmacologically active metabolite norclozapine and pharmacologically inactive metabolite clozapine-*N*-oxide (Fig. **77**) [177].

The antifungals and OCs are CYP450 inhibitors that decrease clozapine metabolism substantially rendering high blood concentration of clozapine and subsequent toxicity. This is why blood level monitoring of clozapine is recommended for patients who are taking clozapine in combination with antifungals and/or OCs [176]. It should also be noted that cigarette smoking is inducer of CYP450 enzymes causing low blood level of clozapine rendering it less active. CF is not a smoker thus has no effect from this.

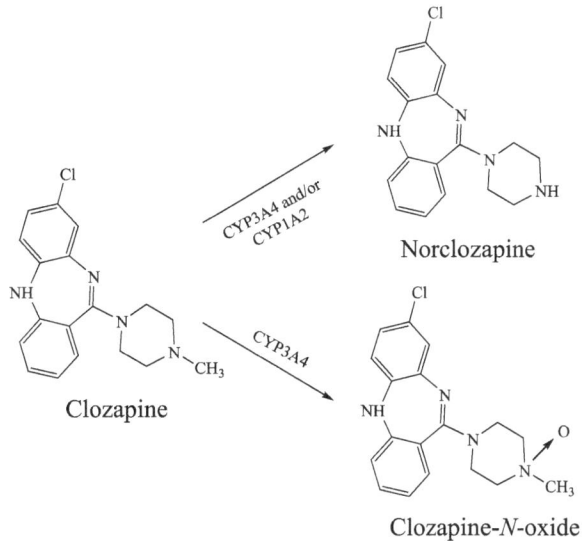

Fig. (77). Metabolism of clozapine.

Case 7: BG, a 61 year-old male, has over 40 years history of smoking. He maintains a moderate exercise routing including swimming. For last three years he has been managing his blood pressure by taking a combination of metoprolol tartrate (50 mg, 2 per day) and amlodipine besylate (10 mg one per day). He decided to quit smoking and thus started to take bupropion hydrochloride (150 mg one per day taper for 5 days and then 2 per day). After starting bupropion, he needed to cut far short his swimming distance due to dyspnea and tiredness. Because he felt otherwise normal, he discussed this with his pharmacist instead of seeking immediate medical attention. The pharmacist advised him to take one-half of metoprolol tablet (25 mg) twice per day. Within 2-3 weeks of this changed dose of metoprolol BG was back to his normal distance of swimming. Later he discussed this incidence with his primary physician who concurred with this reduction in metoprolol dose [178].

Question: Explain clearly the reason for issues after starting bupropion and the reason for reducing the dose of metoprolol. Use the specific metabolic route of metoprolol and enzyme(s) involved.

Answer: Bupropion inhibited the metabolism of metoprolol increasing its blood concentration to toxic level that led to dyspnea and shortage of breathing and tiredness. Bupropion is a strong inhibitor of CY2D6, the major metabolizing enzyme for metoprolol [178]. As shown in Fig. **78**, CYP2D6 plays crucial role in metabolizing metoprolol to α-hydroxymetoprolol and *O*-demethylmetoprolol.

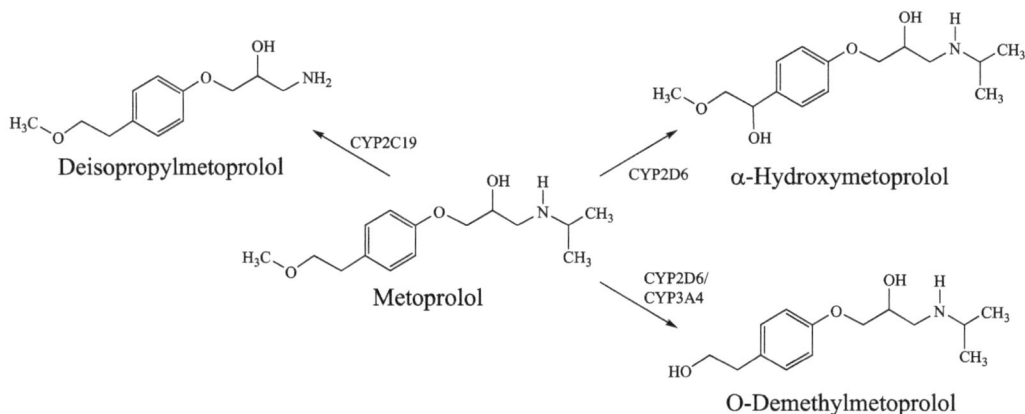

Fig. (78). Metabolic pathway of metoprolol. CYP2D6 is its major metabolizing enzyme that is inhibited by bupropion.

STUDENT SELF-STUDY GUIDE

1. What roles are played by drug metabolism? Identify with specific examples.
2. Explain the role of stereochemistry in metabolism of drugs with example of warfarin, ibuprofen and etomidate.
3. What is first pass effect; enterohepatic circulation? Why and how do they occur? List drug examples.
4. Explain metabolic pathways in the intestinal mucosa.
5. CYP450, Hepatic microsomal flavin containing monooxygenases (MFMO or FMO) Monoamine Oxidase (MAO) and Hydrolases. List drugs metabolized by these enzymes and the active sites of these enzymes. List types of metabolic reactions catalyzed by these enzymes.
6. List specific type of CYP enzymes and the corresponding number of drugs metabolized by each type.
7. List various types of CYP enzymes with their main functions
8. Describe basics of drug interaction with respect to metabolic enzymes
9. List mechanism and routes of aromatic hydroxylation. What are the effects of electron donating and withdrawing groups in aromatic hydroxylation? Give drug examples. What is NIH shift?
10. Explain oxidation of olefins and role of epoxide hydrolase. Can olefinic epoxide be converted to alcohol (aromatic epoxide) by NIH shift?
11. What type of *Carbon* in a drug molecule cannot be hydroxylated?
12. What is allylic and benzylic hydroxylation? Show drug examples.
13. List drug examples where hydroxylation occurs on Cα to C=O and C=N bonds
14. List drug examples where hydroxylation occurs at aliphatic and alicyclic carbon atoms. Which carbons are more easily hydroxylated?
15. What is N-oxidation and N-dealkylation? Which enzymes are involved? How

do you differentiate between N-dealkylation and deamination? Give drug examples. What type of metabolism generates lactams instead of causing dealkylation?

16. What is the difference between mixed function oxidases and amine oxidases?
17. What is the difference between ethanol oxidation and O-dealkylation?
18. Explain S-dealkylation, desulfuration and S-oxidation, with drug examples.
19. How do steric factors influence S- O- and N-dealkylation?
20. Explain oxidative dehalogenation using chloramphenicol as an example. Discuss chloramphenicol toxicity when used in neonates.
21. What is MFMO and its active site? What types of functional groups are metabolized by this enzyme? Give drug examples.
22. Explain MAO, dehydrogenases, xanthene oxidases and their functions with drug examples. Explain the difference between MAO-A and MAO-B.
23. Discuss the functions of coenzymes alcohol and aldehyde dehydrogenase, and the types of drugs they metabolize.
24. Discuss the functions of azo and nitro reductases, and the drugs they metabolize.
25. List various types of hydrolytic enzymes. Compare their rate of hydrolysis of esters, amides, carbonates and carbamates.
26. What are prodrugs and antedrugs? List their advantages, with examples.
27. What are the different types of conjugation reactions?
28. Discuss enzymes and substrates involved in glucuronidation, and sulfate conjugation.
29. Why is acetaminophen toxic to neonates? Describe the mechanism of phenacetin and acetaminophen toxicity.
30. What types of drugs or metabolites form glycine conjugates?
31. What are different mechanisms involved in glutathione conjugation? What is mercapturic acid conjugate? Mercapturic acid conjugate of acetaminophen is a sign of its toxicity – why?
32. Explain the mechanism of acetylation. Explain slow and fast acetylator status?
33. What is COMT? What coenzyme is involved in its action? What types of drugs and/or neurotransmitters are metabolized by COMT?

STUDENT SELF-ASSESSMENT

Part I: Multiple Choice Questions

1. Which of the following metabolites would be the most likely urinary excretion product of nicotine?

Nicotine

A   B   C   D   E

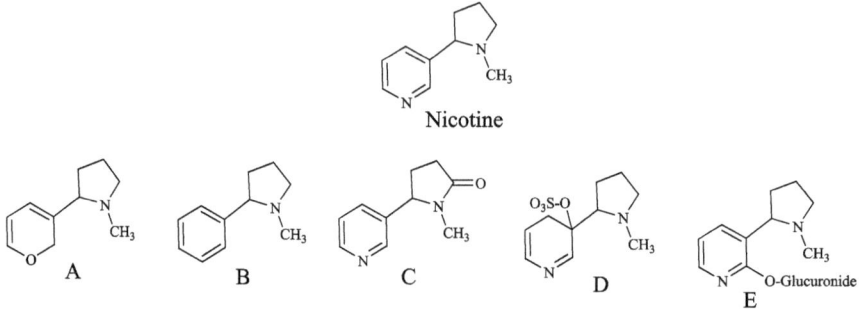

2. All of the following drugs will undergo first pass metabolism except

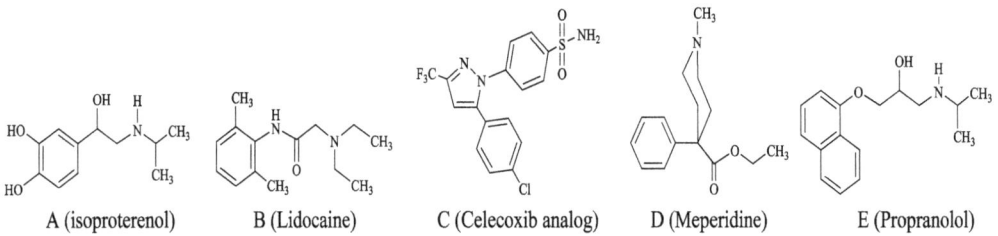

A (isoproterenol)   B (Lidocaine)   C (Celecoxib analog)   D (Meperidine)   E (Propranolol)

3. The active site of drug metabolizing enzyme CYP is
   A. NADPH/NADP$^+$
   B. FADH$_2$/FAD
   C. Hydroquinone/quinone
   D. Fe^{2+}/Fe^{3+}
   E. Fe$^+$/Fe^{2+}

4. Which one is not a metabolite of tolbutamide?

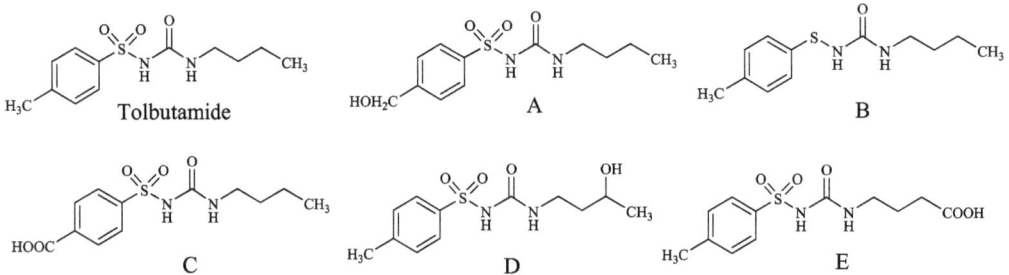

Tolbutamide   A   B

C   D   E

5. The least likely metabolite of indomethacin is

Indomethacin

A   B   C   D   E

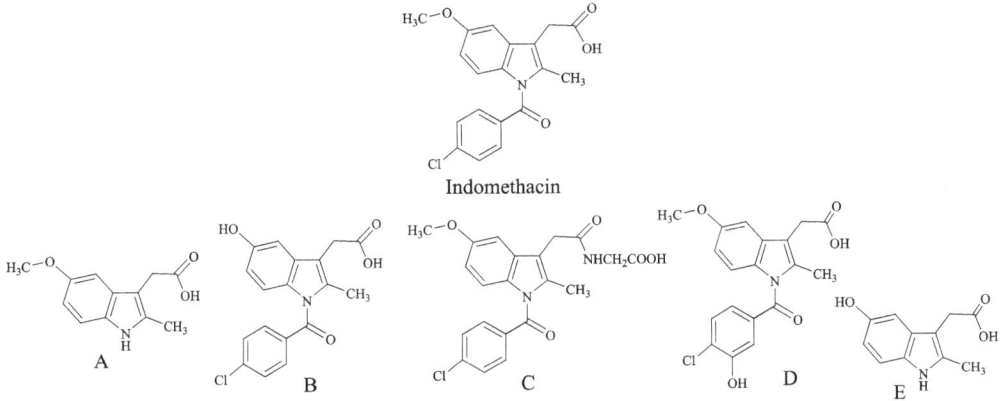

6. The <u>coenzyme</u> involved in the following metabolic reaction is:

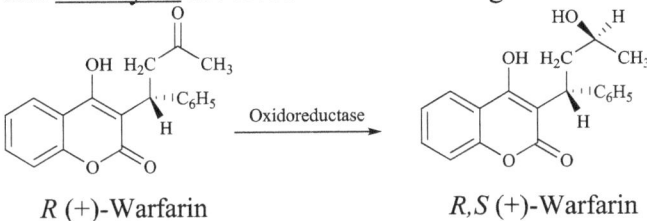

R (+)-Warfarin →(Oxidoreductase)→ *R,S* (+)-Warfarin

A. Iron containing heme
B. FAD
C. FADH$_2$
D. NAD$^+$
E. NADH

7. Which of the following drugs is most susceptible to metabolism by <u>MAO</u>

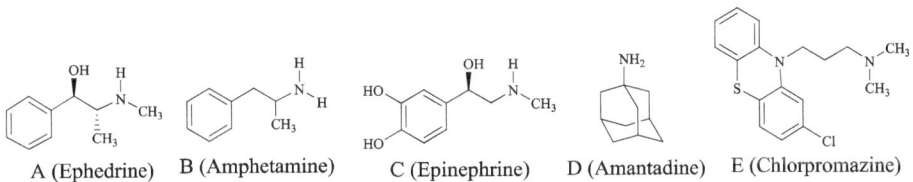

A (Ephedrine) B (Amphetamine) C (Epinephrine) D (Amantadine) E (Chlorpromazine)

8. The correct order (highest to lowest rate)of rate of hydrolytic cleavage is:
    A. Ester → carbonate → amide → carbamate
    B. Carbamate → amide → carbonate → ester
    C. Ester → amide → carbonate → carbamate
    D. Ester → carbamate → carbonate → amide
    E. Ester → carbonate → carbamate → amide

9. Which statement about L-DOPA is <u>not true</u>?

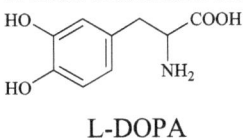

L-DOPA

A. It is highly polar so is not centrally active
B. It is a prodrug
C. It can be delivered into brain by L-amino acid uptake system
D. Dopamine is its active form
E. Decarboxylase is an important metabolizing enzyme into brain for its activation

Part II: K-Type question. Chose the answer

A. If **I only** is correct
B. If **III only** is correct
C. If **I and II** are correct
D. If **II and III** are correct
E. If **I, II, and III** are correct

10. Which of the following statements concerning CYP 450 are correct?
    I. The CYP7, CYP11, and CYP27 subfamilies are involved in cholesterol metabolism.
    II. A single drug may be metabolized by multiple isoforms.
    III. The majority of xenobiotics, or drugs, are metabolized by the CYP4B and CYP1A subfamilies.
11. The chloramphenicol's "Gray baby" syndrome is attributed with the following possible metabolic facts
    I. It is extensively metabolized by glucuronidation in children
    II. It forms toxic metabolites by oxidative metabolism
    III. Its action is prolonged in children
12. Which of the following statements about glutathione are correct?
    I. Glutathione is a tripeptide found virtually in all tissues
    II. Its reduced form contains thiol function which scavenges harmful electrophilic parent drugs or their metabolites
    III. The metabolic reaction types involving glutathione include: electrophilic or aromatic substitution and Michael addition
13. A variety of methyl transferases are involved in drug methylation which may include:
    I. Catechol *O*-methyl transferase
    II. *N*-Methyl transferase
    III. Phenyl *O*-methyl transferase
14. Three possible metabolic pathways for amantadine (structure shown) are:

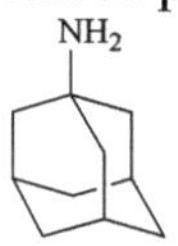

    I. Deamination

    II. *N*-Oxidation

    III. *N*-Glucuronidation

15. Which of the following reactions can be classified as phase II metabolism?

    I.

    II.

    III.

16. Conditions that tend to increase the action of an orally administered drug that undergoes phase II metabolism include

    I. Enterohepatic circulation

    II. Enzyme saturation

    III. First-pass effect

17. Sulfasalazine is metabolically activated by the following enzyme(s):

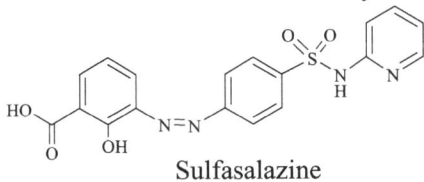

Sulfasalazine

    I. Bacterial reductase in intestine

    II. Microsomal nitro reductase

    III. Hepatic soluble nitro reductase

18. S-Adenosyl methionine is a coenzyme for

    I. COMT

    II. N-Methyl transferase

    III. S-Methyl transferase

19. Advantages of antedrug include

    I. Localization of drug effect

    II. Modification of PK problem

    III. Increase of systemic effect

Part III: Matching questions.

For each drug, select its most likely metabolic pathway.

    A. Ether glucuronidation

    B. Ester glucuronidation

    C. Nitroreduction

  D. Oxidative deamination
  E. Ester hydrolysis

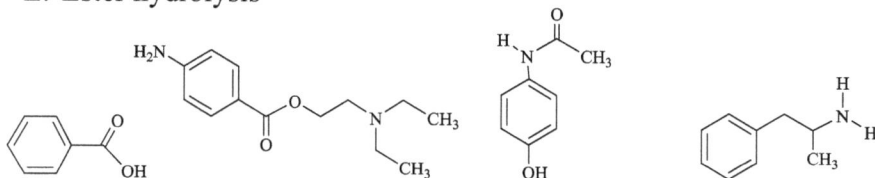

20. Benzoic acid   21. Procaine   22. Acetaminophen   23. Amphetamine

For each pathway shown (24-26), select the correct <u>enzyme</u> from A-E.
  A. Glutathione-S-transferase
  B. Methyl transferase
  C. Acyl-CoA synthetase
  D. Methionine adenosyl transferase
  E. Glycine N-acyltransferase

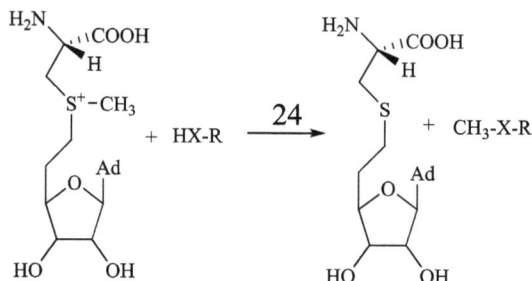

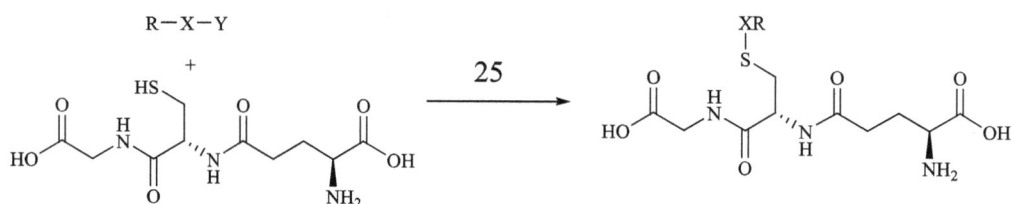

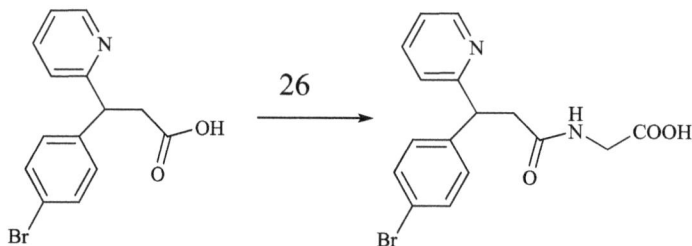

CONSENT FOR PUBLICATION

Not applicable.

CONFLICT OF INTEREST

The authors declare no conflict of interest, financial or otherwise.

ACKNOWLEDGEMENT

Declared none.

REFERENCES

[1]   Danielson PB. The cytochrome P450 superfamily: biochemistry, evolution and drug metabolism in humans. Curr Drug Metab 2002; 3(6): 561-97.
[http://dx.doi.org/10.2174/1389200023337054] [PMID: 12369887]

[2]   King CD, Rios GR, Green MD, Tephly TR. UDP-glucuronosyltransferases. Curr Drug Metab 2000; 1(2): 143-61.
[http://dx.doi.org/10.2174/1389200003339171] [PMID: 11465080]

[3]   Sheehan D, Meade G, Foley VM, Dowd CA. Structure, function and evolution of glutathione transferases: implications for classification of non-mammalian members of an ancient enzyme superfamily. Biochem J 2001; 360(Pt 1): 1-16.
[http://dx.doi.org/10.1042/bj3600001] [PMID: 11695986]

[4]   Müller FO, Van Dyk M, Hundt HK, *et al.* Pharmacokinetics of temazepam after day-time and night-time oral administration. Eur J Clin Pharmacol 1987; 33(2): 211-4.
[http://dx.doi.org/10.1007/BF00544571] [PMID: 2891534]

[5]   Fura A, Shu Y-Z, Zhu M, Hanson RL, Roongta V, Humphreys WG. Discovering drugs through biological transformation: role of pharmacologically active metabolites in drug discovery. J Med Chem 2004; 47(18): 4339-51.
[http://dx.doi.org/10.1021/jm040066v] [PMID: 15317447]

[6]   Maxwell RA, Eckhardt SB. Drug discovery. Humana Press 1990; p. 455.
[http://dx.doi.org/10.1007/978-1-4612-0469-5]

[7]   Nelson SD, Mitchell JR, Timbrell JA, Snodgrass WR, Corcoran GB III. Isoniazid and iproniazid: activation of metabolites to toxic intermediates in man and rat. Science 1976; 193(4256): 901-3.
[http://dx.doi.org/10.1126/science.7838] [PMID: 7838]

[8]   Lin JH, Lu AY. Role of pharmacokinetics and metabolism in drug discovery and development. Pharmacol Rev 1997; 49(4): 403-49.
[PMID: 9443165]

[9]   Guengerich FP. Cytochrome p450 and chemical toxicology. Chem Res Toxicol 2008; 21(1): 70-83.
[http://dx.doi.org/10.1021/tx700079z] [PMID: 18052394]

[10]   Nelson D. Cytochrome P450 Homepage. University of Tennessee. Retrieved 2017-6-18.

[11]   Meunier B, de Visser SP, Shaik S. Mechanism of oxidation reactions catalyzed by cytochrome p450 enzymes. Chem Rev 2004; 104(9): 3947-80.
[http://dx.doi.org/10.1021/cr020443g] [PMID: 15352783]

[12]   Poulos TL, Finzel BC, Howard AJ. High-resolution crystal structure of cytochrome P450cam. J Mol Biol 1987; 195(3): 687-700.
[http://dx.doi.org/10.1016/0022-2836(87)90190-2] [PMID: 3656428]

[13]   Ortiz de Montellano PR, Ed. Cytochrome P450: structure, mechanism, and biochemistry. 2005.
[http://dx.doi.org/10.1007/b139087]

[14]   Sligar SG, Cinti DL, Gibson GG, Schenkman JB. Spin state control of the hepatic cytochrome P450 redox potential. Biochem Biophys Res Commun 1979; 90(3): 925-32.

[http://dx.doi.org/10.1016/0006-291X(79)91916-8] [PMID: 228675]

[15]   Rittle J, Green MT. Cytochrome P450 compound I: capture, characterization, and C-H bond activation kinetics. Science 2010; 330(6006): 933-7.
[http://dx.doi.org/10.1126/science.1193478] [PMID: 21071661]

[16]   Gavhane YN, Yadav AV. Loss of orally administered drugs in GI tract. Saudi Pharm J 2012; 20(4): 331-44.
[http://dx.doi.org/10.1016/j.jsps.2012.03.005] [PMID: 23960808]

[17]   Rowland M. Influence of route of administration on drug availability. J Pharm Sci 1972; 61(1): 70-4.
[http://dx.doi.org/10.1002/jps.2600610111] [PMID: 5019220]

[18]   Gao Y, Shao J, Jiang Z, *et al.* Drug enterohepatic circulation and disposition: constituents of systems pharmacokinetics. Drug Discov Today 2014; 19(3): 326-40.
[http://dx.doi.org/10.1016/j.drudis.2013.11.020] [PMID: 24295642]

[19]   Barry M, Feely J. Enzyme induction and inhibition. Pharmacol Ther 1990; 48(1): 71-94.
[http://dx.doi.org/10.1016/0163-7258(90)90019-X] [PMID: 2274578]

[20]   Peltoniemi M. Effects of cytochrome p450 enzyme inhibitors and inducers on the metabolism of S-ketamine. Annales of Turun Yliopisto, University of Turku, Turku, Finland; 2013.

[21]   Testa B, Jenner P. Novel drug metabolites produced by functionalization reactions: chemistry and toxicology. Drug Metab Rev 1978; 7(2): 325-69.
[http://dx.doi.org/10.3109/03602537808993771] [PMID: 79468]

[22]   White RE, Coon MJ. Oxygen activation by cytochrome P-450. Annu Rev Biochem 1980; 49: 315-56.
[http://dx.doi.org/10.1146/annurev.bi.49.070180.001531] [PMID: 6996566]

[23]   White RE. The involvement of free radicals in the mechanisms of monooxygenases. Pharmacol Ther 1991; 49(1-2): 21-42.
[http://dx.doi.org/10.1016/0163-7258(91)90020-M] [PMID: 1852787]

[24]   Loida PJ, Sligar SG. Molecular recognition in cytochrome P-450: mechanism for the control of uncoupling reactions. Biochemistry 1993; 32(43): 11530-8.
[http://dx.doi.org/10.1021/bi00094a009] [PMID: 8218220]

[25]   Guengerich FP, Macdonald TL. Chemical mechanisms of catalysis by cytochromes P-450: a unified view. Acc Chem Res 1984; 17(1): 9-16.
[http://dx.doi.org/10.1021/ar00097a002]

[26]   Ferrandes B, Eymard P. Metabolism of valproate sodium in rabbit, rat, dog, and man. Epilepsia 1977; 18(2): 169-82.
[http://dx.doi.org/10.1111/j.1528-1157.1977.tb04465.x] [PMID: 326545]

[27]   Rettie AE, Rettenmeier AW, Howald WN, Baillie TA. Cytochrome P-450-catalyzed formation of delta 4-VPA, a toxic metabolite of valproic acid. Science 1987; 235(4791): 890-3.
[http://dx.doi.org/10.1126/science.3101178] [PMID: 3101178]

[28]   Palmer KH, Fowler MS, Wall ME. Metabolism of optically active barbiturates. II. *S* (-)--pentobarbital. J Pharmacol Exp Ther 1970; 175(1): 38-41.
[PMID: 5482417]

[29]   Holtzman JL, Thompson JA. Metabolism of R-(+)-and S-(minus)-pentobarbital by hepatic microsomes from male rats. Drug Metab Dispos 1975; 3(2): 113-7.
[PMID: 236157]

[30]   Kaiser DG, Vangiessen GJ, Reischer RJ, Wechter WJ. Isomeric inversion of ibuprofen (*R*)-enantiomer in humans. J Pharm Sci 1976; 65(2): 269-73.
[http://dx.doi.org/10.1002/jps.2600650222] [PMID: 1255461]

[31]   Hamman MA, Thompson GA, Hall SD. Regioselective and stereoselective metabolism of ibuprofen by human cytochrome P450 2C. Biochem Pharmacol 1997; 54(1): 33-41.

[http://dx.doi.org/10.1016/S0006-2952(97)00143-3] [PMID: 9296349]

[32] Kerr BM, Thummel KE, Wurden CJ, *et al.* Human liver carbamazepine metabolism. Role of CYP3A4 and CYP2C8 in 10,11-epoxide formation. Biochem Pharmacol 1994; 47(11): 1969-79. [http://dx.doi.org/10.1016/0006-2952(94)90071-X] [PMID: 8010982]

[33] Bertilsson L, Tomson T. Clinical pharmacokinetics and pharmacological effects of carbamazepine and carbamazepine-10,11-epoxide. An update. Clin Pharmacokinet 1986; 11(3): 177-98. [http://dx.doi.org/10.2165/00003088-198611030-00001] [PMID: 3524954]

[34] Blobaum AL, Kent UM, Alworth WL, Hollenberg PF. Mechanism-based inactivation of cytochromes P450 2E1 and 2E1 T303A by tert-butyl acetylenes: characterization of reactive intermediate adducts to the heme and apoprotein. Chem Res Toxicol 2002; 15(12): 1561-71. [http://dx.doi.org/10.1021/tx020052x] [PMID: 12482238]

[35] Blobaum AL. Mechanism-based inactivation and reversibility: is there a new trend in the inactivation of cytochrome p450 enzymes? Drug Metab Dispos 2006; 34(1): 1-7. [http://dx.doi.org/10.1124/dmd.105.004747] [PMID: 16369051]

[36] Brodie BB, Gillette JR, Eds. Concepts in Biochemical Pharmacology Part 2. 285.

[37] Daly J. Brodie BB, Gillette JR (ed). Concepts in Biochemical Pharmacology. Part 2. Berlin, Springer-Verlag. 1971; P. 285.

[38] Black AE, Hayes RN, Roth BD, Woo P, Woolf TF. Metabolism and excretion of atorvastatin in rats and dogs. Drug Metab Dispos 1999; 27(8): 916-23. [PMID: 10421619]

[39] Yamazaki H, Komatsu T, Takemoto K, *et al.* Decreases in phenytoin hydroxylation activities catalyzed by liver microsomal cytochrome P450 enzymes in phenytoin-treated rats. Drug Metab Dispos 2001; 29(4 Pt 1): 427-34. [PMID: 11259327]

[40] Yoshimoto K, Echizen H, Chiba K, Tani M, Ishizaki T. Identification of human CYP isoforms involved in the metabolism of propranolol enantiomers--N-desisopropylation is mediated mainly by CYP1A2. Br J Clin Pharmacol 1995; 39(4): 421-31. [http://dx.doi.org/10.1111/j.1365-2125.1995.tb04472.x] [PMID: 7640150]

[41] Freudenthal RI, Carroll FI. Metabolism of certain commonly used barbiturates. Drug Metab Rev 1973; 2(2): 265-78. [PMID: 4604760]

[42] Midgley I, Fowkes AG, Chasseaud LF, Hawkins DR, Girkin R, Kesselring K. Biotransformation of lofexidine in humans. Xenobiotica 1983; 13(2): 87-95. [http://dx.doi.org/10.3109/00498258309052241] [PMID: 6880242]

[43] Cunningham RF, Israili ZH, Dayton PG. Clinical pharmacokinetics of probenecid. Clin Pharmacokinet 1981; 6(2): 135-51. [http://dx.doi.org/10.2165/00003088-198106020-00004] [PMID: 7011657]

[44] Yasumori T, Nagata K, Yang SK, *et al.* Cytochrome P450 mediated metabolism of diazepam in human and rat: involvement of human CYP2C in N-demethylation in the substrate concentration-dependent manner. Pharmacogenetics 1993; 3(6): 291-301. [http://dx.doi.org/10.1097/00008571-199312000-00003] [PMID: 8148870]

[45] Mei Q, Tang C, Assang C, *et al.* Role of a potent inhibitory monoclonal antibody to cytochrome P-450 3A4 in assessment of human drug metabolism. J Pharmacol Exp Ther 1999; 291(2): 749-59. [PMID: 10525096]

[46] Spielberg SP, Gordon GB, Blake DA, Mellits ED, Bross DS. Anticonvulsant toxicity *in vitro*: possible role of arene oxides. J Pharmacol Exp Ther 1981; 217(2): 386-9. [PMID: 7229980]

[47]   Maggs JL, Naisbitt DJ, Tettey JN, Pirmohamed M, Park BK. Metabolism of lamotrigine to a reactive arene oxide intermediate. Chem Res Toxicol 2000; 13(11): 1075-81.
       [http://dx.doi.org/10.1021/tx0000825] [PMID: 11087428]

[48]   Kluwe WM, Maronpot RR, Greenwell A, Harrington F. Interactions between bromobenzene dose, glutathione concentrations, and organ toxicities in single- and multiple-treatment studies. Fundam Appl Toxicol 1984; 4(6): 1019-28.
       [http://dx.doi.org/10.1016/0272-0590(84)90241-0] [PMID: 6542890]

[49]   Kapitulnik J, Levin W, Morecki R, Dansette PM, Jerina DM, Conney AH. Hydration of arene and alkene oxides by epoxide hydrase in human liver microsomes. Clin Pharmacol Ther 1977; 21(2): 158-65.
       [http://dx.doi.org/10.1002/cpt1977212158] [PMID: 65238]

[50]   Madden S, Maggs JL, Park BK. Bioactivation of carbamazepine in the rat *in vivo*. Evidence for the formation of reactive arene oxide(s). Drug Metab Dispos 1996; 24(4): 469-79.
       [PMID: 8801063]

[51]   Fretland AJ, Omiecinski CJ. Epoxide hydrolases: biochemistry and molecular biology. Chem Biol Interact 2000; 129(1-2): 41-59.
       [http://dx.doi.org/10.1016/S0009-2797(00)00197-6] [PMID: 11154734]

[52]   Decker M, Arand M, Cronin A. Mammalian epoxide hydrolases in xenobiotic metabolism and signalling. Arch Toxicol 2009; 83(4): 297-318.
       [http://dx.doi.org/10.1007/s00204-009-0416-0] [PMID: 19340413]

[53]   Jerina DM, Daly JW. Arene oxides: a new aspect of drug metabolism. Science 1974; 185(4151): 573-82.
       [http://dx.doi.org/10.1126/science.185.4151.573] [PMID: 4841570]

[54]   Guroff G, Daly JW, Jerina DM, Renson J, Witkop B, Udenfriend S. Hydroxylation-induced migration: the NIH shift. Recent experiments reveal an unexpected and general result of enzymatic hydroxylation of aromatic compounds. Science 1967; 157(3796): 1524-30.
       [http://dx.doi.org/10.1126/science.157.3796.1524] [PMID: 6038165]

[55]   Hartmann S, Hultschig C, Eisenreich W, Fuchs G, Bacher A, Ghisla S. NIH shift in flavin-dependent monooxygenation: mechanistic studies with 2-aminobenzoyl-CoA monooxygenase/reductase. Proc Natl Acad Sci USA 1999; 96(14): 7831-6.
       [http://dx.doi.org/10.1073/pnas.96.14.7831] [PMID: 10393907]

[56]   Shiiyama S, Soejima-Ohkuma T, Honda S, *et al.* Major role of the CYP2C isozymes in deamination of amphetamine and benzphetamine: evidence for the quinidine-specific inhibition of the reactions catalysed by rabbit enzyme. Xenobiotica 1997; 27(4): 379-87.
       [http://dx.doi.org/10.1080/004982597240532] [PMID: 9149377]

[57]   Binda C, Mattevi A, Edmondson DE. Structure-function relationships in flavoenzyme-dependent amine oxidations: a comparison of polyamine oxidase and monoamine oxidase. J Biol Chem 2002; 277(27): 23973-6.
       [http://dx.doi.org/10.1074/jbc.R200005200] [PMID: 12015330]

[58]   Kaul PN, Whitfield LR, Clark ML. Chlorpromazine metabolism VIII: blood levels of chlorpromazine and its sulfoxide in schizophrenic patients. J Pharm Sci 1976; 65(5): 694-7.
       [http://dx.doi.org/10.1002/jps.2600650516] [PMID: 6774]

[59]   Köppel C, Tenczer J. A revision of the metabolic disposition of amantadine. Biomed Mass Spectrom 1985; 12(9): 499-501.
       [http://dx.doi.org/10.1002/bms.1200120910] [PMID: 2932178]

[60]   Beckett AH, Belanger PM. The disposition of phentermine and its N-oxidized metabolic products in the rabbit. Xenobiotica 1978; 8(1): 55-60.
       [http://dx.doi.org/10.3109/00498257809060383] [PMID: 24306]

[61] Wiśniewska-Knypl JM, Jabłońska JK. The rate of aniline metabolism *in vivo* in rats exposed to aniline and drugs. Xenobiotica 1975; 5(8): 511-9.
[http://dx.doi.org/10.3109/00498257509056121] [PMID: 1166665]

[62] Essien EE, Ogonor JI, Coker HA, Bamisile MM. Metabolic N-oxidation of metronidazole. J Pharm Pharmacol 1987; 39(10): 843-4.
[http://dx.doi.org/10.1111/j.2042-7158.1987.tb05130.x] [PMID: 2891825]

[63] Lemoine A, Gautier JC, Azoulay D, *et al.* Major pathway of imipramine metabolism is catalyzed by cytochromes P-450 1A2 and P-450 3A4 in human liver. Mol Pharmacol 1993; 43(5): 827-32.
[PMID: 8502233]

[64] Akutsu T, Kobayashi K, Sakurada K, Ikegaya H, Furihata T, Chiba K. Identification of human cytochrome p450 isozymes involved in diphenhydramine N-demethylation. Drug Metab Dispos 2007; 35(1): 72-8.
[http://dx.doi.org/10.1124/dmd.106.012088] [PMID: 17020955]

[65] Lim CK, Yuan ZX, Lamb JH, White IN, De Matteis F, Smith LL. A comparative study of tamoxifen metabolism in female rat, mouse and human liver microsomes. Carcinogenesis 1994; 15(4): 589-93.
[http://dx.doi.org/10.1093/carcin/15.4.589] [PMID: 8149466]

[66] Hutsell TC, Stachelski SJ. Determination of disopyramide and its mono-N-dealkylated metabolite in blood serum and urine. J Chromatogr A 1975; 106(1): 151-8.
[http://dx.doi.org/10.1016/S0021-9673(01)81057-8] [PMID: 1150784]

[67] Olesen OV, Linnet K. Identification of the human cytochrome P450 isoforms mediating *in vitro* N-dealkylation of perphenazine. Br J Clin Pharmacol 2000; 50(6): 563-71.
[http://dx.doi.org/10.1046/j.1365-2125.2000.00298.x] [PMID: 11136295]

[68] Valoti M, Frosini M, Palmi M, De Matteis F, Sgaragli G. *N*-Dealkylation of chlorimipramine and chlorpromazine by rat liver microsomal cytochrome P450 isoenzymes. J Pharm Pharmacol 1998; 50(9): 1005-11.
[http://dx.doi.org/10.1111/j.2042-7158.1998.tb06915.x] [PMID: 9811161]

[69] Vickers S, Polsky SL. The biotransformation of nitrogen containing xenobiotics to lactams. Curr Drug Metab 2000; 1(4): 357-89.
[http://dx.doi.org/10.2174/1389200003338929] [PMID: 11465044]

[70] Caldwell J, Dring LG, Williams RT. Metabolism of (^{14} C)methamphetamine in man, the guinea pig and the rat. Biochem J 1972; 129(1): 11-22.
[http://dx.doi.org/10.1042/bj1290011] [PMID: 4646771]

[71] Bach MV, Coutts RT, Baker GB. Involvement of CYP2D6 in the *in vitro* metabolism of amphetamine, two N-alkylamphetamines and their 4-methoxylated derivatives. Xenobiotica 1999; 29(7): 719-32.
[http://dx.doi.org/10.1080/004982599238344] [PMID: 10456690]

[72] Tsang CF, Wilkinson GR. Diazepam disposition in mature and aged rabbits and rats. Drug Metab Dispos 1982; 10(4): 413-6.
[PMID: 6126343]

[73] Andersson T, Miners JO, Veronese ME, Birkett DJ. Diazepam metabolism by human liver microsomes is mediated by both S-mephenytoin hydroxylase and CYP3A isoforms. Br J Clin Pharmacol 1994; 38(2): 131-7.
[http://dx.doi.org/10.1111/j.1365-2125.1994.tb04336.x] [PMID: 7981013]

[74] Calder IC, Hart SJ, Healey K, Ham KN. *N*-hydroxyacetaminophen: a postulated toxic metabolite of acetaminophen. J Med Chem 1981; 24(8): 988-93.
[http://dx.doi.org/10.1021/jm00140a014] [PMID: 7328601]

[75] Chen W, Koenigs LL, Thompson SJ, *et al.* Oxidation of acetaminophen to its toxic quinone imine and nontoxic catechol metabolites by baculovirus-expressed and purified human cytochromes P450 2E1 and 2A6. Chem Res Toxicol 1998; 11(4): 295-301.

[http://dx.doi.org/10.1021/tx9701687] [PMID: 9548799]

[76]   Thummel KE, Lee CA, Kunze KL, Nelson SD, Slattery JT. Oxidation of acetaminophen to N-acety-
       -p-aminobenzoquinone imine by human CYP3A4. Biochem Pharmacol 1993; 45(8): 1563-9.
       [http://dx.doi.org/10.1016/0006-2952(93)90295-8] [PMID: 8387297]

[77]   Laine JE, Auriola S, Pasanen M, Juvonen RO. Acetaminophen bioactivation by human cytochrome
       P450 enzymes and animal microsomes. Xenobiotica 2009; 39(1): 11-21.
       [http://dx.doi.org/10.1080/00498250802512830] [PMID: 19219744]

[78]   Jollow DJ, Mitchell JR, Potter WZ, Davis DC, Gillette JR, Brodie BB. Acetaminophen-induced
       hepatic necrosis. II. Role of covalent binding *in vivo*. J Pharmacol Exp Ther 1973; 187(1): 195-202.
       [PMID: 4746327]

[79]   Yang CS. Prevention of acetaminophen overdose toxicity with organosulfur compounds. WO
       1994008628 A1; PCT/US1993/009692; Publication date: Apr 28, 1994, Filing date: Oct 8, 1993.

[80]   Corcoran GB, Todd EL, Racz WJ, Hughes H, Smith CV, Mitchell JR. Effects of *N*-acetylcysteine on
       the disposition and metabolism of acetaminophen in mice. J Pharmacol Exp Ther 1985; 232(3): 857-
       63.
       [PMID: 3973834]

[81]   Slattery JT, Wilson JM, Kalhorn TF, Nelson SD. Dose-dependent pharmacokinetics of
       acetaminophen: evidence of glutathione depletion in humans. Clin Pharmacol Ther 1987; 41(4): 413-8.
       [http://dx.doi.org/10.1038/clpt.1987.50] [PMID: 3829578]

[82]   Forrest JA, Clements JA, Prescott LF. Clinical pharmacokinetics of paracetamol. Clin Pharmacokinet
       1982; 7(2): 93-107.
       [http://dx.doi.org/10.2165/00003088-198207020-00001] [PMID: 7039926]

[83]   Kunitoh S, Imaoka S, Hiroi T, Yabusaki Y, Monna T, Funae Y. Acetaldehyde as well as ethanol is
       metabolized by human CYP2E1. J Pharmacol Exp Ther 1997; 280(2): 527-32.
       [PMID: 9023260]

[84]   Axelrod J. The enzymatic conversion of codeine to morphine. J Pharmacol Exp Ther 1955; 115(3):
       259-67.
       [PMID: 13272175]

[85]   Duggan DE, Hogans AF, Kwan KC, McMahon FG. The metabolism of indomethacin in man. J
       Pharmacol Exp Ther 1972; 181(3): 563-75.
       [PMID: 4555898]

[86]   Belpaire FM, Wijnant P, Temmerman A, Rasmussen BB, Brøsen K. The oxidative metabolism of
       metoprolol in human liver microsomes: inhibition by the selective serotonin reuptake inhibitors. Eur J
       Clin Pharmacol 1998; 54(3): 261-4.
       [http://dx.doi.org/10.1007/s002280050456] [PMID: 9681670]

[87]   Taylor JA, Twomey TM, von Wittenau MS. The metabolic fate of prazosin. Xenobiotica 1977; 7(6):
       357-64.
       [http://dx.doi.org/10.3109/00498257709035794] [PMID: 610052]

[88]   Gyrd-Hansen N, Friis C, Nielsen P, Rasmussen F. Metabolism of trimethoprim in neonatal and young
       pigs: comparative *in vivo* and *in vitro* studies. Acta Pharmacol Toxicol (Copenh) 1984; 55(5): 402-9.
       [http://dx.doi.org/10.1111/j.1600-0773.1984.tb02002.x] [PMID: 6528810]

[89]   Sigel CW, Grace ME. A new fluorescence assay of trimethoprim and metabolites using quantitative
       thin-layer chromatography. J Chromatogr A 1973; 80(1): 111-6.
       [http://dx.doi.org/10.1016/S0021-9673(01)85355-3] [PMID: 4708398]

[90]   Sarcione EJ, Stutzman L. A comparison of the metabolism of 6-mercaptopurine and its 6-methyl
       analog in the rat. Cancer Res 1960; 20: 387-92.
       [PMID: 14441717]

[91]   Watson WA, Godley PJ, Garriott JC, Bradberry JC, Puckett JD. Blood pentobarbital concentrations during thiopental therapy. Drug Intell Clin Pharm 1986; 20(4): 283-7.
[http://dx.doi.org/10.1177/106002808602000414] [PMID: 3698825]

[92]   Buratti FM, D'Aniello A, Volpe MT, Meneguz A, Testai E. Malathion bioactivation in the human liver: the contribution of different cytochrome p450 isoforms. Drug Metab Dispos 2005; 33(3): 295-302.
[http://dx.doi.org/10.1124/dmd.104.001693] [PMID: 15557345]

[93]   Ng SF, Waxman DJ. Biotransformation of N,N',N''-triethylenethiophosphoramide: oxidative desulfuration to yield N,N',N''-triethylenephosphoramide associated with suicide inactivation of a phenobarbital-inducible hepatic P-450 monooxygenase. Cancer Res 1990; 50(3): 464-71.
[PMID: 2105156]

[94]   Chakraborty BS, Hawes EM, McKay G, *et al.* S-oxidation of thioridazine to psychoactive metabolites: an oral dose-proportionality study in healthy volunteers. Drug Metabol Drug Interact 1988; 6(3-4): 425-37.
[http://dx.doi.org/10.1515/DMDI.1988.6.3-4.425] [PMID: 3271648]

[95]   Anders MW. Bioactivation of halogenated hydrocarbons. J Toxicol Clin Toxicol 1982; 19(6-7): 699-706.
[http://dx.doi.org/10.3109/15563658208990399] [PMID: 7161850]

[96]   Raucy JL, Kraner JC, Lasker JM. Bioactivation of halogenated hydrocarbons by cytochrome P4502E1. Crit Rev Toxicol 1993; 23(1): 1-20.
[http://dx.doi.org/10.3109/10408449309104072] [PMID: 8471158]

[97]   Pohl LR, Nelson SD, Krishna G. Investigation of the mechanism of the metabolic activation of chloramphenicol by rat liver microsomes. Identification of a new metabolite. Biochem Pharmacol 1978; 27(4): 491-6.
[http://dx.doi.org/10.1016/0006-2952(78)90383-0] [PMID: 343786]

[98]   Bories GF, Cravedi JP. Metabolism of chloramphenicol: a story of nearly 50 years. Drug Metab Rev 1994; 26(4): 767-83.
[http://dx.doi.org/10.3109/03602539408998326] [PMID: 7875065]

[99]   Gruenke LD, Konopka K, Koop DR, Waskell LA. Characterization of halothane oxidation by hepatic microsomes and purified cytochromes P-450 using a gas chromatographic mass spectrometric assay. J Pharmacol Exp Ther 1988; 246(2): 454-9.
[PMID: 3404442]

[100]   Huwyler J, Jedlitschky G, Keppler D, Gut J. Halothane metabolism. Impairment of hepatic omega-oxidation of leukotrienes *in vivo* and *in vitro*. Eur J Biochem 1992; 206(3): 869-79.
[http://dx.doi.org/10.1111/j.1432-1033.1992.tb16995.x] [PMID: 1318837]

[101]   Pohl LR, Bhooshan B, Whittaker NF, Krishna G. Phosgene: a metabolite of chloroform. Biochem Biophys Res Commun 1977; 79(3): 684-91.
[http://dx.doi.org/10.1016/0006-291X(77)91166-4] [PMID: 597296]

[102]   Branchflower RV, Nunn DS, Highet RJ, Smith JH, Hook JB, Pohl LR. Nephrotoxicity of chloroform: metabolism to phosgene by the mouse kidney. Toxicol Appl Pharmacol 1984; 72(1): 159-68.
[http://dx.doi.org/10.1016/0041-008X(84)90260-6] [PMID: 6143425]

[103]   Constan AA, Sprankle CS, Peters JM, *et al.* Metabolism of chloroform by cytochrome P450 2E1 is required for induction of toxicity in the liver, kidney, and nose of male mice. Toxicol Appl Pharmacol 1999; 160(2): 120-6.
[http://dx.doi.org/10.1006/taap.1999.8756] [PMID: 10527910]

[104]   Crabb DW, Bosron WF, Li TK. Ethanol metabolism. Pharmacol Ther 1987; 34(1): 59-73.
[http://dx.doi.org/10.1016/0163-7258(87)90092-1] [PMID: 3310044]

[105]   Hawkins RD, Kalant H. The metabolism of ethanol and its metabolic effects. Pharmacol Rev 1972;

24(1): 67-157.
[PMID: 4402043]

[106] Albano E. Alcohol, oxidative stress and free radical damage. Proc Nutr Soc 2006; 65(3): 278-90.
[http://dx.doi.org/10.1079/PNS2006496] [PMID: 16923312]

[107] Hoek JB, Pastorino JG. Ethanol, oxidative stress, and cytokine-induced liver cell injury. Alcohol 2002; 27(1): 63-8.
[http://dx.doi.org/10.1016/S0741-8329(02)00215-X] [PMID: 12062639]

[108] Liesivuori J, Savolainen H. Methanol and formic acid toxicity: biochemical mechanisms. Pharmacol Toxicol 1991; 69(3): 157-63.
[http://dx.doi.org/10.1111/j.1600-0773.1991.tb01290.x] [PMID: 1665561]

[109] Guo C, McMartin KE. The cytotoxicity of oxalate, metabolite of ethylene glycol, is due to calcium oxalate monohydrate formation. Toxicology 2005; 208(3): 347-55.
[http://dx.doi.org/10.1016/j.tox.2004.11.029] [PMID: 15695020]

[110] Jacobsen D, McMartin KE. Methanol and ethylene glycol poisonings. Mechanism of toxicity, clinical course, diagnosis and treatment. Med Toxicol 1986; 1(5): 309-34.
[http://dx.doi.org/10.1007/BF03259846] [PMID: 3537623]

[111] Wall ME, Brine DR, Perez-Reyes M. Metabolism and disposition of naltrexone in man after oral and intravenous administration. Drug Metab Dispos 1981; 9(4): 369-75.
[PMID: 6114837]

[112] Stanczyk FZ, Roy S. Metabolism of levonorgestrel, norethindrone, and structurally related contraceptive steroids. Contraception 1990; 42(1): 67-96.
[http://dx.doi.org/10.1016/0010-7824(90)90093-B] [PMID: 2143719]

[113] Green CE, LeValley SE, Tyson CA. Comparison of amphetamine metabolism using isolated hepatocytes from five species including human. J Pharmacol Exp Ther 1986; 237(3): 931-6.
[PMID: 3712286]

[114] Coutts RT, Prelusky DB, Jones GR. The effects of cofactor and species differences on the *in vitro* metabolism of propiophenone and phenylacetone. Can J Physiol Pharmacol 1981; 59(2): 195-201.
[http://dx.doi.org/10.1139/y81-032] [PMID: 7225947]

[115] Steentoft A, Linnet K. Blood concentrations of clonazepam and 7-aminoclonazepam in forensic cases in Denmark for the period 2002-2007. Forensic Sci Int 2009; 184(1-3): 74-9.
[http://dx.doi.org/10.1016/j.forsciint.2008.12.004] [PMID: 19150586]

[116] Lauritsen K, Hansen J, Ryde M, Rask-Madsen J. Colonic azodisalicylate metabolism determined by *in vivo* dialysis in healthy volunteers and patients with ulcerative colitis. Gastroenterology 1984; 86(6): 1496-500.
[PMID: 6143704]

[117] Pan-Zhou XR, Cretton-Scott E, Zhou XJ, Yang MX, Lasker JM, Sommadossi JP. Role of human liver P450s and cytochrome b5 in the reductive metabolism of 3'-azido-3'-deoxythymidine (AZT) to 3'-amino-3'-deoxythymidine. Biochem Pharmacol 1998; 55(6): 757-66.
[http://dx.doi.org/10.1016/S0006-2952(97)00538-8] [PMID: 9586947]

[118] Cobby J, Mayersohn M, Selliah S. The rapid reduction of disulfiram in blood and plasma. J Pharmacol Exp Ther 1977; 202(3): 724-31.
[PMID: 197231]

[119] Baker MT, Bates JN. Metabolic activation of the halothane metabolite, [14C]2-chloro-1-1-difluoroethene, in hepatic microsomes. Drug Metab Dispos 1988; 16(2): 169-72.
[PMID: 2898328]

[120] Williams FM. Clinical significance of esterases in man. Clin Pharmacokinet 1985; 10(5): 392-403.
[http://dx.doi.org/10.2165/00003088-198510050-00002] [PMID: 3899454]

[121] Jones AW. Profiles in drug metabolism and toxicology: Richard Tecwyn Williams (1909-1979). Drug Metab Rev 2015; 47(4): 401-5.
[http://dx.doi.org/10.3109/03602532.2015.1115516] [PMID: 26610047]

[122] Drug Metabolism and Disposition: Considerations in Clinical Pharmacology, Wilkinson and Rawlins. Boston 1985.

[123] Hodgson E. 2004.http://faculty.ksu.edu.sa/73069/Documents/Toxicology.pdf

[124] Wells PG, Mackenzie PI, Chowdhury JR, *et al.* Glucuronidation and the UDP-glucuronosyltransferases in health and disease. Drug Metab Dispos 2004; 32(3): 281-90.
[http://dx.doi.org/10.1124/dmd.32.3.281] [PMID: 14977861]

[125] Mackenzie PI, Rogers A, Treloar J, Jorgensen BR, Miners JO, Meech R. Identification of UDP glycosyltransferase 3A1 as a UDP N-acetylglucosaminyltransferase. J Biol Chem 2008; 283(52): 36205-10.
[http://dx.doi.org/10.1074/jbc.M807961200] [PMID: 18981171]

[126] Silverman RB, Holladay MW. Drug Metabolism.The Organic Chemistry of Drug Design and Drug Action. 3rd ed. Waltham, MA: Academic Press, Elsevier 2014; pp. 357-422.
[http://dx.doi.org/10.1016/B978-0-12-382030-3.00008-8]

[127] Wang Q, Jia R, Ye C, Garcia M, Li J, Hidalgo IJ. Glucuronidation and sulfation of 7-hydroxycoumarin in liver matrices from human, dog, monkey, rat, and mouse. In vitro Cell Dev Biol Anim 2005; 41(3-4): 97-103.
[http://dx.doi.org/10.1290/0501005.1] [PMID: 16029080]

[128] Shaffer CL, Ryder TF, Venkatakrishnan K, Henne IK, O'Connell TN. Biotransformation of an α4β2 nicotinic acetylcholine receptor partial agonist in sprague-dawley rats and the dispositional characterization of its N-carbamoyl glucuronide metabolite. Drug Metab Dispos 2009; 37(7): 1480-9.
[http://dx.doi.org/10.1124/dmd.109.027037] [PMID: 19339375]

[129] Elvin AT, Keenaghan JB, Byrnes EW, *et al.* Tocainide conjugation in humans: novel biotransformation pathway for a primary amine. J Pharm Sci 1980; 69(1): 47-9.
[http://dx.doi.org/10.1002/jps.2600690113] [PMID: 7354440]

[130] Sakaguchi K, Green M, Stock N, Reger TS, Zunic J, King C. Glucuronidation of carboxylic acid containing compounds by UDP-glucuronosyltransferase isoforms. Arch Biochem Biophys 2004; 424(2): 219-25.
[http://dx.doi.org/10.1016/j.abb.2004.02.004] [PMID: 15047194]

[131] Abolin CR, Tozer TN, Craig JC, Gruenke LD. C-Glucuronidation of the acetylenic moiety of ethchlorvynol in the rabbit. Science 1980; 209(4457): 703-4.
[http://dx.doi.org/10.1126/science.7394528] [PMID: 7394528]

[132] Hosey CM, Broccatelli F, Benet LZ. Predicting when biliary excretion of parent drug is a major route of elimination in humans. AAPS J 2014; 16(5): 1085-96.
[http://dx.doi.org/10.1208/s12248-014-9636-1] [PMID: 25004821]

[133] van Montfoort JE, Hagenbuch B, Groothuis GMM, Koepsell H, Meier PJ, Meijer DKF. Drug uptake systems in liver and kidney. Curr Drug Metab 2003; 4(3): 185-211.
[http://dx.doi.org/10.2174/1389200033489460] [PMID: 12769665]

[134] Kerdpin O, Elliot DJ, Mackenzie PI, Miners JO. Sulfinpyrazone C-glucuronidation is catalyzed selectively by human UDP-glucuronosyltransferase 1A9. Drug Metab Dispos 2006; 34(12): 1950-3.
[http://dx.doi.org/10.1124/dmd.106.012385] [PMID: 16985098]

[135] Kowalczyk I, Hawes EM, McKay G. Stability and enzymatic hydrolysis of quaternary ammonium-linked glucuronide metabolites of drugs with an aliphatic tertiary amine-implications for analysis. J Pharm Biomed Anal 2000; 22(5): 803-11.
[http://dx.doi.org/10.1016/S0731-7085(00)00244-2] [PMID: 10815723]

[136] Markey SP. Chapter 11 - Pathways of Drug Metabolism. In: Atkinson AJJr, Huang S-M, Lertora JJL, Markey SP, (Eds.), Principles of Clinical Pharmacology (Third Edition), Academic Press, 2012; pp. 153-172.

[137] Buhl AE, Waldon DJ, Baker CA, Johnson GA. Minoxidil sulfate is the active metabolite that stimulates hair follicles. J Invest Dermatol 1990; 95(5): 553-7.
[http://dx.doi.org/10.1111/1523-1747.ep12504905] [PMID: 2230218]

[138] Banoglu E. Current status of the cytosolic sulfotransferases in the metabolic activation of promutagens and procarcinogens. Curr Drug Metab 2000; 1(1): 1-30.
[http://dx.doi.org/10.2174/1389200003339234] [PMID: 11467078]

[139] Hinson JA. Reactive metabolites of phenacetin and acetaminophen: a review. Environ Health Perspect 1983; 49: 71-9.
[http://dx.doi.org/10.1289/ehp.834971] [PMID: 6339229]

[140] Flint RB, Roofthooft DW, van Rongen A, *et al.* Exposure to acetaminophen and all its metabolites upon 10, 15, and 20 mg/kg intravenous acetaminophen in very-preterm infants. Pediatr Res 2017; 82(4): 678-84.
[PMID: 28553988]

[141] Ketterer B. Protective role of glutathione and glutathione transferases in mutagenesis and carcinogenesis. Mutat Res 1988; 202(2): 343-61.
[http://dx.doi.org/10.1016/0027-5107(88)90197-2] [PMID: 3057366]

[142] Hinchman CA, Matsumoto H, Simmons TW, Ballatori N. Intrahepatic conversion of a glutathione conjugate to its mercapturic acid. Metabolism of 1-chloro-2,4-dinitrobenzene in isolated perfused rat and guinea pig livers. J Biol Chem 1991; 266(33): 22179-85.
[PMID: 1939239]

[143] Walker K, Ginsberg G, Hattis D, Johns DO, Guyton KZ, Sonawane B. Genetic polymorphism in N-Acetyltransferase (NAT): Population distribution of NAT1 and NAT2 activity. J Toxicol Environ Health B Crit Rev 2009; 12(5-6): 440-72.
[http://dx.doi.org/10.1080/10937400903158383] [PMID: 20183529]

[144] Morgan CD, Sandler M, Davies DS, Conolly M, Paterson JW, Dollery CT. The metabolic fate of DL(7-3H)isoprenaline in man and dog. Biochem J 1969; 114(1): 8P-P.
[http://dx.doi.org/10.1042/bj1140008Pa] [PMID: 5810074]

[145] Crossley R. Chirality and the Biological Activity of Drugs. Boca Raton: CRC Press 1995.

[146] Eichelbaum M, Testa B, Somogyi A. Stereochemical Aspects of Drug Action and Disposition. Berlin: Springer-Verlag 2003.
[http://dx.doi.org/10.1007/978-3-642-55842-9]

[147] Kunze KL, Nelson WL, Kharasch ED, Thummel KE, Isoherranen N. Stereochemical aspects of itraconazole metabolism *in vitro* and *in vivo*. Drug Metab Dispos 2006; 34(4): 583-90.
[http://dx.doi.org/10.1124/dmd.105.008508] [PMID: 16415110]

[148] Tougou K, Gotou H, Ohno Y, Nakamura A. Stereoselective glucuronidation and hydroxylation of etodolac by UGT1A9 and CYP2C9 in man. Xenobiotica 2004; 34(5): 449-61.
[http://dx.doi.org/10.1080/00498250410001691280] [PMID: 15370961]

[149] van Beem H, Manger FW, van Boxtel C, van Bentem N. Etomidate anaesthesia in patients with cirrhosis of the liver: pharmacokinetic data. Anaesthesia 1983; 38 (Suppl.): 61-2.
[http://dx.doi.org/10.1111/j.1365-2044.1983.tb15181.x] [PMID: 6869758]

[150] Lewis RJ, Trager WF, Chan KK, *et al.* Warfarin. Stereochemical aspects of its metabolism and the interaction with phenylbutazone. J Clin Invest 1974; 53(6): 1607-17.
[http://dx.doi.org/10.1172/JCI107711] [PMID: 4830225]

[151] Sten T, Qvisen S, Uutela P, Luukkanen L, Kostiainen R, Finel M. Prominent but reverse

stereoselectivity in propranolol glucuronidation by human UDP-glucuronosyltransferases 1A9 and 1A10. Drug Metab Dispos 2006; 34(9): 1488-94.
[http://dx.doi.org/10.1124/dmd.106.010371] [PMID: 16763014]

[152] Obach RS. Pharmacologically active drug metabolites: impact on drug discovery and pharmacotherapy. Pharmacol Rev 2013; 65(2): 578-640.
[http://dx.doi.org/10.1124/pr.111.005439] [PMID: 23406671]

[153] Albert A. Chemical aspects of selective toxicity. Nature 1958; 182(4633): 421-2.
[http://dx.doi.org/10.1038/182421a0] [PMID: 13577867]

[154] Glossary of terms used in medicinal chemistry (IUPAC recommendations 1998). Pure Appl Chem 1998; 70: 1129-43.
[http://dx.doi.org/10.1351/pac199870051129]

[155] Bodor N. Novel approaches for the design of membrane transport properties of drugs. 1977.

[156] Lee HJ, Soliman MRI. Anti-inflammatory steroids without pituitary-adrenal suppression. Science 1982; 215(4535): 989-91.
[http://dx.doi.org/10.1126/science.6296999] [PMID: 6296999]

[157] Khan MOF, Lee HJ. Synthesis and pharmacology of anti-inflammatory steroidal antedrugs. Chem Rev 2008; 108(12): 5131-45.
[http://dx.doi.org/10.1021/cr068203e] [PMID: 19035773]

[158] Rautio J, Kumpulainen H, Heimbach T, *et al.* Prodrugs: design and clinical applications. Nat Rev Drug Discov 2008; 7(3): 255-70.
[http://dx.doi.org/10.1038/nrd2468] [PMID: 18219308]

[159] Stańczak A, Ferra A. Prodrugs and soft drugs. Pharmacol Rep 2006; 58(5): 599-613.
[PMID: 17085852]

[160] Mandell AI, Stentz F, Kitabchi AE. Dipivalyl epinephrine: a new pro-drug in the treatment of glaucoma. Ophthalmology 1978; 85(3): 268-75.
[http://dx.doi.org/10.1016/S0161-6420(78)35668-2] [PMID: 662280]

[161] Wei CP, Anderson JA, Leopold I. Ocular absorption and metabolism of topically applied epinephrine and a dipivalyl ester of epinephrine. Invest Ophthalmol Vis Sci 1978; 17(4): 315-21.
[PMID: 640779]

[162] Nutt JG, Woodward WR. Levodopa pharmacokinetics and pharmacodynamics in fluctuating parkinsonian patients. Neurology 1986; 36(6): 739-44.
[http://dx.doi.org/10.1212/WNL.36.6.739] [PMID: 3703280]

[163] Nutt JG, Woodward WR, Hammerstad JP, Carter JH, Anderson JL. The "on-off" phenomenon in Parkinson's disease. Relation to levodopa absorption and transport. N Engl J Med 1984; 310(8): 483-8.
[http://dx.doi.org/10.1056/NEJM198402233100802] [PMID: 6694694]

[164] Schwartz PS, Chen CS, Waxman DJ. Sustained P450 expression and prodrug activation in bolus cyclophosphamide-treated cultured tumor cells. Impact of prodrug schedule on P450 gene-directed enzyme prodrug therapy. Cancer Gene Ther 2003; 10(8): 571-82.
[http://dx.doi.org/10.1038/sj.cgt.7700601] [PMID: 12872138]

[165] Ruffolo RR Jr, Bondinell W, Hieble JP. α- and β-adrenoceptors: from the gene to the clinic. 2. Structure-activity relationships and therapeutic applications. J Med Chem 1995; 38(19): 3681-716.
[http://dx.doi.org/10.1021/jm00019a001] [PMID: 7562902]

[166] Turlapaty P, Laddu A, Murthy VS, Singh B, Lee R. Esmolol: a titratable short-acting intravenous beta blocker for acute critical care settings. Am Heart J 1987; 114(4 Pt 1): 866-85.
[http://dx.doi.org/10.1016/0002-8703(87)90797-6] [PMID: 2889341]

[167] Crowley JJ, Cusack BJ, Jue SG, Koup JR, Park BK, Vestal RE. Aging and drug interactions. II. Effect

of phenytoin and smoking on the oxidation of theophylline and cortisol in healthy men. J Pharmacol Exp Ther 1988; 245(2): 513-23.
[PMID: 3367304]

[168] Carrillo JA, Herráiz AG, Ramos SI, Gervasini G, Vizcaíno S, Benítez J. Role of the smoking-induced cytochrome P450 (CYP)1A2 and polymorphic CYP2D6 in steady-state concentration of olanzapine. J Clin Psychopharmacol 2003; 23(2): 119-27.
[http://dx.doi.org/10.1097/00004714-200304000-00003] [PMID: 12640212]

[169] Lucas C, Martin J. Smoking and drug interactions. Aust Prescr 2013; 36(3): 102-4.
[http://dx.doi.org/10.18773/austprescr.2013.037]

[170] Tassaneeyakul W, Guo LQ, Fukuda K, Ohta T, Yamazoe Y. Inhibition selectivity of grapefruit juice components on human cytochromes P450. Arch Biochem Biophys 2000; 378(2): 356-63.
[http://dx.doi.org/10.1006/abbi.2000.1835] [PMID: 10860553]

[171] Rubin B. Grapefruit juice and prescription drug interactions. Geriatrics 2006; 61(11): 12-8.
[PMID: 17112309]

[172] Wang Y, Roy A, Sun L, Lau CE. A double-peak phenomenon in the pharmacokinetics of alprazolam after oral administration. Drug Metab Dispos 1999; 27(8): 855-9.
[PMID: 10421610]

[173] Fowles J, Banton M, Klapacz J, Shen H. A toxicological review of the ethylene glycol series: Commonalities and differences in toxicity and modes of action. Toxicol Lett 2017; 278: 66-83.
[http://dx.doi.org/10.1016/j.toxlet.2017.06.009] [PMID: 28689762]

[174] Barceloux DG, Bond GR, Krenzelok EP, Cooper H, Vale JA. American Academy of Clinical Toxicology practice guidelines on the treatment of methanol poisoning. J Toxicol Clin Toxicol 2002; 40(4): 415-46.
[http://dx.doi.org/10.1081/CLT-120006745] [PMID: 12216995]

[175] Mafukidzea AT, Calnana M, Furinb J. Peripheral neuropathy in persons with tuberculosis. J Clin Tuberc Other Mycobact Dis 2016; 2: 5-11.
[http://dx.doi.org/10.1016/j.jctube.2015.11.002]

[176] Cadeddu G, Deidda A, Stochino ME, Velluti N, Burrai C, Del Zompo M. Clozapine toxicity due to a multiple drug interaction: a case report. J Med Case Reports 2015; 9: 77-82.
[http://dx.doi.org/10.1186/s13256-015-0547-2] [PMID: 25890012]

[177] Tóth K, Csukly G, Sirok D, *et al.* Potential role of patients' CYP3A-status in clozapine pharmacokinetics. Int J Neuropsychopharmacol 2017; 20(7): 529-37.
[http://dx.doi.org/10.1093/ijnp/pyx019] [PMID: 28340122]

[178] Moody DE, Backman RL. Metoprolol and bupropion interaction: A case study. Clin Exp Pharmacol 2004; 4: 165.
[http://dx.doi.org/10.4172/2161-1459.1000165]

CHAPTER 8

Biosynthetic Pathways Frequently Targeted by Pharmaceutical Intervention

Jason L. Johnson[1] and **M. O. Faruk Khan**[2,*]

[1] *Department of Chemistry and Physics, Southwestern Oklahoma State University, Weatherford, Oklahoma, USA*

[2] *Department of Pharmaceutical Sciences and Research, Marshall University School of Pharmacy, Huntington, WV, USA*

Abstract: This chapter provides a brief review of the important biosynthetic pathways frequently targeted by pharmaceutical interventions. The biosynthetic pathways discussed in this chapter include: eicosanoid biosynthesis (prostaglandins, prostacyclins and leukotrienes), epinephrine and norepinephrine biosynthesis, folic acid biosynthesis, steroid biosynthesis (cholesterol, adrenocorticoids and sex hormones) and nucleic acid biosynthesis (purines and pyrimidines anabolism, catabolism and salvages). After studying this chapter, students will be able to:

• Highlight important biosynthetic pathways frequently targeted by pharmaceutical interventions
• Explain detail chemical steps of each of the biosynthetic pathways
• Apply the chemical principles in the regulations of each of the pathways and relevant mechanisms of drug action
• Define the monomer units of the biomolecules and their chemical properties
• Comprehend the significance of biosynthetic pathways in different diseased states

Keywords: Biosynthetic pathways, Eicosanoid, Norepinephrine, Folic acid, Steroids, Nucleic acids, Regulations of biosynthesis.

EICOSANOID BIOSYNTHESIS

Introduction

The *eicosanoids* are a broad group of signaling lipids classified as *paracrine hormones*, meaning that they are *synthesized in virtually all tissues and act immediately to modulate metabolism within or very proximal to the tissues in*

---
* **Corresponding author M.O. Faruk Khan**: Department of Pharmaceutical Sciences and Research, Marshall University School of Pharmacy, Huntington, WV, USA; Tel: 304-696-3094; Fax: 304-696-7309; E-mail: khanmo@marshall.edu

M.O. Faruk Khan & Ashok Philip (Eds.)

which they are synthesized. The precursors for eicosanoid synthesis are polyunsaturated fatty acids that are 20 carbons in length and contain three to five carbon-carbon double bonds; in fact, the Greek origin for the name, *eicosi*, translates directly as "twenty". The "series 2" eicosanoids derived specifically from *arachidonic acid* [20:4 (5, 8, 11, 14)] represent the most abundant in humans, and the pathways leading to their synthesis are collectively referred to as the "*arachidonate cascade*" (Fig. **1**). Arachidonic acid itself is synthesized by the elongation and step-wise reduction of the essential fatty acid *linoleic acid* [18:2(9, 12)]. Nutritional deficiencies in linoleic acid thus lead to symptoms (*e.g.*, elevation of blood pressure, inhibited blood clotting, and poor wound healing) generally reflective of a loss in eicosanoid signaling.

The first and rate-limiting step toward the synthesis of eicosanoids is the ester hydrolysis of membrane-bound phospholipids at position C2 to release arachidonic acid. This reaction is catalyzed in response to stimuli primarily through the actions of *phospholipase A$_2$*. Arachidonic acid in turn serves as the precursor in the synthesis of three eicosanoid classes (Fig. **1**), including: the prostaglandins and thromboxanes, collectively called the *prostanoids*, formed through the "cyclic" or "*cyclooxygenase*" *pathway*; the leukotrienes and lipoxins formed through the "linear" or "*lipoxygenase*" *pathway*; and the epoxyeicosatrienic acids formed through the "*epoxygenase*" *pathway*.

Prostaglandins and Thromboxanes Biosynthesis: The "Cyclic" Pathway

Originally isolated from the prostate gland, *prostaglandins (PG)* represent potent, signaling molecules controlling a wide array of biological activities, owing to their affinity for a number of G-protein receptors. For example, prostaglandins are known to mediate: smooth muscle contractions of the uterus during labor; induction of gastric mucosa and inhibition of acid synthesis in the gastrointestinal tract; relaxation of blood vessels to regulate blood flow; and, most notably, activation of the inflammatory response, including pain and fever. Several functional classes of prostaglandins exist, interchangeable *via* a series of reductions and isomerizations and designated alphabetically as PGA through PGI; subscripts are appended to these designations to indicate the number of double bonds outside the characteristic cyclopentane ring, and Greek subscripts are appended to denote the orientation of ring hydroxyl groups. Structurally, all prostaglandins share a common cyclopentane ring, a trans double bond between C13 and C14, and a hydroxyl group at C15.

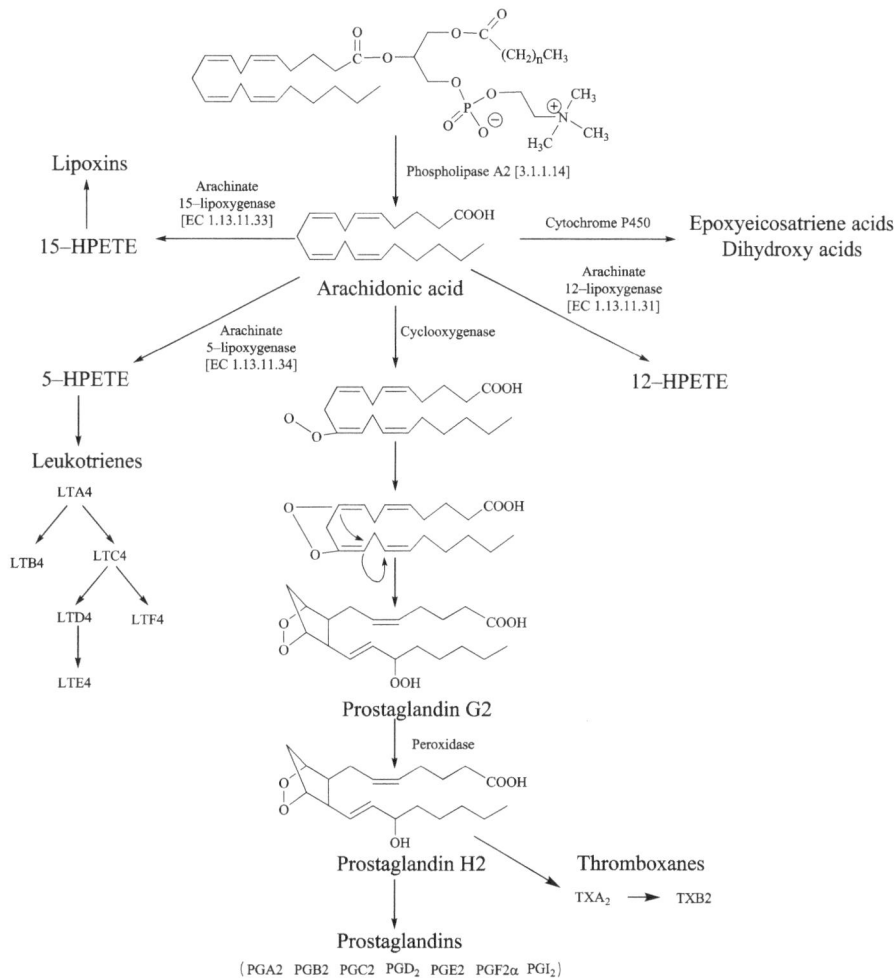

Fig. (1). The arachidonic acid cascade and the cyclic pathway. LT = Leukotriene; PG = Prostaglandin; TX = Thromboxane; HETE = Hydroxyeicosatetraenoic acid; HPETE = Hydroperoxyeicosatetraenoic acid.

Thromboxanes (TX) represent a group of signaling molecules closely related to the prostaglandins, but whose cyclopentane ring has been replaced with a six-membered oxygen-containing ring (termed an *oxane*). The thromboxanes are preferentially synthesized in platelets (*thrombocytes*) and primarily serve to modulate blood flow and clotting in association with tissue damage.

As outlined in Fig. (**1**), the first committed step of the cyclic pathway towards the synthesis of both prostaglandins and thromboxanes from arachidonic acid is catalyzed by the enzyme *cyclooxygenase (COX)* (also known as *PGH₂ synthase* or *prostaglandin endoperoxide synthase*). COX is a membrane-bound glycoprotein of the endoplasmic reticulum that exhibits two activities, each from a separate domain. Within the cyclooxygenase domain, the addition of two molecules of O_2 accompanies the cyclization of C8-C12 to form PGG_2; within the endoperoxidase domain, a glutathioine-dependent reduction of the 15-hydroperoxide to an alcohol yields PGH_2. Two isoenzymes of COX exist, sharing 60% sequence and structural homology; *COX1* is a constitutive (expression without regulation) enzyme of the gastric mucosa, platelets, vascular endothelium and kidney, whereas *COX2* is inducible in response to inflammation. Both isoenzymes are equally inhibited by NSAIDS (non-steroidal anti-inflammatory drugs).

PGH_2 serves as the central precursor for the synthesis of the remaining classes of prostaglandins and thromboxanes (Fig. **1**). The fate of PGH_2 depends largely on the relative activities of competing enzymes, particularly *PGF_2 synthase, PGE_2 synthase, PGI_2 synthase, PGD_2 synthase, and thromboxane A_2 synthase*. The level of activities, in turn, is determined by the tissue-specific expression of the proteins. For example, PGI_2 synthase is very active in vascular endothelial cells, where *PGI_2* serves as a vasodilator and inhibitor of clotting; *thromboxane A_2 synthase*, in contrast, is very active within platelets, forming *thromoboxane A_2* to serve as a vasoconstrictor and stimulator of clotting.

Leukotrienes and Lipoxins Biosynthesis: The "Linear" Pathway

In the "linear" pathway, arachidonic acid undergoes a monodioxygenation (as opposed to the bis-dioxygenation of COX) *via* the addition of a hydroperoxy group by *lipooxygenases (LOs)* to produce *monohydroperoxyeicosatetraienoic acids (HPETEs)* (see Fig. **3** and **4** later). Hydroperoxy substitution may occur at the 5-, 12-, or 15-positions of arachidonic acid. The *leukotrienes (LTs)* are the products of 5-LO, the *hepoxilins* of 12-LO, and the *lipoxins (LXs)* of 15-LO in combination with 5-LO. The functions of these three classes of molecules are distinct. Leukotrienes, for example, constitute a part of the slow-acting substances of anaphylactic shock released after an allergic reaction; they act to stimulate contraction of the smooth muscle lining of the lungs, particularly the bronchi, stimulate mucus secretion within respiratory and intestinal tissue, and generally mediate asthmatic attacks. Lipoxins, in contrast, exhibit anti-inflammatory properties in opposition to the leukotrienes. The function of the hepoxilins is still not clear, although they may be associated with regulating apoptosis.

HPETEs are inherently unstable; they are quickly reduced either by cellular peroxidases to *hydroxyeicosatetraenoic acids (HETEs)* or undergo catalytic hydroperoxy substitution. The first committed step toward leukotriene (named for the minimum of three conjugated double bonds contained within their structures) biosynthesis is catalyzed in two successive steps by *5-LO* (Fig. **2**). First, an iron-catalyzed addition of O_2 to arachidonic acid yields *5-HPETE,* followed by a base-catalyzed elimination of water to form an epoxide within *leukotriene A_4 (LTA_4).* (Note that the subscript once again highlights the number of double bonds present in the molecules, and that this number remains consistent within the leukotriene class despite molecular rearrangements.) LTA4 may follow two synthetic routes: hydrolysis by *LTA_4 epoxide hydrolase* yields *leukotriene B_4*, a chemotactic agent that induces the adhesion of leukocytes on epithelium during inflammation, whereas *LTC_4 synthetase* catalyzes the addition of a sulfhydryl group from the natural reductant glutathione at the 4-epoxide to yield *LTC_4*. Sequential removal of glutamic acid and glycine by the dipeptidases *γ-glutamyl transferase* and *cysteinyl glycinase*, respectively, gives rise to *LTD_4* and *LTE_4*.

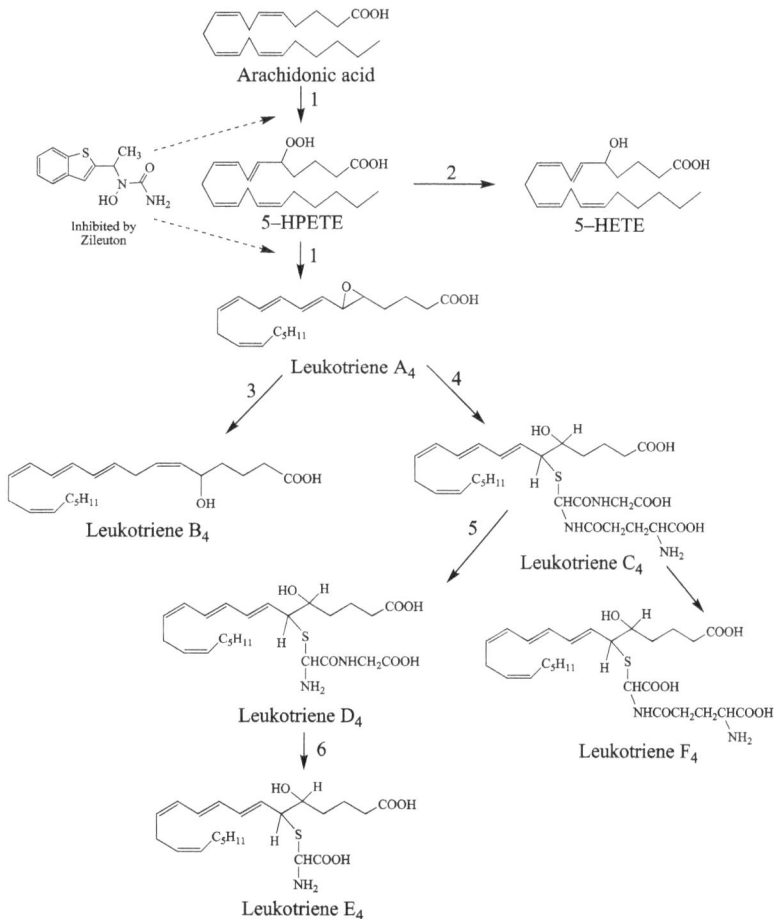

Fig. (2). Leukotriene Biosynthesis: the linear pathway. 1. 5–Lipoxygenase + FLAP [EC 1.13.11.34]; 2. Peroxidase; 3. Leukotriene A4 epoxide hyrolase (LTA4 hydrolase) [EC 3.3.2.6]; 4. LTC4 synthetase (α glutathione–S–transferase) [EC 2.5.1.37]; 5. γ–Glutamyl transferase [EC 2.3.2.2]; 6. Cysteinyl glycinase (an amino dipeptidase).

Lipoxins characteristically contain three hydroxyl groups and the conjugated tetraene. Fig. (**2**) outlines a possible route for the synthesis of *LXA₄*, *LXB₄*, and *LXC₄* proceeding through a centralized 5,15-HPETE intermediate. In general, successive oxidations of arachidonic acid by *15-LO* and *5-LO* coupled with *peroxidase* reductions gives rise to this class of eicosanoid.

Eicosanoid biosynthesis is *regulated* largely *via* the bioavailability of arachidonic acid (AA), as determined by two competing processes: the insertion/reinsertion (*acylation*) of AA into phospholipids *via* the action of *acyl-CoA lysophospholipid acyltransferase*, and the removal (*deacylation*) of AA from stored phospholipids *via* the action of the family of *phospholipase A2* enzymes (*PLA₂s*). PLA2 activity is known to be enhanced *via* a variety of cellular stimuli, including: phosphorylation *via* the mitogen-activation protein (MAP) kinases; the binding of calcium to induce PLA2 translocation into the cellular membrane where phospholipids acylated with AA reside; and stimulation of PLA2 transcription by cytokines and growth factors. Once released, competing, tissue-specific activities of COX, 5-LO, 12-LO, 15-LO, and cytochrome P450 enzymes determine the fate of AA, with each enzyme prone to its own activating stimuli.

EPINEPHRINE AND NOREPINEPHRINE BIOSYNTHESIS

Introduction

Within the adrenal medulla (the center of the adrenal gland surrounded by the cortex), a class of amine-containing derivatives of 1,2-dihydroxybenzene known as *catecholamines* is synthesized. These hormonally active compounds, which include *norepinephrine* (*noradrenaline*) and its methyl derivative *epinephrine* (*adrenaline*), are synthesized from the amino acid L-tyrosine and mediate their effects on cellular function *via* their agonistic or antagonistic interactions with the α-and β-adrenergic receptors. The tissue-specific synthesis of these receptors results in varying functions of the catecholamines around the body. Generally, to meet the extra fuel requirements of a physiological stress, epinephrine serves to stimulate glycogen breakdown in the muscle and liver, *gluconeogenesis* in the liver, and *lipolysis* in adipose tissue, while simultaneously suppressing insulin secretion. Norepinephrine acts generally as a *neurotransmitter* to alter the sympathetic nervous system, serving to increase cardiac output and blood pressure and thus enhancing the delivery of mobilized fuels.

Pathways

Most L-tyrosine is used to support protein synthesis or catabolized to acetoacetate and fumarate. A small amount, however, is diverted into the catecholamine pathway *via* the action of *tyrosine hydroxylase* (also known as *tyrosine- 3-*

monoxygenase) (Fig. **3**). Tyrosine hydroxylase catalyzes the oxidation of the phenyl ring in tyrosine to produce *L-dihydroxyphenylalanine (L-DOPA)*. Tyrosine hydroxylase is classified as a *mixed-function oxidase*, in that it oxidizes both a primary substrate (L-tyrosine) *via* the addition of a hydroxyl group from one oxygen atom of O_2, as well as the oxidation of a co-substrate *via* hydrogen atom donation to the second oxygen atom of O_2 to produce water. In this case, the co-substrate is the organic cofactor *tetrahydrobiopterin*, analogous in structure to both tetrahydrofolate and *flavin coenzymes*. Mechanistically, tetrahydrobiopterin serves to facilitate complexation of O_2 with Fe(II) bound within tyrosine hydroxylase to produce an iron(IV)-oxy species, thought to be the hydroxylating agent. Tetrahydrobiopterin is concomitantly oxidized to *dihydrobiopterin*, requiring the tetrahydro form to be regenerated *via* the action of NADH-dependent *dihydropteridine reductase*. As the rate-limiting step of the pathway, tyrosine hydroxylase is allosterically regulated in part through *feedback* inhibition by accumulating concentrations of the catecholamines.

L-DOPA is decarboxylated to 2-(3,4-dihydroxyphenyl) ethylamine, or simply *dopamine*, by *DOPA decarboxylase*, also known as *aromatic L-amino acid decarboxylase* to reflect its nonspecific affinity for substrates that include a variety of amines. DOPA decarboxylase, like many enzymes catalyzing amino acid decarboxylation, requires the organic cofactor *pyridoxal phosphate* (PLP, derived from Vitamin B6) to stabilize the resulting carbanion upon decarboxylation.

In dopaminergic neurons and some parts of the brain (*e.g.*, the *substania nigra*), dopamine represents the end product of the pathway. However, within the adrenal medulla, dopamine enters secretary granules where copper-containing and membrane-bound *dopamine β-hydroxylase* catalyzes the further oxidation of dopamine into *norepinephrine*. *Dopamine β-hydroxylase*, like tyrosine hydroxylase, is a *mixed-function oxidase*; in this case, the co-substrate that supplies the hydrogen atoms during oxygen reduction to water is *ascorbic acid* (Vitamin C). Reduction of ascorbic acid produces *dehydroascorbic acid* as a byproduct.

After diffusing into the cytosol, norepinephrine is methylated by the enzyme Phenylethanolamine *N-methyltransferase* to produce the final catecholamine, *epinephrine*. The alkylating agent in this biological reaction is *S-adenosyl methionine (SAM)*. SAM contains a methyl group destabilized *via* its attachment to a methionine sulfur atom (a sulfonium ion); thus SN2 nucleophilic attack by nucleophiles, such as the amine in norepinephrine, is readily supported. Loss of the activated methyl group by SAM results in the formation of homocysteine (loosely translated as one more carbon than cysteine), which is recycled into SAM through a pathway known as the activated methyl cycle.

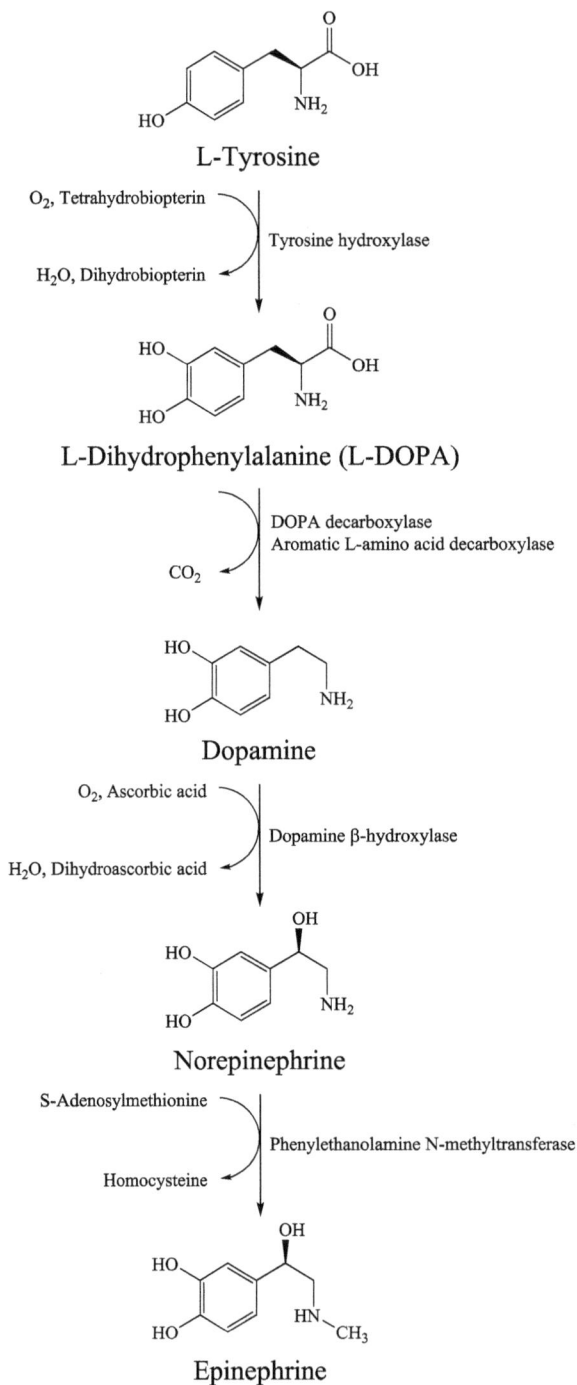

Fig. (3). Biosynthetic pathway of epinephrine and norepinephrine.

Epinephrine, thus formed, remains in the granules until activated for release. The signal for release is mediated by stress-induced transmissions of nerve impulses from the hypothalamus that stimulate the release of acetylcholine. Acetylcholine in turn acts to depolarize the plasma membranes of the adrenomedullary cells, resulting in a calcium ion influx. The presence of Ca^{+2} stimulates the release of epinephrine and norepinephrine from the chromaffin granules into the extracellular space.

Catecholamine biosynthesis is *regulated* by the *sympathetic nervous system*. The biosynthesis is triggered by stimuli (*e.g.*, pain, fright, cold, exercise, hypoglycemia, hypoxia, infection, trauma, *etc.*) that are processed by the sympathetic nervous system. In fact, the adrenal medulla, where synthesis occurs, is often considered a modified *sympathetic* ganglion. The mechanism for up-regulation focuses on the polypeptide hormone *corticotropin-releasing factor (CRF)* manufactured and released from the paraventricular neuroendocrine cells of the *hypothalamus* during times of physiological stress. Note that release factors like CRF, by definition, serve to stimulate the release of their corresponding trophic hormones, and trophic hormones in turn stimulate their target endocrine tissues to secrete the hormones they synthesize. In this particular case, CRH is carried to the *anterior lobe* of the pituitary, where it stimulates the secretion of *adrenocorticotropic hormone (ACTH)*. ACTH binds to its G-protein receptor in target tissues, initiating the activation of cAMP-dependent *protein kinase A*. While ACTH is perhaps best known for stimulating the secretion of steroid hormones such as cortisol from the adrenal cortex (see later Section below), ACTH also enhances the activity of epinephrine precursors *via* the activation of both *tyrosine hydroxylase* and *dopamine-β-hydroxylase* (Fig. **3**). Reinforcing this activation and exemplifying the functional relationship between the adrenal medulla and cortex, cortisol secondarily acts to increase the transcription of *phenylethanolamine N-methyltransferase*, the enzyme responsible for converting norepinephrine to epinephrine. By contrast, during times of little or no stress, cellular catecholamines are inactivated through a combination of oxidative deamination carried out by *monoamine oxidase* and O-methylation carried out by *catechol-O-methyl-transferase*. The products of these reactions are excreted in the urine.

FOLIC ACID BIOSYNTHESIS

Introduction

Pteroylglutamic acid (PteGlu), also known as *folic acid*, is a compound composed of 6-methyl pterin, p-amino benzoic acid (PABA), and glutamic acid (Fig. **4**). (*Note that a pteridine residue is a heterobicyclic system containing two nitrogens* *per ring*). The doubly oxidized form, folate, is considered a vitamin (Vitamin B9) for humans, since only bacteria and plants are capable of synthesizing PABA and linking pteric acid to glutamate. While many sources of folic acid exist in the diet, leafy vegetables are among the richest (Latin: *folium*, leaf); in addition, intestinal bacteria contribute in small amounts to human uptake. The most common dietary form of folic acid is *pteroylheptaglutamate,* oligoglutamyl conjugates in which the peptide linkages occur between the γ-carboxyl group of one glutamate and the α-amino group of the next. Once in the intestine, the oligomers are hydrolyzed by *glutamate conjugase* into pteroylmonoglutamate for absorption. Subsequent distribution of the folate in the blood is mediated by *reduced folate carrier (RFC)* originating from the *intestinal mucosa.*

The biosynthesis of most molecules involves some form of carbon addition to precursor molecules. The source of the donated carbon depends in large part on its oxidation state. Fully *reduced* methyl groups are most often provided by *S-adenosyl-methionine (SAM)*; for example, SAM is a required cofactor in the methylation of norepinephrine into epinephrine (Fig. **3**). In contrast, fully *oxidized* carbon, *i.e.* carbon dioxide (or bicarbonate), is most often activated for biological addition by the organic cofactor *biotin* (Vitamin B7); its utility is demonstrated in gluconeogenesis during the conversion of three-carbon pyruvate into four-carbon oxaloacetate. Many biological reactions, however, require the addition of an intermediary oxidation state of carbon; in these cases, fully reduced folate, known as *tetrahydrofolate ($H_4PteGlu$)*, serves as the C1-donating cofactor.

Examples of biosynthetic reactions dependent on tetrahydrofolate include: (1) the *phosphoribosylglycinamide formyltransferase* catalyzed step of purine ring synthesis, utilizing the N^{10}-formyl group; (2) the *thymidylate synthase* step in the conversion of deoxynucleotide dUMP to dTMP to support DNA synthesis, utilizing the N^5,N^{10}-methylene group; (3) the synthesis of N-formylmethiony--tRNA by *transformylase* to initiate translation in bacteria, utilizing the N^{10}-formyl group; and (4) the Vitamin B12-dependent, *methionine synthase* catalyzed step of the activated *methyl cycle*, in which an N^5-methyl group is used to convert homocysteine into methionine for subsequent formation of SAM. The latter example explains the tight, metabolic relationship between folic acid, SAM, and Vitamin B12 (*cobalamin*).

Deficiencies in folic acid intake, absorption, or metabolism are commonly characterized by a condition known as *megloblastic anemia*. Without adequate supplies of purines and dTMP, DNA synthesis is readily inhibited, arresting cells in the S phase of mitosis. Tissue types that generally undergo frequent and rapid cell division are particularly affected. Slower maturation of erythrocytes, for example, lead to abnormally large cellular structure with weakened cellular

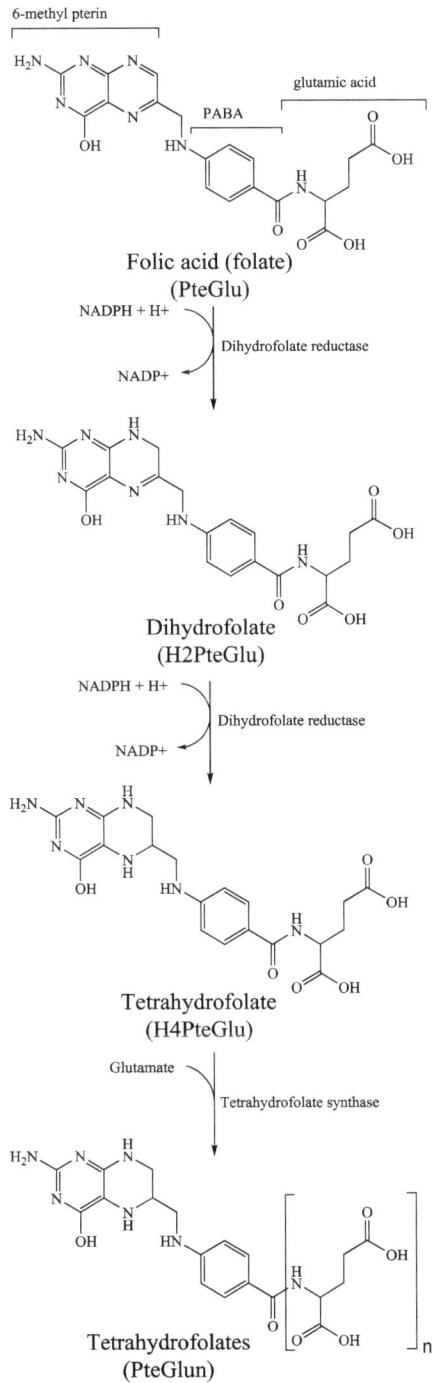

Fig. (4). Biosynthesis of tetrahydrofolates from folic acid.

membranes and a resulting anemic state for the organism. Neural tube birth defects are also commonly associated with folic acid deficiencies in women, a condition that has led most industrialized countries to fortify grains and cereals with the vitamin. Finally, because folic acid supports rapid cellular division, those enzymes that activate or regenerate folic acid in the human body are themselves targets for pharmaceuticals to control rapidly proliferating cancers and parasitic/viral infections. Methotrexate, for example, is a well-known chemotherapeutic agent that is a competitive inhibitor of dihydrofolate reductase.

Pathways

Once absorbed, folic acid must be doubly reduced in the human body to convert it into an active coenzyme. Both reductions are catalyzed by *dihydrofolate reductase (DHFR)* (Fig. **4**). Folic acid is first reduced to *7,8-dihydrofolate ($H_2PteGlu$)* through the transfer of a hydride ion from NADPH to position N-8 of the pteridine ring. In a second step catalyzed by the same enzyme, $H_2PteGlu$ is further reduced by NADPH-dependent hydride ion transfer to position N5 to produce 5,6,7,8-tetrahydrofolate. In its activated form, tetrahydrofolate is converted to polyglutamyl forms by an ATP-dependent *tetrahydrofolate synthase.*

Reduced polyglutamyl forms of tetrahydrofolate serve as the coenzyme of folate-dependent biosynthetic reactions. The C1 group to be donated may be bound as N^5,N^{10}-methylene (-CH_2-), N^5-formimino(-C=NH), or N^5-formyl (-C=O). The most common entry of the C1 group within the $H_4PteGlu$ pool is as the N^5,N^{10}-methylene group coming from the conversion of serine to glycine, catalyzed by *serine hydroxymethyltransferase,* and from the *glycine cleavage enzyme system*; histidine degradation most often provides the N^5-formimino group. Once the single carbon has been incorporated into tetrahydrofolate, it can undergo additional interconversions involving change in oxidation state prior to being used in biosynthesis.

A complex array of transcription factors are known to *regulate* the dihydrofolate reductase (DHFR) gene, serving to modulate mRNA synthesis in accordance with cell cycle requirements. However, studies designed to understand cellular resistance to drugs that target DHFR have led to an awareness of other levels of DHFR regulation. For example, cancer patients treated with methotrexate to inhibit DHFR activity may resultantly experience an increase in cellular levels of the targeted enzyme, thereby counteracting the drug. Recent studies have shown that the source for such resistance may relate, in part, to the natural mechanism of *feedback*, translational inhibition of DHFR. High concentrations of the enzyme result in DHFR binding to the mRNA which encodes it, thereby inhibiting its own translation. When methotrexate binds to DHFR, the enzyme appears to undergo an allosteric change; the new conformer lacks the ability to bind mRNA, and DHFR translation thus increases. As studies continue, it is becoming apparent that DHFR activity is controlled at multiple levels.

STEROID BIOSYNTHESIS

Introduction

Steroids are organic compounds that are characterized by or derived from a four, fused-ring structure known as *gonane*, or that at least contain the repeating five-carbon *isoprenoid* unit serving as a building block for the steroid nucleus. The structure of gonane consists of three fused cyclohexane rings arranged in the form of phenanthrene (in which rings are labeled A, B, and C), fused with a fourth ring (D) of cyclopentane. While gonane contains 20 carbons, biological steroids often have an alkyl side chain of varying length that extends from C17, as well as multiple methyl, hydroxyl, or other functional groups added at one or more of the remaining positions.

The biological functions of steroids are somewhat diverse. *Bile acids*, for example, serve in the emulsification, digestion, and absorption of dietary lipids. Fat-soluble, *steroid vitamins*, such as Vitamins A, D, E, and K, are associated with mediating vision, calcium uptake and metabolism for bone growth, antioxidant properties, and the blood coagulation cascade, respectively. *Cholesterol* serves as one of the primary membrane components in eukaryotic cells, along with the non-steroidal phosphoglycerides and sphingolipids. Finally, five classes of *steroid hormones* exist that are differentiated based upon the nature of their target receptors but that all exist as structural modifications of cholesterol, including the progestins, glucocorticoids and mineralocorticoids (collectively referred to as the *adrenocorticoids*), and the androgens and estrogens (collectively referred to as the *sex hormones*).

Cholesterol Biosynthesis

The most abundant steroid and essential component for moderating the fluidity of all biological membranes is cholesterol. Free cholesterol also serves as the precursor for a number of steroid hormones and Vitamin D, and its primary route of excretion from the body is as bile acids in the feces. As a large nonpolar compound, the solubility of cholesterol in aqueous environments is quite low, requiring that its distribution in the body occur as part of micelle-like, *apoprotein* particles such as LDL and VLDL. Within the apoproteins, cholesterol esters are sequestered on the interior while the polar *lipoproteins* and phospholipids are on the outside.

Structurally, cholesterol contains a gonane ring that is functionalized at multiple positions: a hydroxyl group exists at position C3, responsible for the *amphipathic* nature of the otherwise hydrophobic compound; a double bond exists between C5 and C6; single methyl groups occur at positions C10 and C13; and an eight-membered, branched hydrocarbon chain is extended from C17. At 27 total carbons, its *de novo* synthesis is thus rather complicated, but is carried out cytoplasmically in all cells. Although various tissues have the ability to synthesize cholesterol, the liver is the chief organ.

Generally, the biosynthetic steps of cholesterol can be divided into *four broad stages* (Fig. **5**). In the *first stage*, three acetate moieties activated through thioester formation with Coenzyme-A are condensed into the six-carbon intermediate *mevalonate*. Ready sources of acetyl-CoA include the *β-oxidation of fatty acids*, the oxidation of ketogenic amino acids, and the *pyruvate dehydrogenase* reaction. Conversion of mevalonate to activated, five-carbon isoprenoid units such as *isopentenyl diphosphate* and *3,3-dimethylallyl diphosphate* constitutes *stage two*, followed by the *polymerization* (*stage three*) of six of these isoprenoid units into the thirty-carbon, linear molecule of *squalene*. In the *fourth and final stage*, oxygenation and cyclization of squalene accompanied by a series of oxidations and transmethylations give rise to the final product.

Mevalonate synthesis in stage one consists of three successive reactions (Fig. **5A**). Initially, *acetyl-CoA acetyltransferase*, better known as *thiolase*, catalyzes a *Claisen condensation* between two acetyl-CoA molecules to yield four-carbon *acetoacetyl-CoA*. This represents the same enzyme that drives the *retro-Claisen condensation* as the last step in the β-oxidation of fatty acids in the mitochondria. Condensation of acetoacetyl-CoA with a third activated acetate group through the enzyme *β-hydroxy-β-methylglutaryl-CoA (HMG-CoA) synthase)* (also known as *3-hydroxy-3-methylglutaryl-CoA synthetase*) gives rise to a branched-chain intermediate known as *HMG-CoA*. The thioester group of HMG-CoA is subsequently reduced to an alcohol in an NADPH-dependent reaction driven by *HMG-CoA reductase* (3-hydroxy-3-methylglutaryl-CoA reductase) to produce *mevalonate*, which is *the rate limiting step* in the biosynthesis of cholesterol.

The biosynthesis of activated isoprenoid units proceeds through two, step-wise phosphorylation reactions, a reduction, and an isomerization. First, the transfer of a gamma-phosphate of ATP to the newly generated hydroxy group of HMG-CoA by *mevalonate-5-phosphotransferase* gives rise to *mevalonic acid--monophosphate*. *Phosphomevalonate kinase* transfers a second equivalent of phosphate from ATP to generate the 5-diphospho or pyrophospho group within *mevalonic acid-5-diphosphate*. A decarboxylation / dehydration reaction catalyzed by *diphosphomevalonate decarboxylase* proceeds through the intermediate 3-phosphomevalonate-5-pyrophosphate prior to phosphate release and formation of *Δ³-isopentenyl diphosphate*. Finally, the protonation/deprotonation equilibration between the two isoprenoid units (isopentenyl diphosphate and its allylic isomer *3,3-dimethylallyldiphosphate*) is catalyzed by *isopentyldiphosphate delta-isomerase*.

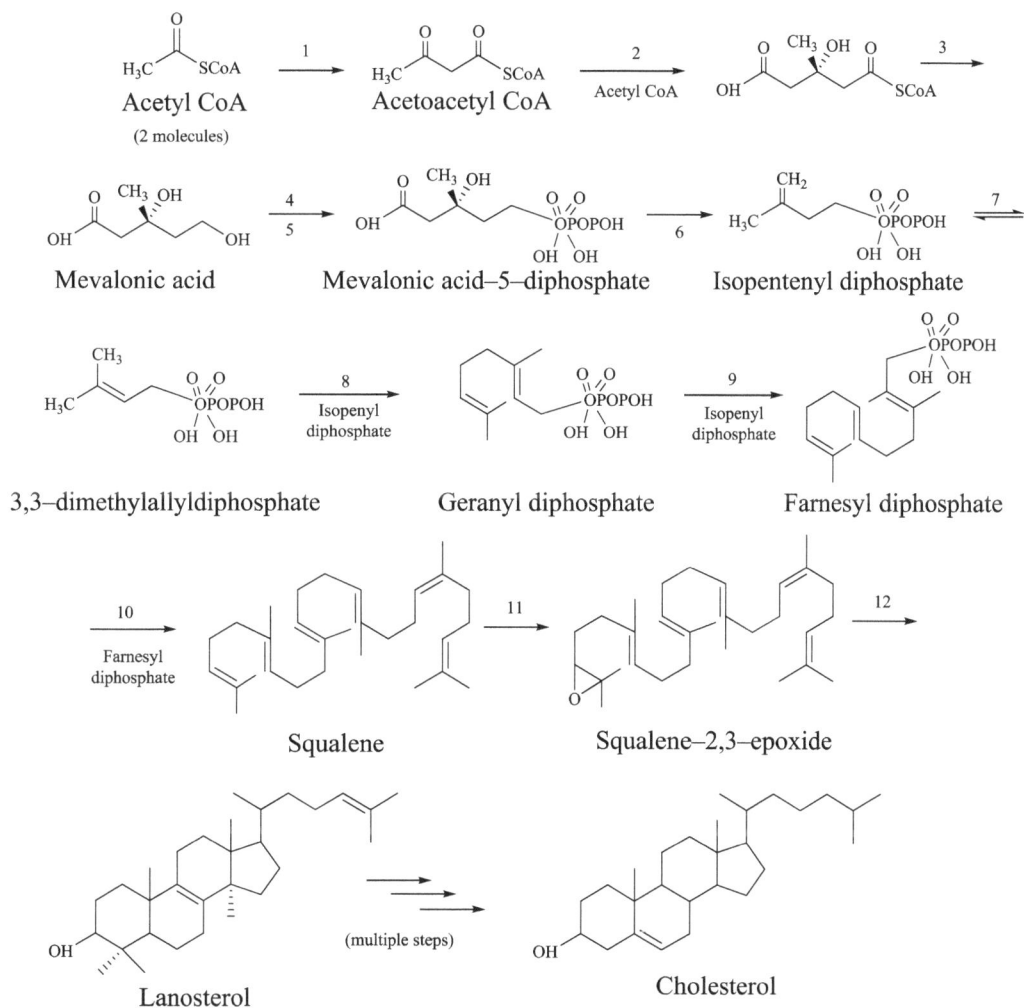

Fig. (5A). *De novo* biosynthesis of cholesterol from acetyl Co A. 1. Acetyl–CoA acetyltransferase [EC 2.3.1.9]; 2. 3–Hydroxy–3–methylglutaryl–CoA synthetase [EC 4.1.3.5]; 3. 3–Hydroxy–3– methylglutaryl–CoA reductase [EC 1.1.1.34]; 4. Mevalonate–5–phosphotranferase [EC 2.7.1.36]; 5. Phosphomevalonate kinase [EC 2.7.4.2]; 6. Diphosphomevalonate decarboxylate [EC 4.1.1.33]; 7. Isopentyldiphosphate delta–isomease [EC 5.3.3.2]; 8. Dimethylallyl transferase [EC 2.5.1.1]; 9. Geranyl transferase [EC 2.5.1.10]; 10. Farnesyl–diphosphate farnesyl transferase [EC 2.5.1.21] (Squalene synthetase); 11. Squalene 1–monooxygenase [EC 1.14.99.7]; 12. Lanosterol synthase [EC 5.4.99.7].

The "head-to-tail" (1'-4) polymerization of resulting isoprenoids proceeds through the general outline, C5→ C10 → C15 → C30. The sequence begins with *dimethylallyl transferase* (often abbreviated *prenyltransferase*) catalyzing a condensation reaction between the carbocation formed upon pyrophosphate release from the "head" (C1) of dimethylallyl diphosphate and the nucleophilic "tail" (C4 of the double bond) of isopentyldiphosphate to form 10-carbon *geranyl diphosphate*. *Geranyl transferase* catalyzes an identical condensation between geranyl diphosphate and a second equivalent of isopentyldiphosphate to produce 15-carbon *farnesyl diphosphate*. (Note that the odd names of these aromatic compounds are derived from their initial sources; for example, geraniol is the oil of geraniums, and farnesol of the flowers of the Farnese acacia tree, both sources of floral essence in perfumes.) Finally, *squalene synthetase*, anchored in the membrane of the endoplasmic reticulum, catalyzes the head-to-head condensation of two molecules of farnesyl diphosphate, accompanied by the elimination of both pyrophosphate groups and an NADPH-dependent reduction. The result is 30-carbon *squalene*.

The conversion of squalene to cholesterol proceeds through multiple steps, but the positioning of the C3 hydroxyl group and the initial cyclization of the characteristic four-fused ring structure to give rise to the primary intermediate *lanosterol* stand out prominently. *Squalene 1-monooxygenase* represents another example of a mixed-function oxidase (a class shared by tyrosine hydroxylase and dopamine β-hydroxylase, enzymes encountered earlier in this chapter); it catalyzes the conversion of squalene to its *2,3-epoxide* form. Molecular O_2 provides the source of the epoxide oxygen, with the remaining oxygen atom reduced to water in an NADPH-dependent manner. The next reaction, catalyzed by *lanosterol synthase*, involves a multifaceted molecular transformation—four rings, six carbon-carbon bonds, and seven chirality centers are simultaneously generated. The enzyme literally folds the squalene molecule so as to make contact between the various double bonds, which then engage in successive electrophilic addition reactions that are accompanied by a series of 1,2-hydride and methyl shifts.

There are 19 remaining steps in the synthesis of cholesterol from lanosterol, as outlined in Fig. (**5B**). While these reactions will not be discussed in detail, and are in many cases still poorly understood, the structural transformations that occur can be generalized as: (1) oxidation reactions that involve removal of three methyl groups (one removed from C14 in the form of formate by *lanosterol-14a-methyl demethylase,* and removal of two from C4 in the form of carbon dioxide by separate reactions catalyzed by *sterol-4a-carboxylate 3-dehydrogenase*); (2) the reduction of a double bond between C24 and C25 in the side chain, catalyzed by *sterol delta 24 reductase*; and (3) the relocation of a second double bond from C8 to C5, catalyzed by an interplay of *cholestenol delta-isomerase, lathosterol oxidase*, and *7-dehydrocholesterol reductase*. Note that the enzymes associated with these reaction often take on multiple roles, and that the sequence of reactions can be variable.

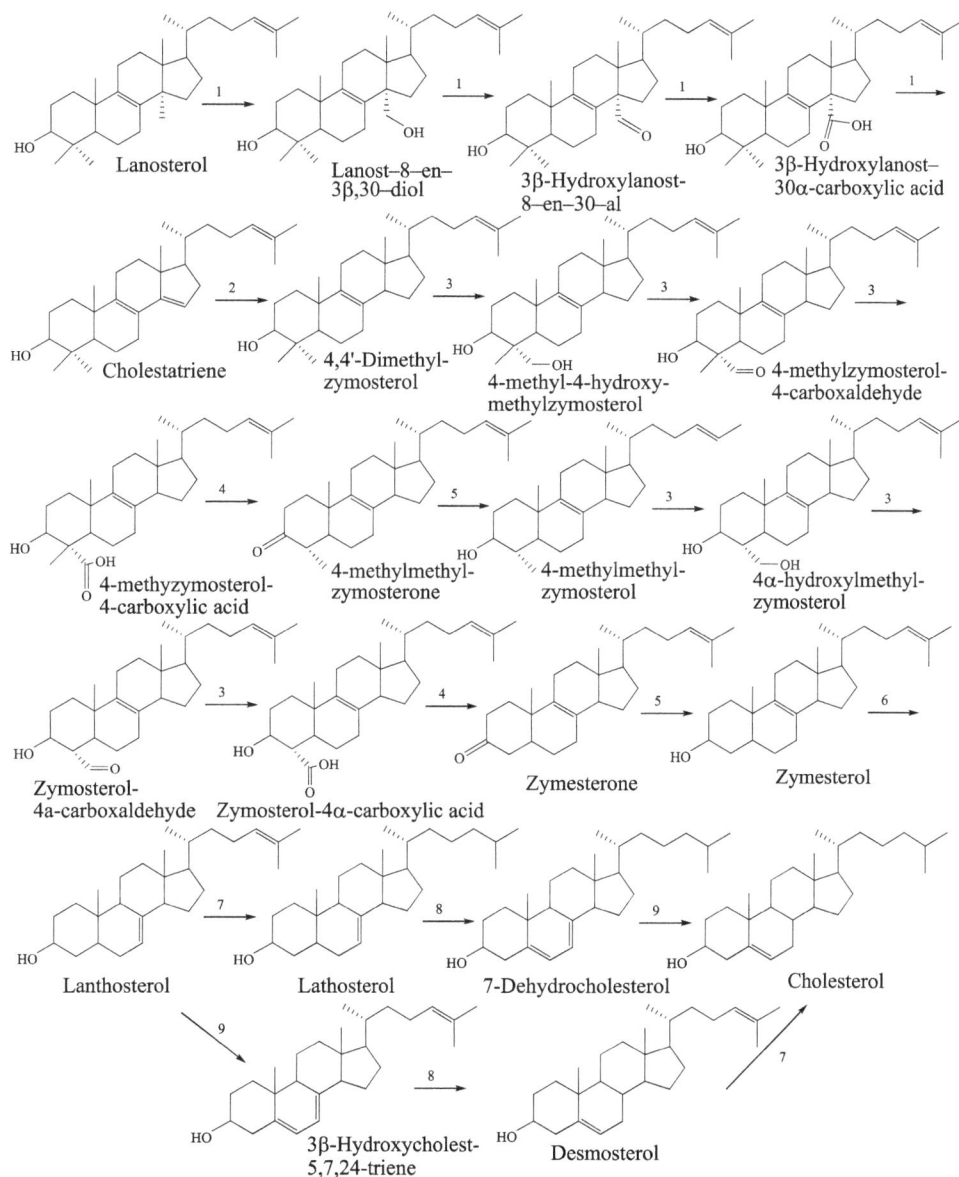

Fig. (5B). Biosynthesis of cholesterol from lanosterol. 1 Lanosterol-14a-methyl demethylase [1.14.13.70]; 2 Steroid-14-reductase; 3 Methylsterol monooxygenase [1.14.13.72]; 4 Sterol-4a-carboxylate 3-dehydrogenase; (decarboxylating) [1.1.1.170]; 5 Steroid 3-keto reductase; 6 Cholestenol delta-isomerase [5.3.3.5]; 7 Sterol delta 24 reductase [1.3.1.72]; 8 Lathosterol oxidase [1.3.3.2]; 9 7-Dehydrocholesterol reductase [1.3.1.21].

Cholesterol biosynthesis is a complex and energy-consuming process, and hence *tightly controlled via* transcriptional regulation, hormone-induced phosphorylation cascades, limited proteolysis, and rates of receptor-mediated endocytosis. The reaction catalyzed by HMG-CoA reductase represents the *rate-limiting step* for all of cholesterol synthesis and therefore is the target of each of these regulatory mechanisms, as well as the target for statin drugs designed to lower cholesterol levels in patients. For example, in response to decreasing intracellular cholesterol concentrations, the amino termini of *sterol regulatory element-binding proteins (SREBPs)*, normally integrated into the ER, are excised and allowed to pass into the nucleus, where they activate transcription of the HMG-CoA reductase gene. Peptide hormones also act to control HMG-CoA reductase activity; *insulin*, secreted by the pancreas when ample supplies of glucose are available to support biosynthesis, leads to the dephosphorylation and activation of the enzyme, whereas *glucagon* initiates a phosphorylation cascade to ultimately inhibit the enzyme. The accumulation of as yet unidentified intermediates of cholesterol synthesis also appears to stimulate the enzymatic breakdown of HMG-CoA reductase, reducing enzyme levels when the synthetic pathways are over-stimulated. Finally, levels of transcription of the LDL receptor, responsible for bringing cholesterol into the cell, are also dictated by the relative concentration of cholesterol; specifically, cholesterol inhibits the synthesis of *LDL receptors* by down-regulating their synthesis. If not brought into the cells, cholesterol disposal *via* the bile salts are enhanced. Together, these regulatory mechanisms maintain a rate of synthesis proportional to intracellular cholesterol levels.

Adrenocorticoid Biosynthesis

As the name implies, adrenocorticoid hormones are characteristically synthesized in the adrenal cortex. Three classes of adrenocorticoid hormones exist. The *progestins*, exemplified by *progesterone*, generally control events during pregnancy and serve as intermediates toward the formation of other adrenocorticoid and sex hormones. (Since they are also synthesized in regions other than the adrenal gland, such as testes and ovaries, many texts do not strictly classify the progestins as adrenocorticoid hormones.) The *glucocorticoids*, such as hydrocortisone, mediate effects opposite to that of insulin, such as stimulating gluconeogenesis, inhibiting glucose uptake by muscle and fat, and stimulating fat mobilization *via* transcriptional regulation of PEP *carboxykinase*, all designed toward assisting an organism in reacting to external stress. As immunological mediators, they also up-regulate the transcription of anti-inflammatory proteins. Finally, the *mineralocorticoids*, such as aldosterone, modulate the ion balance by promoting reabsorption of potassium, sodium, chloride and bicarbonate ions by the kidney.

The activation of steroid hormone synthesis, originating from central nervous system signals that trigger intermediary hormones, initially involves stimulation of the hydrolysis of cholesterol esters through *cholesterol esterase* (Fig. **6**) and the uptake of the cholesterol into the mitochondria of cells in the target tissue. Since adrenocorticoids characteristically exhibit 21 carbons, whereas cholesterol has 27, the next stage in their synthesis involves removal of a six-carbon unit from the C17 side chain of cholesterol as isocaproic acid to form the central intermediate, *pregnenolone*. The *rate limiting* P450-enzyme *cholesterol monooxygenase* (also referred to as *cholesterol 20, 22-desmolase* or *cholesterol side-chain cleaving enzyme*) carries out this transformation, by hydroxylating both C20 and C22 and cleaving the bond between the two carbons in an NADPH-dependent oxidation of C20 into a ketone. In this oxidation, the cofactors *adrenodoxin reductase* and adrenodoxin shuttle six-electrons to the P450 complex. Generally, the presence or absence of the enzyme cholesterol monooxygenase largely defines whether or not a given cell type is capable of synthesizing steroid hormones.

Pregnenolone serves as the precursor of all other steroid hormones. Progesterone is synthesized from pregnenolone in two steps—the 3-hydroxyl group is oxidized by *β-hydroxysteroid Δ^5 dehydrogenase*, and the Δ^5 double bond is *isomerized* from the B to the A ring by *steroid Δ^5 isomerase*. Hydrocortisone is synthesized in *zona fasciculata* from progesterone *via* cytochrome P450-directed hydroxylation at C17 (catalyzed by *steroid 17α-monoxygenase; CYP17*), present in *zona fasciculata*, C21 (catalyzed by *steroid 21-monooxygnease*), and C11 (catalyzed by *steroid 11β-monooxygenase*). The pattern of modifications can be variable, with hydroxylation of C17 occurring within either pregnenolone to produce *17α-hydroxypregnenolone* or within progesterone to produce *17α-hydroxyprogesterone* (Fig. **6**). The intermediate *11-deoxycortisol* represents hydroxylation of progesterone at only C17 and C21, with the name implying that the structure is a C11-hydroxyl group shy of becoming *hydrocortisone*.

The *bioregulation* of hydrocortisone involves the hypothalamus, pituitary and adrenal cortex, known as the *HPA axis*. The hypothalamus communicates with the posterior pituitary by nerves to release the polypeptide hormones. The hypothalamus communicates with the anterior pituitary by releasing polypeptide releasing factors. These travel *via* blood vessels to the anterior pituitary to cause release of the polypeptide hormones. With respect to glucocorticoids the hypothalamus secretes *Corticotropin–releasing factor* (*CRF*) which travels in the blood to the anterior pituitary where it initiates synthesis and release of corticotropin (*ACTH*). Recall that the precursor for ACTH is POMC (Pro-opiomelanocortin is a precursor polypeptide with 241 amino acid). ACTH travels to the adrenal cortex where it stimulates the rate limiting step in the biosynthesis, the hydroxylation of C20 and C22. The adrenal gland subsequently releases

Hydrocortisone into the blood where it travels to nearly all cells in the body. Circulating Hydrocortisone also closes a negative biofeedback loop by binding to receptors in the hypothalamus and pituitary. Exogenous glucocorticoids bind these receptors and thus suppress the HPA axis resulting in decreased production of endogenous Hydrocortisone.

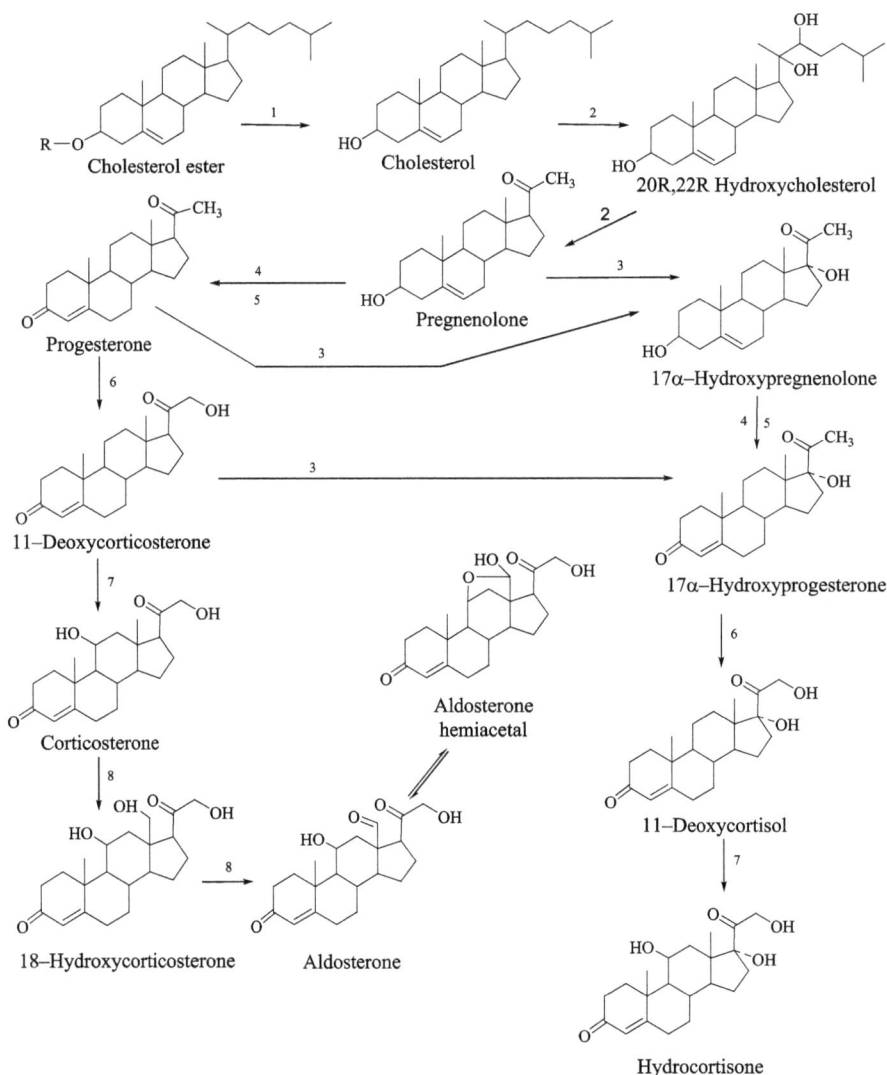

Fig. (6). Biosynthesis of adrenocorticoids from cholesterol. 1 Cholesterol esterase [EC 3.1.1.13]; 2 Cholesterol monooxygenase (side-chain-cleaving) [EC 1.14.15.6], C20–22 Lyase, Desmolase, CYP11A1; 3 Steroid 17α–monooxygenase [EC1.14.99.9], CYP17; 4 3β–Hydroxysteroid Δ5 dehydrogenase [EC 1.1.1.145], CYP11B1; 5 Steroid Δ5 isomerase [EC 5.3.3.1]; 6 Steroid 21–monooxygenase [EC1.14.99.10], CYP21A2; 7 Steroid 11β–monooxygenase [EC1.14.15.4], CYP11B1; 8 Corticosterone 18–monooxygenase [EC 1.14.15.5], CYP11B2.

Mineralocorticoid biosynthesis takes place in *zona glomerulosa* and does not involve C17 hydroxylation; instead, progesterone is directly hydroxylated at C21 to produce *11-deoxycorticosterone* and subsequently at C11 to produce *corticosterone*. The enzyme *corticosterone 18-monooxygenase*, present in *zona glomerulosa*, successively oxidizes the C18 angular methyl group to a hydroxyl group within 18-hydroxycorticosterone and finally an aldehyde within aldosterone. The resulting C18-aldehyde is prone to intra-molecular nucleophilic attack by the C11 hydroxyl group, resulting in a majority of aldosterone existing in the cyclized hemiacetal form.

The *bioregulation* of aldosterone involves the kidneys, liver, and adrenal glands. The two most significant regulators of aldosterone secretion are: (i) concentrations of potassium ion in extracellular fluid and (ii) angiotensin II.

Sex Hormone Biosynthesis

Androgens, known collectively as the *male sex hormones*, include the entire collection of 19-carbon intermediates leading up to the formation of testosterone, as outlined in Fig. (**7**). Androgen synthesis builds once again upon the structures of pregnenolone and progesterone. The hydroxylation of progesterone at C17 by *steroid 17α-monooxygenase* and subsequent aldol cleavage of the side chain consisting of C20 and C21 by *17α-hydroxyprogesterone aldolase/lyase* yield *17α-hydroxyprogesterone* and *androstenedione*, respectively. Reduction of androstenedione by *17β-hydroxysteroid dehydrogenase* produces *testosterone*. A slight alternation in the route to testosterone provides additional androgens (Fig. **7**). For example, if pregnenolone is directly hydroxylated at C17 and cleaved at C17-C30, *dehydroepiandrosterone* is produced—note that this compound maintains the Δ^5 double bond within the B ring and is reduced at C3 relative to androstenedione. Further reduction at C17 by *17β-hydroxysteroid dehydrogenase* produces androstenediol, while further oxidation at C3 and Δ^5 isomerization produces testosterone once again.

Finally, estrogens, *female sex hormones* characterized by 18 total carbons, are themselves synthesized from the androgens. The estrogens include *estrone*, derived from androstenedione, and *estradiol*, derived from testosterone (Fig. **8**). The key enzyme in this transformation is *aromatase*, a cytochrome P450 enzyme that catalyzes the formation of an aromatic A ring proceeding through the oxidation and elimination of the angular C19 methyl group. *Estradiol 17α-dehydrogenase* interconverts estrone and estradiol *via* C17 oxidation. Interestingly, since aromatase in essence synthesizes female sex hormones from male sex hormones, male athletes involved in steroid supplementation often take aromatase inhibitors to avoid the development of female sexual characteristics, such as enlarged breasts.

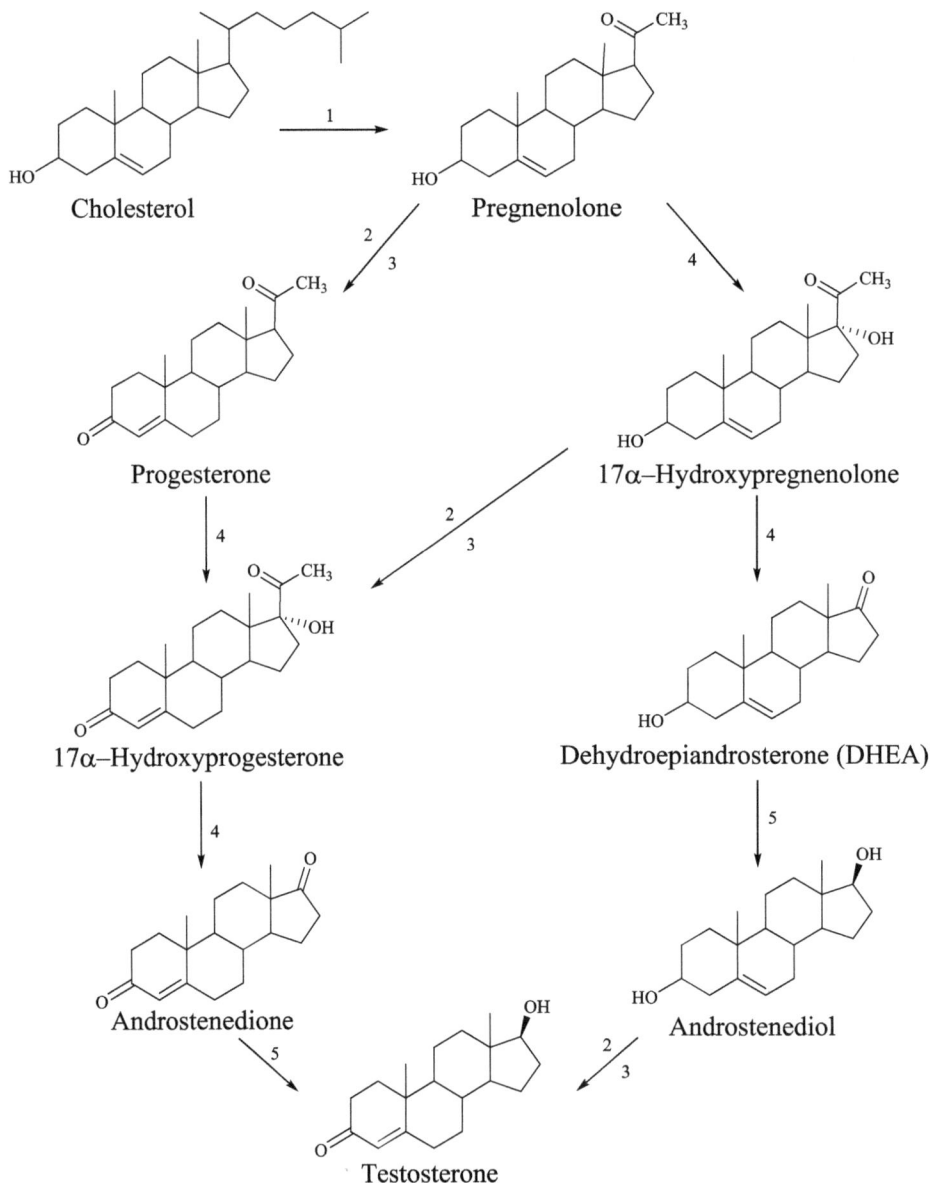

Fig. (7). Biosynthesis of testosterone. 1. Cholesterol monooxygenase (side-chain-cleaving) [EC 1.14.15.6], C20–22 Lyase, Desmolase, CYP11A1; 2. 3β–Hydroxysteroid Δ5 dehydrogenase [EC 1.1.1.145]; 3. Steroid Δ5 isomerase [EC 5.3.3.1]; 4. Steroid 17α–monooxygenase [EC 1.14.99.9], 17α–Hydroxyprogesterone aldolase [EC 4.1.2.30], C17,20 Lyase; 5. 17β–Hydroxysteroid dehydrogenase [EC 1.1.1.51].

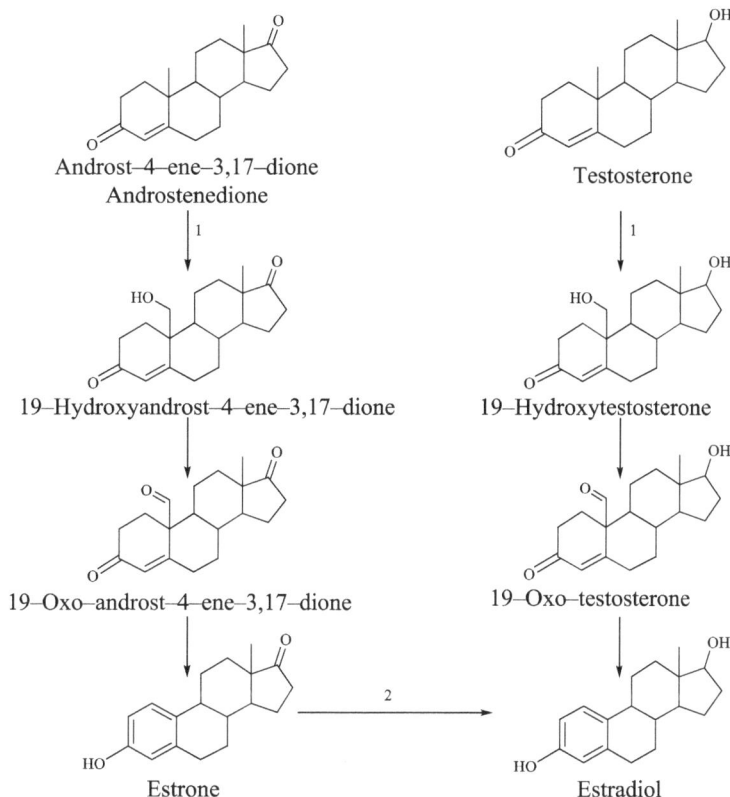

Fig. (8). Biosynthesis of estrogens. 1 Aromatase [EC 1.14.14.1], Estrogen synthetase, CYP19; 2 Estradiol 17β–dehydrogenase.

Steroid hormones are not stored for ready use; instead, their rate of synthesis directly *controls* the concentration and availability of hormones within tissues. As described above for the regulation of catecholamine synthesis and release, the hypothalamus' response to stress stimuli initiates the hormonal cascade that first invokes the release of CRF, which activates the release of ACTH, which in turn activates the release of adrenocortical steroids. ACTH is the primary regulator of glucocorticoid synthesis, while is only of secondary importance to the mineralocorticoids. For glucocorticoid synthesis, the ACTH-induced, cAMP-dependent phosphorylation cascade *stimulates*: (a) the expression of the LDL-receptor, allowing adrenal tissue to uptake cholesterol as a precursor for steroidogenesis; (b) the activity of cholesterol esterase, to increase the hydrolysis of cholesterol esters within vesicles entering cells *via* the LDL-receptor; (c) the transcription of steroidogenic acute regulatory protein for translocating cholesterol into the mitochondria, where initial steps in steroidogenesis occur; and (d) the activity of cholesterol monooxygenase, controlling the synthesis of the pregnenolone, the central intermediate in the manufacture of the adrenocortical steroids. A system of feedback inhibition also modulates the cascade, in which ACTH inhibits the release of CRF, while the adrenocortical steroids inhibit the release of both CRF and ACTH.

NUCLEIC ACID BIOSYNTHESIS

Introduction

Nucleotide biosynthesis supports numerous biological processes, including providing: the building blocks for the synthesis of DNA and RNA; the energy currency of the cell in the form of ATP and GTP, whose concentrations modulate a variety of metabolic pathways; second messengers for hormonal action, such as cAMP; and components of coenzymes such as NAD^+, FAD, and Coenzyme A. Thus, it is not surprising that every cell type in the human body participates in the elaborate biosynthetic pathways manufacturing the nucleotides, and that these pathways are under tight regulatory control.

The synthesis of nucleotides requires enzymes belonging to a variety of different classes. However, one class of enzyme is particularly associated with nucleotide synthesis and deserves special mention—the *glutamine amidotransferases* (*GATs*). These enzymes characteristically exhibit two domains or subunits. In the amidotransferase domain/subunit, substrate glutamine associates and is hydrolyzed to release free ammonia. This hydrolysis generally involves nucleophilic attack by an active-site cysteine on the delta carbonyl group of glutamine. Upon release, ammonia travels through a molecular tunnel to a second domain/subunit of the protein known as the ligase or synthetase component. Here, a receptor substrate binds and, having been activated through phosphoryl transfer, is attacked by the incoming ammonia in a nucleophilic displacement reaction of the phosphate moiety. Because nucleotide synthesis (and hence rapid cell division) is so dependent on the action of the GATs, a class of chemotherapeutic drugs exist that arrest tumor growth by forming covalent adducts within the glutamine binding site—such pharmaceuticals include acivicin, azaserine, and diazooxonorleucine.

Purine Biosynthesis

The purine ring system within nucleotides is manufactured from five different precursor molecules. These include the amine of *aspartate* at position N1, the single carbon of two *formate* molecules at C2 and C8, the amide nitrogen of two *glutamine* molecules at N3 and N9, the *bicarbonate* carbon at C6, and the entire *glycine* molecule at C4, C5, and N7. Interestingly, the purine ring is *not* manufactured first as a free base and later attached to the ribose and phosphate units to give rise to a nucleotide structure, but rather the nitrogenous base system is built directly upon the sugar-phosphate base. The first purine nucleotide constructed in the pathway is *inosine-5'-monophosphate* (*IMP*) containing the hypoxanthine ring; from the shared intermediate IMP, *guanosine-5--monophosphate* (*GMP*) and *adenine-5'-monophosphate* (*AMP*) are synthesized through separate reactions. Once the monophosphate nucleotides are constructed, various ubiquitous enzymes such as adenylate kinase and nucleoside-monophosphate and –diphosphate kinases equilibrate cellular pools between the diphosphate (ADP and GDP) and triphosphate (ATP and GTP) forms.

The synthesis of IMP is somewhat complicated, involving at least eleven distinct enzymatic steps (Fig. **9**). *Ribose-5-phosphate*, deriving directly from the oxidative phase of the pentose phosphate pathway, is first activated by ATP to give rise to *5-phosphoribosyl-α-pyrophosphate* (*PRPP*). The enzyme *ribose phosphate diphosphokinase* (or *PRPP synthetase*) catalyzes this step, providing PRPP as an activated precursor (meaning it is activated toward nucleophilic substitution *via* the positioning of a stable pyrophosphate leaving group at C1') for not only purine synthesis, but also for pyrimidine, tryptophan, and histidine synthesis. This enzyme is generally down-regulated by ADP and GDP, which would reflect insufficient energy reserves for continued nucleotide manufacture.

Ribose-5-phosphate thus serves as the base molecule for the manufacture of a purine ring system to be attached at C1' of the ribose sugar. The first atom of the purine ring to be positioned is N9, provided by *glutamine* in a reaction catalyzed by *amidophosphoribosyl transferase*. This enzyme is the first of many examples of the GAT class of enzyme described above, in which ammonia is hydrolyzed from glutamine and serves as a nucleophile in attacking the activated C1' position of PRPP (the receptor substrate). The product is *5-phosphoribosylamine* (*PRA*).

The remaining seven steps toward the synthesis of IMP are summarized in Fig. (**9**); while mechanistically interesting, these reactions are generally not the focus of *in vivo* regulatory mechanisms. The next molecule incorporated into PRA is *glycine*, which contributes all of its atoms at C4, C5, and N7 of the eventual purine ring system. Glycine forms an amide bond with the amino group of PRA in a reaction catalyzed by *glycinamide ribotide (GAR) synthetase* and driven by the hydrolysis of ATP. The product GAR is very unstable and likely channeled into the active site of the next enzyme, *GAR formyltransferase*. The extended α-amino group of GAR is then formylated (addition of -C=O) using N^{10}-*formyltetrahydrofolate* (N^{10}-formyl-THF) as the donor, adding the eventual C8 atom of the ring and producing *formylglycinamide ribotide (FGAR)*. The oxygen of FGAR reacts with ATP to form a phosphoryl ester, which is then attacked by nucleophilic ammonia released from glutamine hydrolysis to form *formylglycinamidine ribotide (FGAM)*. FGAM thus acquires the purine atom N3 in this reaction catalyzed by *FGAM synthase*, an amidotransferase protein similar in function to amidophosphoribosyltransferase.

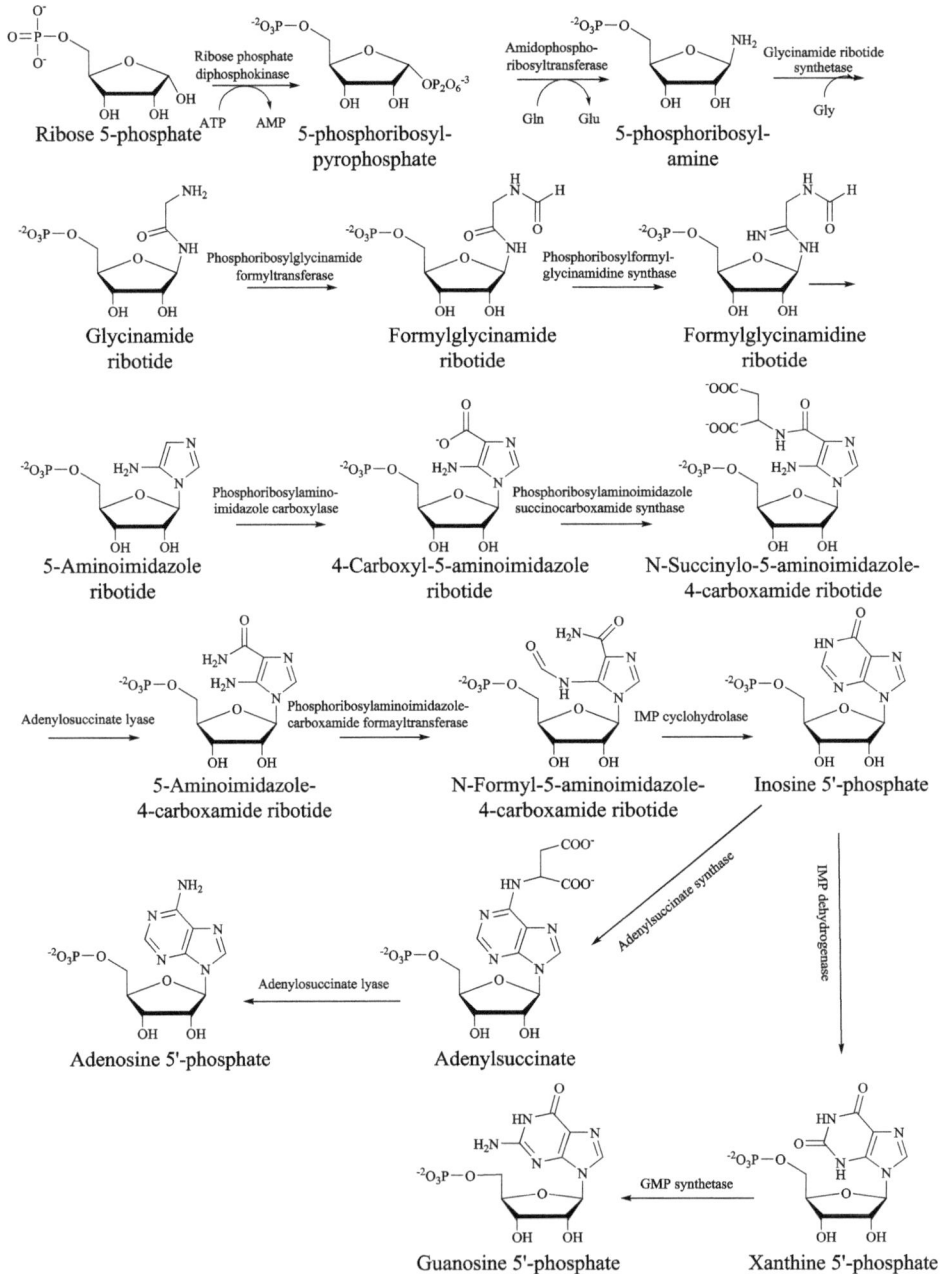

Fig. (9). Purine biosynthesis.

FGAM may spontaneously undergo an intramolecular condensation reaction followed by an aromatization of the imidazole ring to produce *5-aminoimidazole ribotide (AIR);* however, the reaction is more often catalyzed in an ATP-dependent manner by *AIR cyclase* or *synthetase. AIR carboxylase* (systematically known as *phosphoribosylaminoimidazole carboxylase*) next introduces the C6 atom of the purine ring by reaction of bicarbonate (activated as carboxyphosphate by reaction with ATP). The product of AIR carboxylase is *4-caroboxyl-5-aminoimidazole ribotide,* or *CAIR.* In a pair of reactions analogous to those of the urea cycle, N1 of the purine ring is acquired first by condensation of CAIR with *asparate* to yield *N-succinylo-5-aminoimidazole-4-carboxamide ribotide (SAICAR)*, followed by the elimination of *fumarate* by *adenylosuccinate lyase* to yield *5-aminoimidazole-4-carboxamide ribotide (AICAR)*. The last atom of the purine ring system to be added is C2, deriving once again from N^{10}-formyl-THF in a nucleophilic addition reaction catalyzed by *phosphoribosylaminioimidazole-carboxamideformyltransferase*, or simply *AIRCAR transformylase*. Finally, *IMP cyclohydrolase* catalyzes the elimination of water and resultant ring closure to give rise to IMP.

Once IMP is formed, separate pathways lead directly to the biosynthesis of AMP and GMP; IMP itself is not allowed to accumulate to appreciable concentrations in the cell. GMP differs from IMP in that an amino group exists exocyclic at C2. The amino group is added through two reactions: *IMP dehydrogenase* oxidizes C2 in a NAD$^+$-dependent reaction to produce *xanthine-5'-monophosphate (XMP)*, followed by *GMP synthetase* (a GAT enzyme also known as *XMP-glutamine amidotransferase*) catalyzing the replacement of the 2-keto group by an amino group deriving from glutamine hydrolysis. Interestingly, prior to reaction with ammonia, the 2-keto group is first activated by adenylation, meaning that GMP formation relies on a ready source of ATP. Such an inter-dependence between GTP formation and the availability of ATP helps to ensure an equilibration of purine pools in the cell to support nucleic acid synthesis.

AMP differs from IMP in that an amino group, rather than keto group, exists exocyclic at C6. In reactions nearly identical to those converting CAIR to AICAR above, *adenylosuccinate synthetase* catalyzes a GTP-dependent condensation of the C6-keto group with *asparate*, and *adenylosuccinate lyase* eliminates oxidized *fumarate* to leave behind the C6-amino group. Once again, the requirement for GTP to form AMP serves to equilibrate purine pools.

Pyrimidine Biosynthesis

By comparison to purine synthesis, the pathways by which pyrimidine nucleotides are formed are much more direct. Only three molecules contribute to the pyrimidine ring structure: *aspartate* provides N1, C4, C5, and C6; *bicarbonate provides C2; and glutamine provides N3*. Moreover, the pyrimidine ring system is synthesized in its entirety as a free base, and only then coupled with *PRPP* to give rise to nucleotide structure, in direct opposition to the mode of purine biosynthesis.

Step one toward the pyrimidine biosynthetic pathway is the manufacture of carbamoyl phosphate by the GAT enzyme *carbamoyl phosphate synthetase II (CPSII)*, utilizing glutamine, bicarbonate, and two equivalents of ATP as substrates. The δ-amino group of glutamine will become N3 of the purine, and the bicarbonate ion will become C2. This enzyme should not be confused with CPSI, which also forms carbamoyl phosphate from free ammonia (as opposed to glutamine) as an entry point for waste ammonia into the urea cycle.

CPSII actually exists as one component of a trifunctional protein that also includes the next two enzymes of the pyrimidine synthetic pathway (Fig. **10**), *asparate carbamoyl transferase* (better known as *aspartate transcarbamylase)* and *dihydroorotase*. Aspartate transcarbamylase catalyzes the condensation of activated carbamoyl phosphate with aspartate to yield N-carbamoyl asparate; this reaction links together all atoms that will eventually constitute the pyrimidine ring. Dihyrooratase catalyzes a simple intra-molecular condensation reaction that results in ring closure and the formation of *dihydroorotate*. *Dihydroorotate oxidase* utilizes the FMN-flavin to irreversibly oxidize the C5 and C6 carbons into a double bond, forming orotate. Note that orotate, not uracil or cytosine, represents the only *free* pyrmidine base generated through this *de novo* biosynthetic pathway. In a reaction catalyzed by *orotate phoshoribosyltransferase*, PRPP reacts with the nucleophilic N1 of orotate to release pyrophosphate and form the first full nucleotide structure, *orotidine-5'-monophosphate (OMP)*. Decarboxylation of the C6 carboxyl group of OMP leads directly to *uridine-5--monophosphate* (*UMP*); the enzyme for this reaction, *orotidine-5'-phosphate decarboxylase*, is the most catalytically proficient enzyme known, yet uses no organic cofactors to stabilize the decarboxylation products. Ubiquitous kinases then exist to readily phosphorylate UMP into UDP and finally UTP.

CTP synthetase, another example of a GAT enzyme, then utilizes glutamine as an amino donor to UTP. The mechanism of CTP synthetase consists of: (a) ATP-dependent *phosphorylation* of UTP at the #4 exocyclic oxygen; (b) the *hydrolysis* of glutamine into glutamate and release of ammonia; (c) and a *nucleophilic displacement* reaction of ammonia on the activated #4 oxygen of UTP to produce CTP. Since the relative activity of CTP synthetase directly sets the *in vivo* concentrations of CTP, the enzyme is a prime target of anti-cancer, -parasitic, and –viral pharmaceuticals designed to arrest the proliferation of cells. Moreover, the dependence of CTP synthetase on ATP to activate the reaction represents another strategy for the cell to equilibrate pools of purines and pyrmidines to support DNA and RNA synthesis.

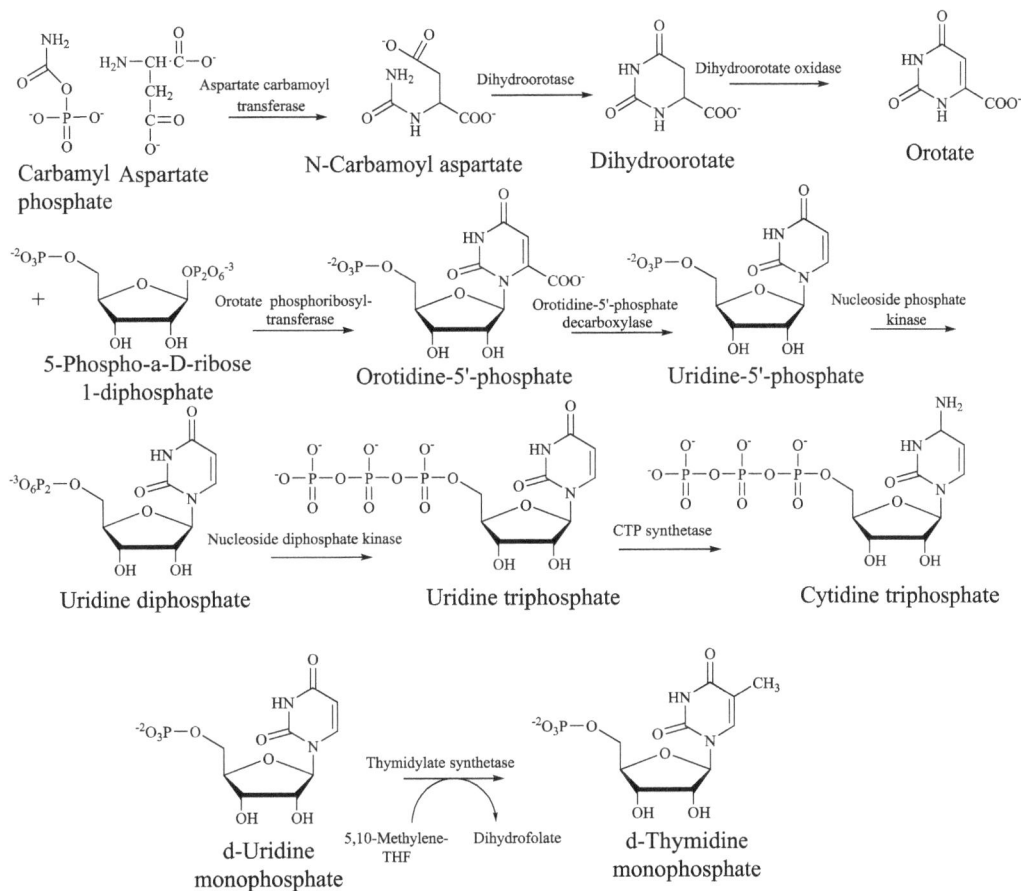

Fig. (10). Pyrimidine biosynthesis.

Deoxyribonucleotide Biosynthesis

Recall that, unlike RNA, DNA's composite nucleotides contain a reduced, 2' deoxy group on ribose, and that T's rather than U's exist in sequence with the remaining three nucleotides. Thus far, we have only seen the biosynthetic pathways for ATP, GTP, UTP, and CTP. What, then, is the origin of the dNTPs, including dTTP?

The deoxyribonucleotides are synthesized from their corresponding ribonucleotides through the direct reduction of the 2' position (Fig. **11**). In eukaryotes, the diphosphate form of the ribonucleotides (NDPs) serves as the substrate for this reduction, catalyzed by ribonucleotide diphosphoreductase, or simply *ribonucleotide reductase*. Important residues at the catalytic site of this enzyme include a tyrosine and two cysteines. Reduction occurs *via* the elimination of the 2'-OH group as water, with a tyrosine free radical indirectly stabilizing the resulting carbocation. Hydride ion transfer from cysteine to the positively charged 2' position generates the dNDP, and the now electrophilic cysteine engages in disulfide bond formation with a second cysteine residue. Eventually, ADP, GDP, UDP, and CDP all serve as substrates in this reaction; however, the order of reduction is allosterically synchronized by substrate specificity sites occupied by the product dNDPs, ensuring that equimolar concentrations of DNA's building blocks exist to support cell division.

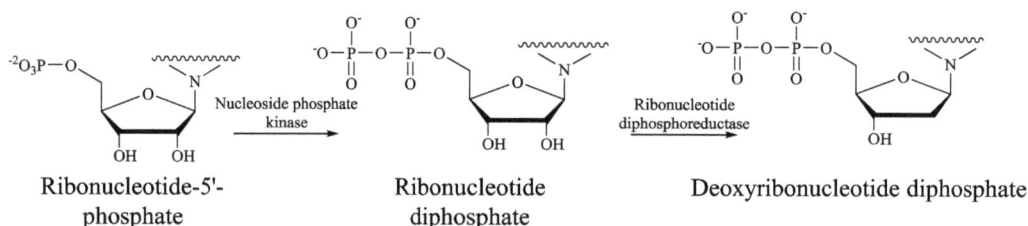

Fig. (11). Deoxyribonucleotide biosynthesis from corresponding ribonucleotide.

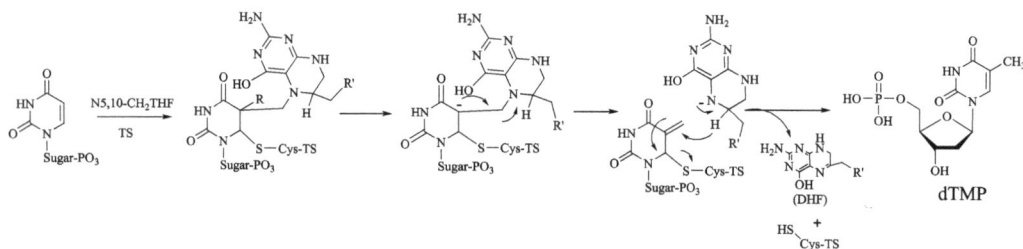

Fig. (12). Biosynthesis of dTMP from dUMP. Thymidylate Synthase: HS-Cys-TS.

Interestingly, although U's do not exist in strands of DNA, ribonucleotide reductase recognizes and reduces UDP into dUDP. Why should the deoxy form of uridine nucleotides exist in the cell? The answer highlights the second distinction between RNA and DNA—dUMP serves as the substrate for making dTMP (Fig. **12**). *Thymidylate synthetase* catalyzes the transfer of a methyl group from N^5,N^{10}-*methylene-tetrahydrofolate (THF)* to C5 of the pyrimidine ring of dUMP to make dTMP. Note that the methylene group of N^5,N^{10}-methylene-THF is reduced to a methyl group during this transfer, requiring that the organic cofactor be concomitantly oxidized to *dihydrofolate (DHF)*. N^5,N^{10}-methylene-THF must be regenerated from DHF to maintain thymidylate synthetase activity and other biosynthetic reactions dependent upon methyl transfer. Two reactions achieve this regeneration: DHF is reduced to *tetrahydrofolate (THF)* by NADPH-dependent *dihydrofolate reductase,* and THF in turn gains a methylene group from a serine residue to become N^5,N^{10}-methylene-THF in a reaction catalyzed by *serine hydroxymethyltransferase*.

Purine Catabolism

Within primates, almost all dietary purine nucleotides, as well as excess purine nucleotides deriving from nucleic acid turnover that are not reused as part of the purine salvage pathway (see below), are oxidized into a common excretory product uric acid. While the pathways vary slightly between AMP and GMP catabolism (Fig. **13**), both involve the actions of a *5'-nucleotidase* enzyme to hydrolyze away phosphate and convert the nucleotides into nucleosides, and a *nucleosidase* enzyme to hydrolyze away the ribose moiety and convert the nucleosides into free bases. Neither of these enzymes is specific for ribose or deoxyribose. AMP is deaminated into inosine-5'-monophoshate (IMP) *via* the action of AMP deaminase; nucleotidase and nucleosidase activities then release the free base, hypoxanthine.

GMP, in contrast, is first converted to its free base guanine before its amino group is hydrolytically removed by guanine deaminase to yield xanthine. Note that the hypoxanthine base arising through AMP degradation and the xanthine base arising from GMP degradation are both ultimately oxidized by *xanthine oxidase* to produce *uric acid*. The ultimate fate of uric acid is excretion through the urine; however, because of its inherently low solubility limit, excess uric acid deriving from metabolic disorder and/or high purine content in the diet can result in the deposition of sodium urate crystals in the joints, a condition commonly known as gout. Interestingly, mammals other than primates process uric acid further into *allantoin*, in a reaction catalyzed by the enzyme *urate oxidase*.

Pyrimidine Catabolism

The degradation of pyrimidine nucleotides is outlined in Fig. (**14**). Initially, cytidine is hydrolyzed to uridine by cytidine deaminase. As was true for the purine nucleotides, the nonspecific phosphatase enzyme *nucleotidase* then transforms all pyrimidine nucleotides into their nucleosides, and the phosphorolysis activity of *nucleosidase* hydrolyzes the glycosidic bond within the nucleosides to release the corresponding free bases. Uracil and thymine are ultimately degraded into different products, but follow a nearly identical reaction scheme and utilize enzymes that show no preference between the two bases. The first committed and rate-limiting step of pyrimidine catabolism involves the reduction of the in-ring double bond catalyzed by *dihydropyrimidine dehydrogenase*. The dihydro-products react with the enzyme *dihydropyrmidinase* to hydrolyze the carbon-nitrogen bond of the remaining ring structures, producing *β-ureidopropionic acid* from uracil and *α-methyl-β-ureidopropionic acid* from

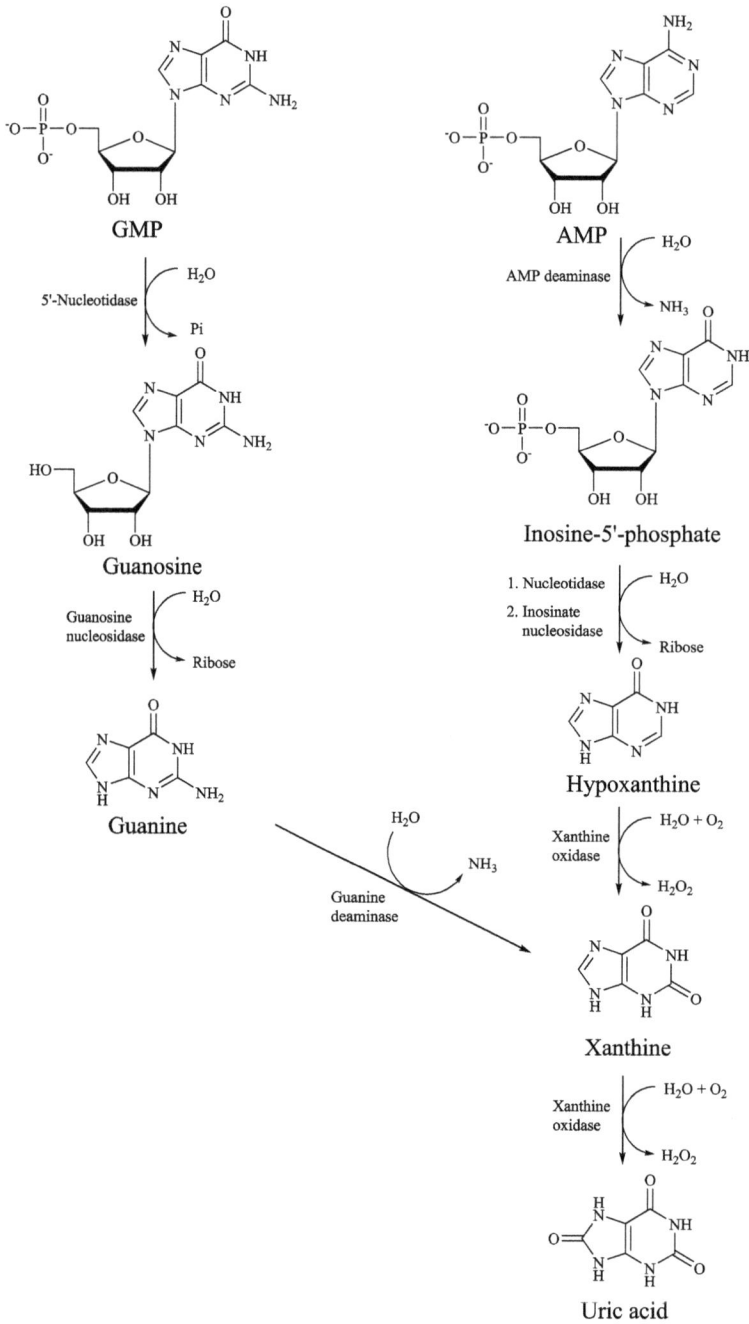

Fig. (13). Purine catabolism.

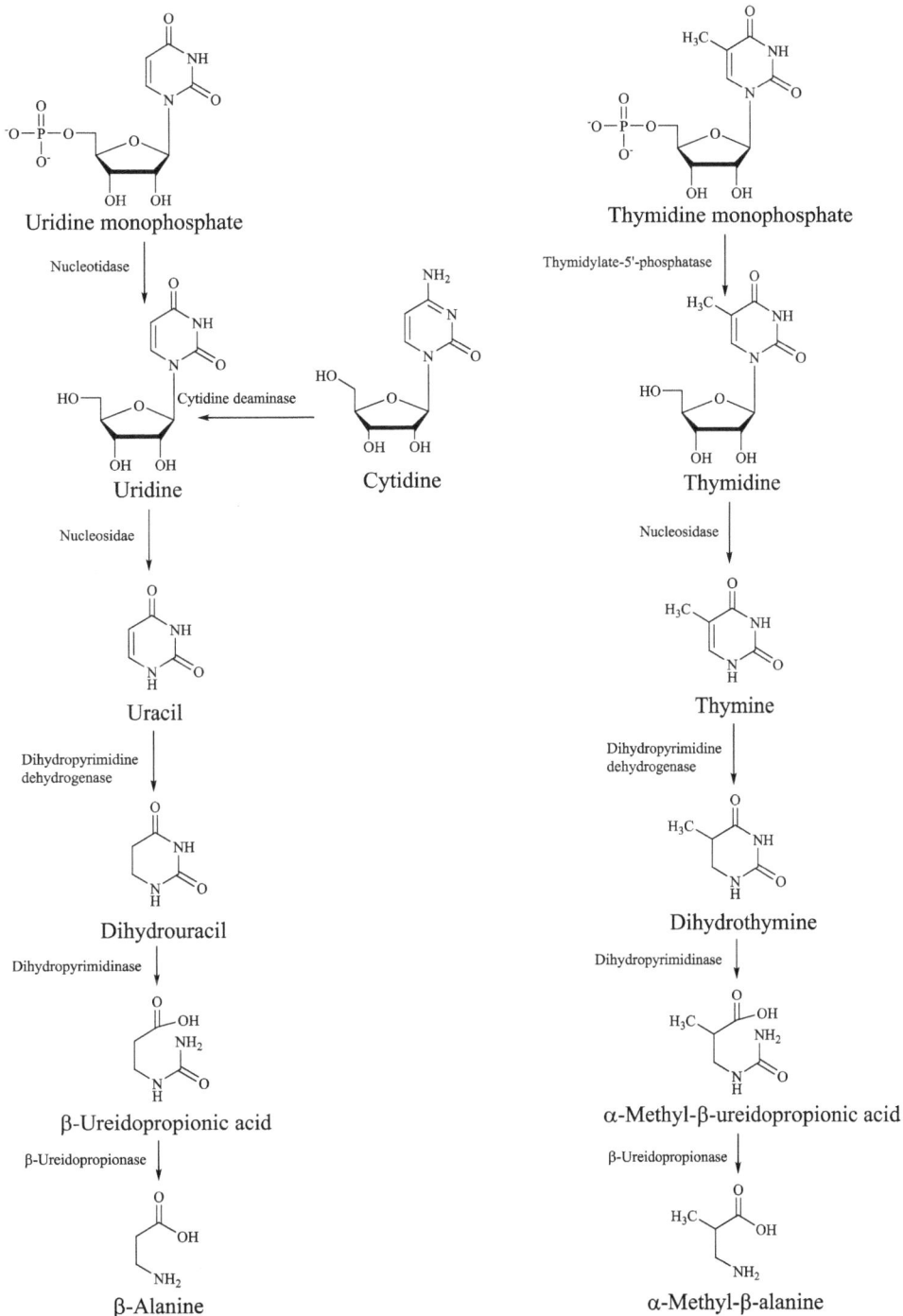

Fig. (14). Pyrimidine catabolism.

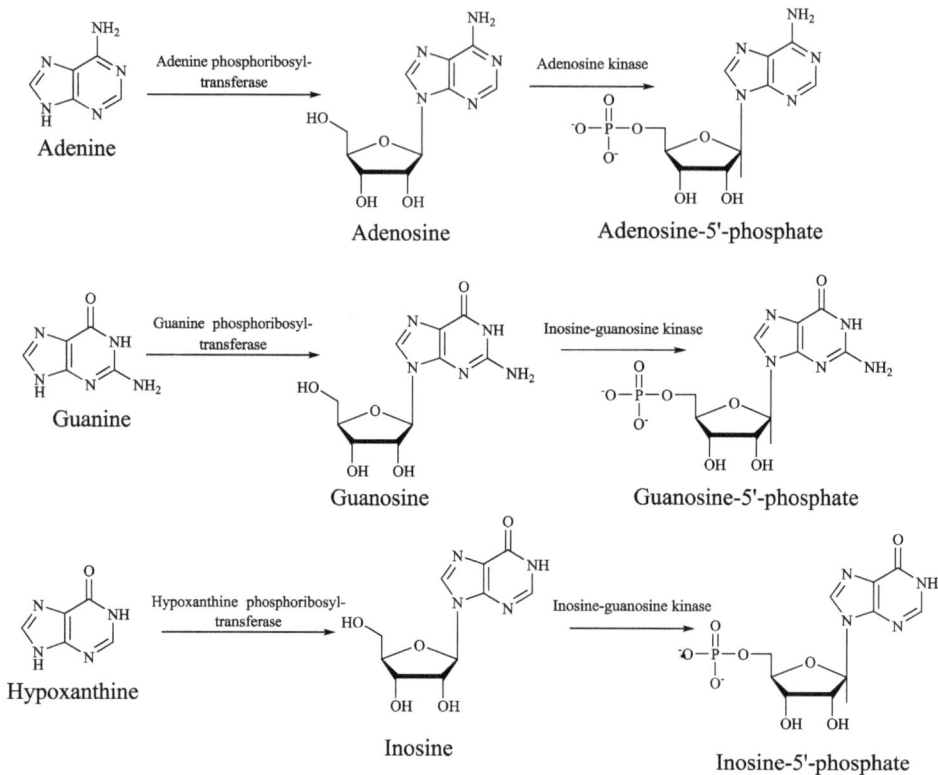

Fig. (15). Purine salvage pathway.

thymine. Finally, carbon dioxide and ammonia are hydrolytically released through the action of *β-ureidopropionase* to produce *β-alanine* and *α-methyl-β-alanine*, respectively. Product ammonia is expelled through the urea cycle, whereas β-alanine and α-methyl-β-alanine are nonstandard amino acids that are converted, *via* transamination and activation reactions, into methylmalonyl-CoA, malonyl-CoA, and eventually the *Krebs Cycle intermediate succinyl-CoA*. Hence, there are no discreet excretory products resulting from pyrimidine catabolism, as there were from purine catabolism.

Purine Salvage Pathway

The *de novo* synthesis of nucleotides supports the majority of cellular requirements. However, nucleic acids released upon cellular degradation can be, to a small degree, recycled through reactions described as the *salvage pathways*. In these reactions, free purine bases are salvaged and reused to make nucleotides directly (Fig. **15**). The pathway is actually a *single-step synthesis* in which the formation of the N-glycosidic bond between each nitrogenous base and the activated form of ribose, PRPP, yields the nucleoside-monophosphate. Phosphoribosyl transferases catalyze these reactions; *adenine phosphoribosyltransferase* is specific for the manufacture of AMP, whereas *hypoxanthine-guanine phosphoribosyltransferase* generates either GMP or IMP. Nucleotides may be further phosphorylated into their corresponding nucleotides by specific kinases, and ribonucleotide reductase once again reduces the 2'-OH of the nucleotide diphosphates to support DNA synthesis. Note then that there is a fundamental distinction between the abbreviated salvage pathways and the much more complex *de novo* pathways for purines; salvage pathways scavenge intact nitrogenous base systems and react them with ribose, whereas *de novo* pathways construct the nitrogenous bases upon the ribose platform.

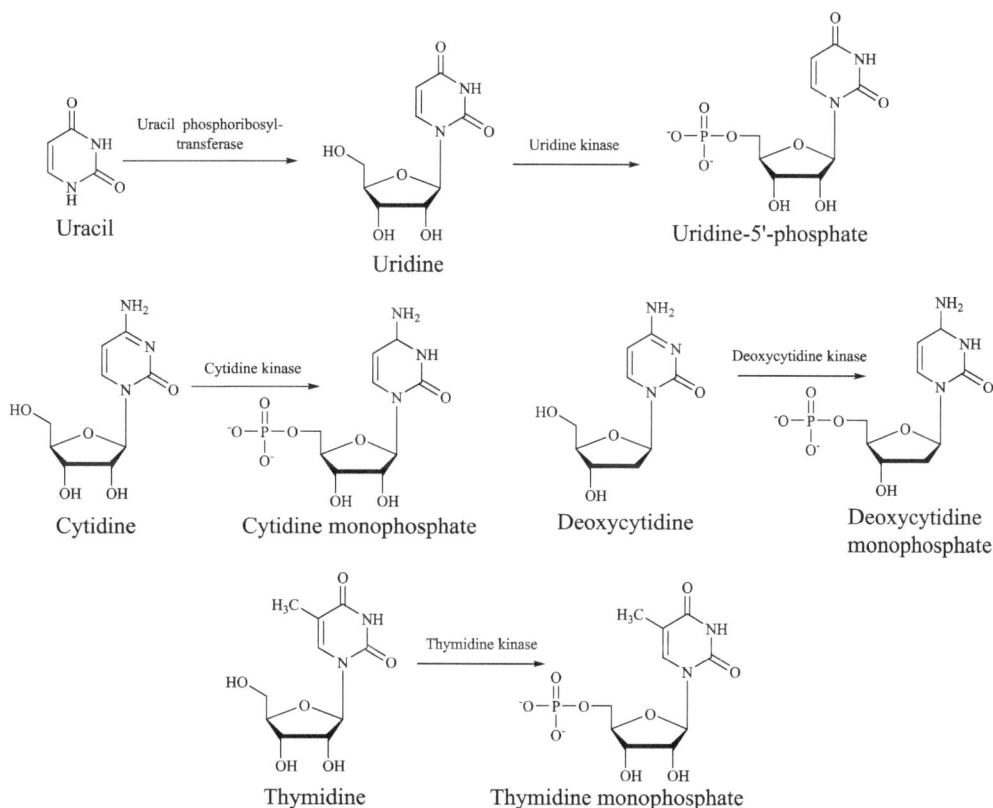

Fig. (16). Pyrimidine salvage pathways.

Pyrimidine Salvage Pathway

Unlike the purines, most pyrimidines are recycled primarily from *nucleosides* rather than free bases (Fig. **16**). For example, *kinases* specific for uridine, cytidine, deoxycytidine, or thymidine directly phosphorylate the nucleosides to generate the corresponding *nucleotide monophosphates*. Only uracil can be salvaged from the free base form by *uracil phosphoribosyltransferase* in a reaction analogous to that of the purine salvage pathways.

Regulation of Nucleotide Metabolism

De novo purine biosynthesis is regulated at three important metabolic junctions *via allosteric, feedback inhibition*. Amidophosphoribosyltransferase, responsible for activating ribose-5-phosphate *via* PRPP formation, catalyzes the first reaction unique to the synthesis of purine nucleotides. Consequently, this enzyme is subject to sequential *feedback inhibition* by all of the eventual purine nucleotide products, including IMP, GMP, and AMP. The next several steps of the *de novo* pathway lead to the formation of IMP, from which GMP is uniquely synthesized beginning with IMP dehydrogenase and AMP is uniquely synthesized beginning with adenylosuccinate synthetase. Thus, in high concentrations, GMP serves as a *feedback inhibitor* to IMP dehydrogenase, and accumulating AMP functions likewise with adenylosuccinate synthetase.

The regulation of pyrimidine biosynthesis also involves allosteric feedback loops, both at the beginning and at the end of the pathway. The activity of CPSII, forming an activated carbamoyl group for condensation with aspartate, is considered the first committed step to pyrimidine synthesis in mammals. Thus, CPSII is tightly *regulated* by the *allosteric inhibitors UDP and UTP*, the first "standard" pyrimidine nucleotide manufactured within the pathway. For additional control, PRPP and ATP function as *allosteric activators* of CPSII, accumulating when sufficient energy reservoirs exist in the cell to support biosynthesis. Finally, the activity of CTP synthetase, responsible for conversion of UTP into CTP, is also tightly *regulated via feedback, allosteric inhibition* by its direct product CTP. Interestingly, the gamma-phosphate of CTP, when bound as an inhibitor, overlaps with that of UTP, making it a competitive inhibitor of the reaction. Pyrimidine analogs, such as cyclopentenylcytosine, 3-deazauridine, and cytosine arabinoside, serve as anti-cancer, anti-viral, and anti-parasitic drugs by binding in the cytosine-specific pocket and mimicking CTPS's mechanism of *feedback inhibition*.

STUDENT SELF-STUDY GUIDE

1. What are cyclooxygenase and peroxydase? What do you mean by prostaglandin endoperoxide H2 synthase (PGHS)?
2. What are COX-1 and COX-2? What physiologically important prostaglandins and thromboxanes are synthesized by them blocking of which gives clinical effects of different NSAIDs?
3. What do you mean by arachidonic acid cascade? What is its significance?

4. What are cyclic and linear pathways? What are their relations with cyclooxygenase and lipoxygenase pathways?

5. "The fate of PGH$_2$ depends largely on the relative activities of competing enzymes" – What are these competing enzymes and their products? Which prostaglandins or thromboxane is present in vascular endothelial cells and platelets, respectively and what are their functions?

6. What are the functions of leukotrienes, hepoxilins and lipoxins?

7. How the eicosanoids biosynthesis are regulated biologically?

8. What are two biologically significant catecholamines? What are their functions? Which amino acid are these biosynthesized from?

9. Discuss the enzyme class, details role and regulations of tyrosine hydroxylase and its cofactor.

10. Which enzyme in the biosynthetic pathway of NE requires a cofactor derived from vitamin B6? What is the end product in this pathway in the substania nigra and in the adrenal medulla, respectively?

11. Which enzyme is critical in the adrenal medulla? Explain its function and the role of vitamin C in its activity.

12. Where and how epinephrine is produced? What enzyme, coenzymes and other factors are involved in this step? Discuss the key mechanisms involved in this step.

13. What is the role of Ca^{2+} in the function of epinephrine and norepinephrine?

14. How the biosynthesis of catecholamines is regulated?

15. Why folic acid is considered a vitamin in human? Which vitamin is this?

16. What is RFC and where does it originate from?

17. What is the role of tetrahydrofolate? Which biological reactions are dependent on this chemical, and in what role?

18. Show with structure the biosynthetic pathway of tetrahydrofolate from folic acid and discuss its bioregulation.

19. What is the site of cholesterol biosynthesis? Discuss the stages involved in this biosynthesis. Show the rate limiting step with structures.

20. In cholesterol biosynthesis what is the first precursor steroid that is produced by the cyclization of squalene? What enzymes are involved in these steps?

21. What are the general transformation reactions involved in the conversion of lanosterol to cholesterol?

22. What is the precursor for the biosynthesis of all the steroid hormones? Show the biosynthetic pathway from cholesterol acetate of this precursor.

23. What are the enzymes common to and unique to zona glomerulosa and zona fasciculate? What is the site for hydrocortisone and aldosterone biosynthesis, respectively? How are these adrenocorticoids bioregulated?

24. Outline the biosynthetic pathway for testosterone and estradiol from pregnenolone with structures. Why male athletes involved in steroid

supplements also take aromatase inhibitor?

25. Discuss the subunits with their functions of the glutamine amidotransferases (GATs). Which chemotherapeutic drugs arrest tumor growth by forming covalent adducts within the glutamine binding site of GATs?

26. What is PRPP and its significance? Show its formation with structures.

27. Show the step in purine biosynthesis that may spontaneously undergo an intramolecular condensation reaction but is more often catalyzed in an ATP-dependent cyclase or synthetase.

28. What is the first purine nucleotide constructed in the biosynthetic pathway. Show the pathways to GMP and AMP from that nucleotide.

29. What are the three molecules that contribute to the pyrimidine ring structure? Also specify which components of pyrimidine are contributed by each of these molecules. How the pyrimidine nucleotide formation differs from the purine nucleotide formation?

30. Which molecule in the *de novo* biosynthetic pathway of pyrmidine represents the only *free* base? Show its synthetic pathway. Also show the first pyrimidine nucleotide formation.

31. Discuss the mechanism of CTP synthetase and show the reactions catalyzed by this enzyme.

32. Which enzyme is the key in the conversion of ribonucleotides into corresponding deoxyribonucleotides? What are the key amino acid residues in its activity and what roles do these residues play? Is dTMP formed in the same mechanism? Show the mechanism of formation of dTMP.

33. What is the catabolic end product of purine nucleotides? What is the reason of gout?

34. What are the end products of pyrimidine catabolism? Is there any discrete excretory product from pyrimidine catabolism like purine catabolism?

35. Discuss the purine and pyrimidine salvage pathways.

36. How the nucleotides are bioregulated?

CONSENT FOR PUBLICATION

Not applicable.

CONFLICT OF INTEREST

The authors declare no conflict of interest, financial or otherwise.

ACKNOWLEDGEMENT

Declared none.

FURTHER READING

General

[1]   Berg JM, Tymoczko JL, Stryer L, Eds. Biochemistry. 6th ed., New York: W. H. Freeman & Company 2007.

[2]   Devlin TM, Ed. Textbook of biochemistry with clinical correlations. 7th ed., New Jersey: John Wiley & Sons, Inc. 2011.

[3]   McMurry J, Begley T, Eds. The organic chemistry of biological pathways. Greenwood Village, CO: Roberts and Company Publishers 2005.

[4]   Nelson DL, Cox MM, Eds. Lehninger: Principles of Biochemistry. 5th ed., New York: W. H. Freeman & Company 2008.

[5]   Voet D, Voet JG, Eds. Biochemistry. 4th ed., New Jersey: John Wiley & Sons, Inc. 2011.

Eicosanoids: Prostaglandins, Thromboxanes, and Leukotrienes

[6]   Balsinde J, Balboa MA, Insel PA, Dennis EA. Regulation and inhibition of phospholipase A2. Annu Rev Pharmacol Toxicol 1999; 39: 175-89.
[http://dx.doi.org/10.1146/annurev.pharmtox.39.1.175]

[7]   De Caterina R, Zampolli A. From asthma to atherosclerosis—5-lipoxygenase, leukotrienes, and inflammation. N Engl J Med 2004; 350: 4-7.
[http://dx.doi.org/10.1056/NEJMp038190]

[8]   Halushka PV, Mais DE, Mayeux PR, Morninelli TA. Thromboxane, prostaglandin, and leukotriene receptors. Annu Rev Pharmacol Toxicol 1989; 29: 213-39.
[http://dx.doi.org/10.1146/annurev.pa.29.040189.001241]

[9]   Kiefer JR, Pawlitz JL, Moreland KT, *et al.* Structural insights into the stereochemistry of the cyclooxygenase reaction. Nature 2000; 405: 97-101.
[http://dx.doi.org/10.1038/35011103]

[10]   Kurumbail RG, Kiefer JR, Marnett LJ. Cyclooxygenase enzymes: catalysis and inhibition. Curr Opin Struct Biol 2001; 11: 752-60.
[http://dx.doi.org/10.1016/S0959-440X(01)00277-9]

[11]   Lands WEM. Biosynthesis of prostaglandins. Annu Rev Nutr 1991; 11: 41-60.
[http://dx.doi.org/10.1146/annurev.nu.11.070191.000353]

[12]   Leslie CC. Regulation of arachidonic acid availability for eicosanoid production. Biochem Cell Biol 2004; 82: 1-17.
[http://dx.doi.org/10.1139/o03-080]

[13]   Needleman P, Truk J, Jakschik BA, Morrison AR, Lefkowith JB. Arachidonic acid metabolism. Annu Rev Biochem 1986; 55: 69-102.
[http://dx.doi.org/10.1146/annurev.bi.55.070186.000441]

[14]   Samuelsson B, Goldyne M, Granstrom E, Hamberg M, Hammarstrom S, Malmsten C. Prostaglandins and thromboxanes. Annu Rev Biochem 1978; 47: 997-1029.
[http://dx.doi.org/10.1146/annurev.bi.47.070178.005025]

[15]   Samuelsson B, Funk CD. Enzymes involved in the biosynthesis of leukotriene B4. J Biol Chem 1989; 264: 19469-72.

[16]   Sigal E. The molecular biology of mammalian arachidonic acid metabolism. Am J Physiol 1991; 260: L13-8.

[17]   Smith WL, DeWitt DL, Garavito RM. Cyclooxygenases: structural, cellular, and molecular biology. Annu Rev Biochem 2000; 69: 145-82.
[http://dx.doi.org/10.1146/annurev.biochem.69.1.145]

[18]   Snyder DW, Fleisch JH. Leukotriene receptor antagonists as potential therapeutic agents. Annu Rev Pharmacol Toxicol 1989; 29: 123-43.
[http://dx.doi.org/10.1146/annurev.pa.29.040189.001011]

[19]   van der Don WA, Tsai A-L, Kulmacz RJ. The cyclooxygenase reaction mechanism. Biochem 2002; 41: 15451-8.
[http://dx.doi.org/10.1021/bi026938h]

[20]   Vane JR, Bakhle YS, Botting RM. Cyclooxygenases 1 and 2. Annu Rev Pharmacol Toxicol 1998; 38: 97-120.
[http://dx.doi.org/10.1146/annurev.pharmtox.38.1.97]

Catecholamines

[21]   Fitzpatrick PF. Tetrahydropterin-dependent amino acid hydroxylases. Annu Rev Biochem 1999; 68: 355-81.
[http://dx.doi.org/10.1146/annurev.biochem.68.1.355]

[22]   Gilman AG. G proteins: transducers of receptor-generated signals. Annu Rev Biochem 1987; 56: 615-49.
[http://dx.doi.org/10.1146/annurev.bi.56.070187.003151]

[23]   Laurent D, Peterson KF, Russell RR, Cline GW, Shulman GI. Effect of epinephrine on muscle glycogenolysis and insulin-stimulated muscle glycogen synthesis in humans. Am J Physiol 1998; 274: E130-8.

[24]   Masserano JM, Weiner N. Tyrosine hydroxylase regulation in the central nervous system. Mol Cell Biochem 1983; 53-54: 129-52.
[http://dx.doi.org/10.1007/BF00225250]

[25]   Moore RY, Bloom FE. Central catecholamine neuron systems: anatomy and physiology of the norepinephrine and epinephrine systems. Annu Rev Neurosci 1979; 2: 113-68.
[http://dx.doi.org/10.1146/annurev.ne.02.030179.000553]

[26]   Wong DL, Lesage A, Siddall B, Funder JW. Glucocorticoid regulation of phenylethanolamine N-methyltransferase. FASEB J 1992; 6: 3310-5.
[http://dx.doi.org/10.1096/fasebj.6.14.1426768]

[27]   Yanagihara N, Futoshi I, Eishichi M, Oka M. Factors affecting adrenal medullary catecholamine biosynthesis and release. Adv Pharmacol 2008; 42: 18-21.
[http://dx.doi.org/10.1016/S1054-3589(08)60684-5]

Folic Acid

[28]   Abali EE, Skacel NE, Celikkaya H, Hsieh YC. Regulation of human dihydrofolate reductase activity and expression. Vitam Horm 2008; 79: 267-92.
[http://dx.doi.org/10.1016/S0083-6729(08)00409-3]

[29]   Banerjee D, Mayer-Kuckuk P, Capiaux G, Budak-Alpdogan T, Gorlick R, Bertino JR. Novel aspects of resistance to drugs targeted to dihydrofolate reductase and thymidylate synthase. Biochim Biophys Acta 2002; 1587: 164-73.
[http://dx.doi.org/10.1016/S0925-4439(02)00079-0]

[30]   Blakley RL. Eukaryotic dihydrofolate reductase. Adv Enzymol Relat Areas Mol Biol 1995; 70: 23-102.

[31]   McGuire JJ. Anticancer antifolates: current status and future directions. Curr Pharm Des 2003; 9: 2593-613.
[http://dx.doi.org/10.2174/1381612033453712]

[32]   Schnell JR, Dyson HJ, Wright PE. Structure, dynamics, and catalytic function of dihydrofolate reductase. Annu Rev Biophys Biomol Struct 2004; 33: 119-40.
[http://dx.doi.org/10.1146/annurev.biophys.33.110502.133613]

[33] Schweitzer BI, Dicker AP, Bertino JR. () "Dihydrofolate reductase as a therapeutic target. FASEB J 1990; 4: 2441-52.
[http://dx.doi.org/10.1096/fasebj.4.8.2185970]

[34] Shane B, Stokstad ELR. Vitamin B12-Folate Interrelationships. Annu Rev Nutr 1985; 5: 115-41.
[http://dx.doi.org/10.1146/annurev.nu.05.070185.000555]

[35] Suh JR, Herbig AK, Stover PJ. New persepectives on folate catabolism. Annu Rev Nutr 2001; 21: 255-82.
[http://dx.doi.org/10.1146/annurev.nutr.21.1.255]

Cholesterol

[36] Bochar DA, Friesen JA, Stauffacher CV, Rodwell VW. Biosynthesis of mevalonic acid from acetyl CoA. Compr Nat Prod Chem 1999; 2: 15-44.
[http://dx.doi.org/10.1016/B978-0-08-091283-7.00035-7]

[37] Brown MS, Goldstein JL. Cholesterol feedback: from Schoenheimer's bottle to Scap's MELADL. J Lipid Res 2009; 50: S15-27.
[http://dx.doi.org/10.1194/jlr.R800054-JLR200]

[38] Charlton-Menys V, Durrington PN. Human cholesterol metabolism and therapeutic molecules. Exp Physiol 2008; 93: 27-42.
[http://dx.doi.org/10.1113/expphysiol.2006.035147]

[39] Goldstein JL, Brown MS. Regulation of the mevalonate pathway. Nature 1990; 343: 425-30.
[http://dx.doi.org/10.1038/343425a0]

[40] Gu P, Ishii Y, Spencer TA, Schechter I. Function-structure studies and identification of three enzyme domains involved in the catalytic activity in rat hepatic squalene synthase. J Biol Chem 1998; 273: 12515-25.
[http://dx.doi.org/10.1074/jbc.273.20.12515]

[41] Istvan ES, Deisenhofer J. Structural mechanism for statin inhibition of HMG-CoA reductase. Science 2001; 292: 1160-4.
[http://dx.doi.org/10.1126/science.1059344]

[42] Jarstfer MB, Blagg BSJ, Rogers DH, Poulter CD. Biosynthesis of squalene. Evidence for a tertiary cyclopropylcarbinyl cationic intermediate in the rearrangement of presqualene disphosphte to squalene. J Am Chem Soc 1996; 118: 13089-90.
[http://dx.doi.org/10.1021/ja963308s]

[43] Risley JM. Cholesterol biosynthesis: lanosterol to cholesterol. J Chem Educ 2002; 79: 377-84.
[http://dx.doi.org/10.1021/ed079p377]

[44] Rudney H, Sexton RC. Regulation of cholesterol biosynthesis. Annu Rev Nutr 1986; 6: 245-72.
[http://dx.doi.org/10.1146/annurev.nu.06.070186.001333]

Adrenocortical Steroids

[45] Connelly MA. SR-BI-mediated HDL cholesteryl ester delivery in the adrenal gland. Mol Cell Endocrinol 2009; 300: 83-8.
[http://dx.doi.org/10.1016/j.mce.2008.09.011]

[46] Edwards PA, Ericsson J. Sterols and isoprenoids: signaling molecules derived from the cholesterol biosynthetic pathway. Annu Rev Biochem 1999; 68: 157-85.
[http://dx.doi.org/10.1146/annurev.biochem.68.1.157]

[47] Falkenstein E, Tillmann H-C, Christ M, Feuring M, Wehling M. Multiple actions of steroid hormones: A focus on rapid, non-genomic effects. Pharmacol Rev 2000; 52: 513-56.

[48] Fardella CE, Miller WL. Molecular biology of mineralocorticoid metabolism. Annu Rev Nutr 1996; 16: 443-70.

[http://dx.doi.org/10.1146/annurev.nu.16.070196.002303]

[49]   Nebert DW, Gonzalez FJ. P450 genes: Structure, evolatuion, and regulation. Annu Rev Biochem 1987; 56: 945-93.
[http://dx.doi.org/10.1146/annurev.bi.56.070187.004501]

[50]   Simpson ER, Waterman MR. Regulation of the synthesis of steroidogenic enzymes in adrenal cortical cells by ACTH. Annu Rev Physiol 1988; 50: 427-40.
[http://dx.doi.org/10.1146/annurev.ph.50.030188.002235]

[51]   Simpson ER, Clyne C, Rubin G, *et al.* Aromatase - a brief overview. Annu Rev Physiol 2002; 64: 93-127.
[http://dx.doi.org/10.1146/annurev.physiol.64.081601.142703]

[52]   Wong LL. Cytochrome P450 monooxygenases. Curr Opin Chem Biol 1998; 2: 263-8.
[http://dx.doi.org/10.1016/S1367-5931(98)80068-9]

Nitrogenous Base Synthesis

[53]   Carreras CW, Santi DV. The catalytic mechanism and structure of thymidylate synthase. Annu Rev Biochem 1995; 64: 721-62.
[http://dx.doi.org/10.1146/annurev.bi.64.070195.003445]

[54]   Eklund H, Uhlin U, Farnegardh M, Logan DT, Nordlund P. Structure and function of the radical enzyme ribonucleotide reductase. Prog Biophys Mol Biol 2001; 77: 177-268.
[http://dx.doi.org/10.1016/S0079-6107(01)00014-1]

[55]   Endrizzi JA, Hanseong K, Anderson PM, Baldwin EP. Crystal structure of *Escherichia coli* cytidine triphosphate synthetase, a nucleotide-regulated glutamine amidotransferase/ATP-dependent amidoligase fusion protein and homologue of anticancer and antiparasitic drug targets. Biochem 2004; 43: 6447-63.
[http://dx.doi.org/10.1021/bi0496945]

[56]   Evans DR, Guy HI. Mammalian pyrimidine biosynthesis: Fresh insights into an ancient pathway. J Biol Chem 2004; 279: 33035-8.
[http://dx.doi.org/10.1074/jbc.R400007200]

[57]   Huang X, Holden HM, Raushel FM. Channeling of substrates and intermediates in enzyme-catalyzed reactions. Annu Rev Biochem 2001; 70: 149-80.
[http://dx.doi.org/10.1146/annurev.biochem.70.1.149]

[58]   Kappock TJ, Ealick SE, Stubbe J. Modular evolution of the purine biosynthetic pathway. Curr Opin Chem Biol 2000; 4: 567-72.
[http://dx.doi.org/10.1016/S1367-5931(00)00133-2]

[59]   Kashlan OB, Scott CP, Lear JD, Cooperman BS. A comprehensive model for the allosteric regulation of mammalian ribonuclease reductase. Functional consequences of ATP- and dATP-induced oligomerization of the large subunit. Biochem 2002; 41: 462-72.
[http://dx.doi.org/10.1021/bi011653a]

[60]   Lee L, Kelly RE, Pastra-Landis SC, Evans DR. Oligomeric structure of the multifunctional protein CAD that initiates pyrimidine biosynthesis in mammalian cells. Proc Natl Acad Sci USA 1985; 82: 6802-6.
[http://dx.doi.org/10.1073/pnas.82.20.6802]

[61]   Loffler M, Fairbanks LD, Zameitat E, Marinaki AM, Simmonds HA. Pyrimidine pathways in health and disease. Trends Mol Med 2005; 11: 430-7.
[http://dx.doi.org/10.1016/j.molmed.2005.07.003]

[62]   Mouilleron S, Golinelli-Pimpaneau B. Conformational changes in ammonia-channeling glutamine amidotransferases. Curr Opin Struct Biol 2007; 17: 653-64.
[http://dx.doi.org/10.1016/j.sbi.2007.09.003]

[63]   Nordlund P, Reichard P. Ribonucleotide reductases. Annu Rev Biochem 2006; 75: 681-706.
[http://dx.doi.org/10.1146/annurev.biochem.75.103004.142443]

[64]   Weber G, Naga M, Natsumeda Y, *et al.* Regulation of de novo and salvage pathways in chemotherapy.
Adv Enzyme Regul 1991; 31: 45-67.
[http://dx.doi.org/10.1016/0065-2571(91)90008-A]

[65]   Zalkin H, Dixon JE. De novo purine nucleotide biosynthesis. Prog Nucleic Acid Res Mol Biol 1992;
42: 259-85.
[http://dx.doi.org/10.1016/S0079-6603(08)60578-4]

SUBJECT INDEX

www.ingramcontent.com/pod-product-compliance
Lightning Source LLC
Chambersburg PA
CBHW050806220326
41598CB00006B/134